Springer Complexity

Springer Complexity is an interdisciplinary program publishing the best research and academic-level teaching on both fundamental and applied aspects of complex systems – cutting across all traditional disciplines of the natural and life sciences, engineering, economics, medicine, neuroscience, social and computer science.

Complex Systems are systems that comprise many interacting parts with the ability to generate a new quality of macroscopic collective behavior the manifestations of which are the spontaneous formation of distinctive temporal, spatial or functional structures. Models of such systems can be successfully mapped onto quite diverse "real-life" situations like the climate, the coherent emission of light from lasers, chemical reaction-diffusion systems, biological cellular networks, the dynamics of stock markets and of the internet, earthquake statistics and prediction, freeway traffic, the human brain, or the formation of opinions in social systems, to name just some of the popular applications.

Although their scope and methodologies overlap somewhat, one can distinguish the following main concepts and tools: self-organization, nonlinear dynamics, synergetics, turbulence, dynamical systems, catastrophes, instabilities, stochastic processes, chaos, graphs and networks, cellular automata, adaptive systems, genetic algorithms and computational intelligence.

The two major book publication platforms of the Springer Complexity program are the monograph series "Understanding Complex Systems" focusing on the various applications of complexity, and the "Springer Series in Synergetics", which is devoted to the quantitative theoretical and methodological foundations. In addition to the books in these two core series, the program also incorporates individual titles ranging from textbooks to major reference works.

Editorial and Programme Advisory Board

W0235199

Péter Érdi
Center for Complex Systems Studies, Kalamazoo College, USA
and Hungarian Academy of Sciences, Budapest, Hungary

Karl Friston
National Hospital, Institute for Neurology, Wellcome Dept. Cogn. Neurology, London, UK

Hermann Haken
Center of Synergetics, University of Stuttgart, Stuttgart, Germany

Janusz Kacprzyk
System Research, Polish Academy of Sciences, Warsaw, Poland

Scott Kelso
Center for Complex Systems and Brain Sciences, Florida Atlantic University, Boca Raton, USA

Jürgen Kurths
Nonlinear Dynamics Group, University of Potsdam, Potsdam, Germany

Linda Reichl
Department of Physics, Prigogine Center for Statistical Mechanics, University of Texas, Austin, USA

Peter Schuster
Theoretical Chemistry and Structural Biology, University of Vienna, Vienna, Austria

Frank Schweitzer
System Design, ETH Zürich, Zürich, Switzerland

Didier Sornette
Entrepreneurial Risk, ETH Zürich, Zürich, Switzerland

Understanding Complex Systems

Founding Editor: J.A. Scott Kelso

Future scientific and technological developments in many fields will necessarily depend upon coming to grips with complex systems. Such systems are complex in both their composition – typically many different kinds of components interacting simultaneously and nonlinearly with each other and their environments on multiple levels – and in the rich diversity of behavior of which they are capable.

The Springer Series in Understanding Complex Systems series (UCS) promotes new strategies and paradigms for understanding and realizing applications of complex systems research in a wide variety of fields and endeavors. UCS is explicitly transdisciplinary. It has three main goals: First, to elaborate the concepts, methods and tools of complex systems at all levels of description and in all scientific fields, especially newly emerging areas within the life, social, behavioral, economic, neuro- and cognitive sciences (and derivatives thereof); second, to encourage novel applications of these ideas in various fields of engineering and computation such as robotics, nano-technology and informatics; third, to provide a single forum within which commonalities and differences in the workings of complex systems may be discerned, hence leading to deeper insight and understanding.

UCS will publish monographs, lecture notes and selected edited contributions aimed at communicating new findings to a large multidisciplinary audience.

Peter beim Graben · Changsong Zhou ·
Marco Thiel · Jürgen Kurths (Eds.)

Lectures in Supercomputational Neuroscience

Dynamics in Complex Brain Networks

With 179 Figures and 18 Tables

 Springer

Dr. Peter beim Graben
University of Reading
School of Psychology and Clinical Language
Sciences
Whiteknights, PO Box 217
Reading RG6 6AH
United Kingdom

Prof. Dr. Changsong Zhou
Hong Kong Baptist University
Department of Physics
224 Waterloo Road
Kowloon Tong, Hong Kong
China

Dr. Marco Thiel
University of Aberdeen
School of Engineering and Physical Sciences
Aberdeen AB24 3UE
United Kingdom

Prof. Dr. Jürgen Kurths
Universität Potsdam
Institut für Physik
LS Theoretische Physik
Am Neuen Palais 10
14469 Potsdam, Germany

ISSN 1860-0832
ISBN 978-3-642-09216-9   e-ISBN 978-3-540-73159-7

Springer is a part of Springer Science+Business Media
springer.com
© Springer-Verlag Berlin Heidelberg 2008
Softcover reprint of the hardcover 1st edition 2008

Cover design: WMX Design, Heidelberg

Preface

Computational neuroscience has become a very active field of research in the last decades. Improved experimental facilities, new mathematical techniques and especially the exponential increase of computational power have lead to stunning new insights into the functioning of the brain. Scientists begin to endeavor simulating the brain from the bottom level of single neurons to the top-level of cognitive behavior. The "Blue Brain Project" (http://bluebrainproject.epfl.ch/) for example, is a hallmark for this approach.

Many scientists are attracted to this highly interdisciplinary field of research, in which only the combined efforts of neuroscientists, biologists, psychologists, physicists and mathematicians, computer scientists, engineers and other specialists, e.g. from anthropology, linguistics, or medicine, seem to be able to shift the limits of our knowledge. However, one of the most common problems of interdisciplinary work is to find a "common language", i.e., an effective way to discuss problems with colleagues with a different scientific background. Therefore, an introduction into this field has to familiarize the reader with aspects from various relevant fields in an intelligible way.

This book is an introduction to the field of computational neuroscience from a physicist's perspective, regarded as *neurophysics*, with in depth contributions of systems neuroscientists. It is based upon the lectures delivered during the *5th Helmholtz Summer School on Supercomputational Physics*:

"Complex Networks in Brain Dynamics"

held in September 2005 at the University of Potsdam.

The book-title *Lectures in Supercomputational Neuroscience: Dynamics in Complex Brain Networks* is motivated by the methods and outcomes of the Summer School: A conceptual model for complex networks of neurons is introduced, which incorporates many important features of the "real" brain, such as different types of neurons, various brain areas, inhibitory and excitatory coupling and plasticity of the network. The model is then implemented in an MPI (message-passing interface)-based parallel computer code, running at appropriate supercomputers, that is introduced and discussed in detail in this book. But beyond the mere presentation of the C-program, the text will enable the reader to modify and adapt the algorithm for his/her own research.

The first part of the book (**Neurophysiology**) gives an introduction to the physiology of the brain on different levels, ranging from the rather large areas of the brain down to individual neurons. Various models for individual neurons are discussed as well as models for the "communication" among these neuronal oscillators. An outlook on cognition and learning is also given in this first part.

The second part (**Complex Networks**) outlines the dynamics of ensembles of neurons forming different types of networks. Recently developed new approaches based on complex networks with special emphasis on the relationships between structure and function of complex systems are presented. The topology of such a network, i.e., how the neurons are coupled, plays an important role for the behavior of the ensemble. Even though the network of the 10^{10} neurons in a human brain is much too complex to be modeled with our present knowledge, the conceptual models presented here are a promising starting point and allow gaining insight into the principles of complex networks in brain dynamics. This part covers all aspects from the basics of networks, their topology and how to quantify them, to the structure and function of complex cortical networks up to collective behavior of large networks such as clustered synchronization. New techniques for the analysis of data of complex networks are also introduced. They allow not only to study large populations of neurons but also to study (neural) oscillators with more than one time scale, e.g. spiking and bursting neurons.

The third part (**Cognition and Higher Perception**) presents results about how structural units of the brain (columns) can be described and how networks of neurons can be used to model cognition and perception as measured by the electroencephalogram. It is shown how networks of simple neuronal models can be used to model, e.g., reaction times from psychological experiments.

The forth part (**Implementations**) discusses the implementation of a model of a network of networks of neurons in an MPI-based C-code. The code is modular in the sense that the model(s) for the neurons, the topology of the network, the coupling and many further parameters can easily be changed and adapted. The main point is to outline how in principle many different features can be implemented in a computer code, rather than presenting a cutting-edge algorithm. The computer code is available for download (http://www.agnld.uni-potsdam.de). In this part we also discuss that computational neuroscience is not simply about parallelizing normal computer code. A very important component of it is the implementation of specially adapted algorithms. An example of such a computer code will be given in this chapter.

In the fifth and final part (**Applications**), three groups of students of the Summer School, discuss the results they obtained running the code on supercomputers. After studying the parameter space for large networks of Morris-Lecar neurons, they use a map of cortical connections from a cat's brain, which was obtained based on experimental studies. In their simulations they consider multiple spatio-temporal scales and study the patterns for synchronized firing of neurons in different brain areas. Results of simulations for different network topologies and neuronal models are also summarized here. These chapters will be helpful to those who are planning to apply the parallel code for their own research, as they give a very practical account of how to actually perform simulations. They point at crucial problems and show how to overcome pitfalls when simulating based on the MPI code.

We hope that this book will help graduate students and researchers to access the field of computational neuroscience and to develop and improve high-end, parallel computer codes for the simulation of large networks of neurons.

Last, but not least, we wish to thank all lecturers and the coordinators of the 5th Helmholtz Summer School on Supercomputational Physics; Mamen Romano, Lucia Zemanová, and Gorka Zamora-López for their assistance; the Land Brandenburg for main funding, EU, NoE, and EU-Network BioSim (contract No. LSHB–CT–2004–005137) for further support; the University of Potsdam for making access to its supercomputer cluster available and also for logistics. Finally, we thank James Ong for his careful proof-reading of the complete book.

Nonlinear Dynamics Group   Peter beim Graben
University of Potsdam   Changsong Zhou
                                                                 Marco Thiel
                                                               Jürgen Kurths

Contents

Part III Cognition and Higher Perception

Part IV Implementations

Part V Applications

Part I

Neurophysiology

1

Foundations of Neurophysics

Peter beim Graben[1,2]

[1] School of Psychology and Clinical Language Sciences,
University of Reading, United Kingdom
p.r.beimgraben@reading.ac.uk
[2] Institute of Physics, Nonlinear Dynamics Group, Universität Potsdam,
Germany

Summary. This chapter presents an introductory course to the biophysics of neurons, comprising a discussion of ion channels, active and passive membranes, action
potentials and postsynaptic potentials. It reviews several conductance-based and
reduced neuron models, neural networks and neural field theories. Finally, the basic
principles of the neuroelectrodynamics of mass potentials, i.e. dendritic fields, local field potentials, and the electroencephalogram are elucidated and their putative
functional role as a mean field is discussed.

1.1 Introduction

Metaphorically, the brain is often compared with a digital computer [1, 2]
that runs *software* algorithms in order to perform cognitive computations.
In spite of its usefulness as a working hypothesis in the cognitive [3–6] and
computational [7–18] neurosciences, this metaphor does obviously not apply
to the *hardware* level. Digital computers consist of circuit boards equipped
with chips, transistors, resistors, capacitances, power supplies, and other electronic components wired together. Digital computation is essentially based on
controlled switching processes in semiconductors which are nonlinear physical
systems. On the other hand, brains consist to 80% of water contained in cells
and also surrounding cells. How can this physical *wet-ware* substrate support
computational dynamics? This question should be addressed in the present
chapter. Starting from the physiological facts about neurons, their cell membranes, electrolytes, and ions [19–21], I shall outline the biophysical principles
of neural computation [12, 13, 15, 18, 22–25] in parallel to those of computation in electronic circuits. Thus, the interesting physiological properties will
be described by electric "equivalent circuits" providing a construction kit of
building blocks that allow the modeling of membranes, single neurons, and
eventually neural networks. This field of research is broadly covered by *computational neuroscience*. However, since this discipline also deals with more
abstract approximations of real neurons (see Sect. 1.4.3) and with artificial

neural networks, I prefer to speak about *neurophysics*, i.e. the biophysics of real neurons.

The chapter is organized as a journey along a characteristic neuron where the stages are Sects. 1.2–1.4. Looking at Fig. 8.1 in Chap. 8, the reader recognizes the *cell bodies*, or *somata*, of three cortical neurons as the triangular knobs. Here, our journey will start by describing the microscopically observable membrane potentials. Membranes separating electrolytes with different ion concentrations exhibit a characteristic resting potential. In a corresponding equivalent circuit, this voltage can be thought of being supplied by a battery. Moreover, passive membranes act as a capacitance while their semipermeability with respect to particular kinds of ions leads to an approximately ohmic resistance. This property is due to the existence of leaky ion channels embedded in the cell membrane. At the neuron's axon hillock (trigger zone), situated at the base of the soma, the composition of the cell membrane changes. Here and along the axon, voltage-gated sodium and potassium channels appear in addition to the leakage channels, both making the membrane active and excitable. As we shall see, the equivalent circuit of the membrane allows for the derivation of the famous *Hodgkin-Huxley equations* of the *action potentials* which are the basic of neural *conductance models*. Traveling along the axon, we reach the presynaptic terminals, where the Hodgkin-Huxley equations have to be supplemented by additional terms describing the dynamics of voltage-gated calcium channels. Calcium flowing into the terminal causes the release of transmitter vesicles that pour their content of neurotransmitter into the synaptic cleft of a chemical synapse. Then, at the postsynapse, transmitter molecules dock onto receptor molecules, which indirectly open other ion channels. The kinetics of these reactions give rise to the *impulse response functions* of the postsynaptic membranes. Because these membranes behave almost passively, a linear differential equation describes the emergence of *postsynaptic potentials* by the convolution product of the postsynaptic pulse response with the *spike train*, i.e. the sequence of action potentials. Postsynaptic potentials propagate along the *dendrites* and the soma of the neuron and superimpose linearly to a resulting signal that eventually arrives at the axon hillock, where our journey ends.

In Sect. 1.5, we shall change our perspective from the microscopic to the macroscopic. Here, the emergence of mass potentials such as the local field potential (LFP) and the electroencephalogram (EEG) will be discussed.

1.2 Passive Membranes

Neurons are cells specialized for the purpose of fast transfer and computation of information in an organism. Like almost every other cell, they posses a cell body containing a nucleus and other organelles and they are surrounded by a membrane separating their interior from the extracellular space. In order to collect information from their environment, the soma of a characteristic neuron branches out into a *dendritic tree* while another thin process, the *axon*, provides an output connection to other neurons [19–21]. The cell plasma in the interior as well as the liquid in the extracellular space are electrolytes, i.e. solutions of different kinds of ions such as sodium (Na^+), potassium (K^+), calcium (Ca^{2+}), chloride (Cl^-), and large organic ions. However, the concentrations of these ions (denoted by $[Na^+], [K^+], [Ca^{2+}]$, etc.) can differ drastically from one side of the cell membrane to the other (see Fig. 2.1 of Chap. 2). Therefore, the membrane is subjected to two competing forces: the osmotic force aiming at a compensation of these concentration gradients on the one hand, and the Coulomb force aiming at a compensation of the electric potential gradient. Biochemically, cell membranes are lipid bi-layers swimming like fat blobs in the plasma soup [19, 20], which makes them perfect electric isolators. Putting such a dielectric between two opposite electric charges yields a capacitance of capacity

$$C_m = \frac{Q}{U}, \tag{1.1}$$

where Q is the total charge stored in the capacitance and U is the voltage needed for that storage. Hence, a membrane patch of a fixed area A that separates different ion concentrations can be represented by a single capacitance $C_m = 1\,\mu F\,cm^{-1} \times A$ in an equivalent "circuit" shown in Fig. 1.1 [19, 20].

Generally, we interpret such equivalent circuits in the following way: The upper clamp refers to the extracellular space whereas the clamp at the bottom measures the potential within the cell. Due to its higher conductance, the extracellular space is usually assumed to be equipotential, which can be designated as $U = 0\,mV$ without loss of generality.

1.2.1 Ion Channels

If neuron membranes were simply lipid bi-layers, there would be nothing more to say. Of course, they are not. All the dynamical richness and computational complexity of neurons is due to the presence of particular proteins, called *ion channels*, embedded in the cell membranes. These molecules form tubes traversing the membrane that are permeable to certain kinds of ions [19–25]. The "zoo" of ion channels is comparable with that of elementary particles. There are channels whose pores are always open (*leakage channels*) but permeable only for sodium or potassium or chloride. Others possess *gates* situated in their pores which are controlled by the membrane potential, or the presence of certain substances or even both. We shall refer to the first kind of channels as to *voltage-gated channels*, and to the second kind as to *ligand-gated*

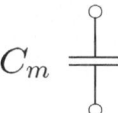

Fig. 1.1. Equivalent "circuit" for the capacitance C_m of a membrane patch

channels. Furthermore, the permeability of a channel can depend on the direction of the ionic current such that it behaves as a rectifier whose equivalent "circuit" would be a diode [19, 20]. Eventually, the permeability could be a function of the concentration of particular reagents either in the cell plasma or in the extracellular space, which holds not only for ligand-gated channels. Such substances are used for classifying ion channels. Generally, there are two types of substances. Those from the first class facilitate the functioning of a channel and are therefore called *agonists*. The members of the second class are named *antagonists* as they impede channel function.

Omitting these complications for a while, we assume that a single ion channel of kind k behaves as an ohmic resistor with conductance

$$\gamma_k = \frac{1}{\rho_k}, \qquad (1.2)$$

where ρ_k is the resistivity of the channel. A typical value (for the gramicidin-A channel) is $\gamma_{\text{GRAMA}} \approx 12\,\text{pS}$. Figure 1.2 displays the corresponding equivalent "circuit".

In the remainder of this chapter, we will always consider membrane patches of a fixed area A. In such a patch, many ion channels are embedded, forming the parallel circuit shown in Fig. 1.3(a).

According to Kirchhoff's First Law, the total conductance of the parallel circuit is

$$g_k = N_k \gamma_k \qquad (1.3)$$

when N_k channels are embedded in the patch, or, equivalently, expressed by the channel concentration $[k] = N_k/A$,

$$g_k = [k]A\gamma_k .$$

1.2.2 Resting Potentials

By embedding leakage channels into the cell membrane, it becomes *semipermeable*, i.e. permeable for certain kinds of ions while impenetrable for others. If there is a concentration gradient of a permeable ion across a semipermeable membrane, a diffusion current I_{diff} through the membrane patch A is created, whose density obeys Fick's Law

$$j_{\text{diff}} = -D\,q\frac{\mathrm{d}[\mathrm{I}]}{\mathrm{d}x}, \qquad (1.4)$$

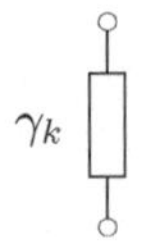

Fig. 1.2. Equivalent "circuit" for a single ohmic ion channel with conductance γ_k

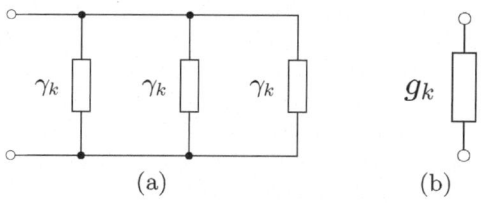

Fig. 1.3. Equivalent circuits (**a**) for ion channels of one kind k connected in parallel; (**b**) Substituted by a single resistor of conductance $g_k = 3\gamma_k$

where $d[I]/dx$ denotes the concentration gradient for ion I, q its charge, and $D = k_B T/\mu$ is the diffusion constant given by Einstein's relation [26] (k_B is Boltzmann's constant, T is the temperature and μ is the viscosity of the electrolyte) [22–25]. This diffusion current can be described by an equivalent "circuit" given by a current source I_{diff} (Fig. 1.4).

The separation of charges by the diffusion current leads to an increasing potential gradient dU/dx across the membrane. Therefore, a compensating ohmic current

$$j_{\mathrm{ohm}} = -\sigma \frac{dU}{dx} \tag{1.5}$$

flows back through the leakage channels ($\sigma = q^2[I]/\mu$ is the conductance of the electrolyte expressed by the ion concentration and its charge). Then the total current $j = j_{\mathrm{diff}} + j_{\mathrm{ohm}}$ (visualized by the circuit in Fig. 1.5) is described by the *Nernst-Planck equation*

$$j = -D\,q\frac{d[I]}{dx} - [I]\,\frac{q^2}{\mu}\frac{dU}{dx}. \tag{1.6}$$

The Nernst Equation

The general quasi-stationary solution of (1.6), the Goldman-Hodgkin-Katz equation ((2.4) in Chap. 2), clearly exhibits a nonlinear dependence of the ionic current on the membrane voltage [22–25]. However, for only small deviations from the stationary solution — given by the *Nernst equation*

$$E_I = \frac{k_B T}{q}\,\ln\frac{[I]_{\mathrm{out}}}{[I]_{\mathrm{int}}}, \tag{1.7}$$

Fig. 1.4. Equivalent "circuit" either for the diffusion currents through the cell membrane or for the active ion pumps

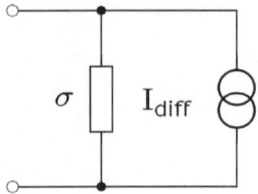

Fig. 1.5. Equivalent circuit for the derivation of the Nernst-Planck equation (1.6)

where $[I]_{out}$ is the ion concentration in the extracellular space and $[I]_{int}$ within the cell — the current can be regarded as being ohmic.

For room temperature, the factor $k_B T/q \approx 25\,mV$. With the concentrations from Fig. 2.1, Chap. 2, this leads to the characteristic resting potentials; e.g. $U_{K+} = -101\,mV$, and $U_{Na+} = +56\,mV$.

Each sort of ion possesses its own Nernst equilibrium potential. We express this fact by a battery in an equivalent "circuit" shown in Fig. 1.6.

Now, we are able to combine different ion channels k all selective for one sort of ions I with their corresponding power supplies. This is achieved by a serial circuit as shown in Fig. 1.7. This equivalent circuit will be our basic building block for all other subsequent membrane models.

If the clamp voltage of this circuit has the value U, we have to distribute this voltage according to Kirchhoff's Second Law as

$$U = \frac{I_k}{g_k} + E_I\,,$$

leading to the fundamental equation

$$I_k = g_k(U - E_I)\,. \tag{1.8}$$

The Goldman Equation

As an example, we assume that three types of ion channels are embedded in the membrane patch, one pervious for sodium with the conductance g_{Na+}, another pervious for potassium with the conductance g_{K+}, and the third pervious for chloride with the conductance g_{Cl-}, respectively. Figure 1.8 displays the corresponding equivalent circuit.

Interpreting the top of the circuit as the extracellular space and the bottom as the interior of the neuron, we see that the resting potential for potassium

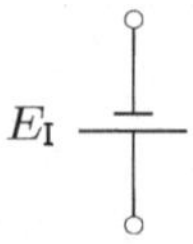

$$E_I \;\; \vdash$$

Fig. 1.6. Equivalent circuit for the Nernst equilibrium potential (1.7)

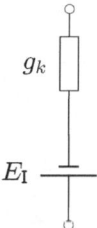

Fig. 1.7. Equivalent circuit for a population of ion channels of kind k selective for the ion sort I embedded in a membrane with resting potential E_{I}

and chloride is negative (denoted by the short tongue of the battery symbol) while the sodium equilibrium potential is positive in comparison to the extracellular space.

According to Kirchhoff's First Law, the total current through the circuit is

$$I = I_{\mathrm{Na}^+} + I_{\mathrm{K}^+} + I_{\mathrm{Cl}^-}\,. \tag{1.9}$$

To obtain the stationary equilibrium, we have to set $I = 0$. Using the fundamental equation (1.8), we get the equation

$$0 = g_{\mathrm{Na}^+}(U - E_{\mathrm{Na}^+}) + g_{\mathrm{K}^+}(U - E_{\mathrm{K}^+}) + g_{\mathrm{Cl}^-}(U - E_{\mathrm{Cl}^-})\,,$$

whose resolution entails the equilibrium potential

$$U = \frac{g_{\mathrm{Na}^+}E_{\mathrm{Na}^+} + g_{\mathrm{K}^+}E_{\mathrm{K}^+} + g_{\mathrm{Cl}^-}E_{\mathrm{Cl}^-}}{g_{\mathrm{Na}^+} + g_{\mathrm{K}^+} + g_{\mathrm{Cl}^-}}\,. \tag{1.10}$$

Equation (1.10) is closely related to the Goldman equation that can be derived from the Goldman-Hodgkin-Katz equation [24]. It describes the net effect of all leakage channels. Therefore, the circuit in Fig. 1.8 can be replaced by the simplification found in Fig. 1.9.

Accordingly, the leakage current is again given by (1.8)

$$I_l = g_l(U - E_l)\,. \tag{1.11}$$

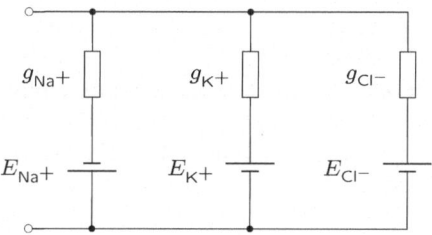

Fig. 1.8. Equivalent circuit for three populations of ion channels permeable for sodium, potassium and chloride with their respective Nernst potentials

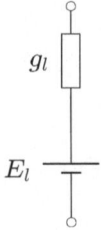

Fig. 1.9. Equivalent circuit for the total leakage current and its corresponding leakage potential E_l

Characteristic values are $g_l = 13\,\mu\text{S}$ for the leakage conductance and $E_l = -69\,\text{mV}$ for the leakage potential as the solution of (1.10) [19, 20].

While the Nernst potential for one kind of ions denotes a stationary state, the Goldman equilibrium potential results from a continuous in- and outflow of ions that would cease when all concentration gradients had been balanced. To stabilize the leakage potential the cell exploits active *ion pumps* modeled by a current source as displayed in Fig. 1.4. These ion pumps are proteins embedded in the cell membrane that transfer ions against their diffusion gradients by consuming energy. Maintaining resting potentials is one of the energetically most expensive processes in the nervous system [27]. This consumption of energy is, though rather indirectly, measurable by neuroimaging techniques such as positron emission tomography (PET) or functional magnetic resonance imaging (fMRI) [19, 20, 28, 29].

1.3 Active Membranes

The resting potentials we have discussed so far are very sensitive to changes in the conductances of the ion channels. While these are almost constant for the leakage channels, there are other types of channels whose conductances are functions of certain parameters such as the membrane potential or the occurrence of particular reagents. These channels make membranes active and dynamic. The former are called voltage-gated whereas the latter are referred to as ligand-gated. Basically, these channels occur in two dynamical states: their pore may be open (O) or closed (C). The conductance of closed channels is zero, while that of an open channel assumes a particular value γ_k. Therefore, a single gated channel can be represented by a serial circuit of a resistor with conductance γ_k and a switch S, as depicted in Fig. 1.10.

Let N_k be the number of gated channels of brand k embedded in our membrane patch of area A, and let O_k and C_k the number of momentarily open and closed channels of this kind, respectively. As argued in Sect. 1.2.1, the total conductance of all open channels is given by Kirchhoff's First Law as

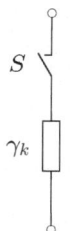

Fig. 1.10. Equivalent circuit for a single gated channel with open-conductance γ_k

$$g_k = O_k\, \gamma_k\,, \tag{1.12}$$

while

$$\bar{g}_k = N_k\, \gamma_k \tag{1.13}$$

is now the maximal conductance of these channels.

1.3.1 Action Potentials

Signals propagate mainly passively along the dendritic and somatic membranes until they reach the axon hillock, or trigger zone of the neuron. Here, the composition of the membrane changes significantly and voltage-gated sodium and potassium channels supplement the all-pervasive leakage channels. Above, we have modeled these channels by switches connected serially with ohmic resistors. Now, the crucial question arises: Who opens the switches?

Here, for the first time, a stochastic account is required. Ion channels are macro-molecules and hence quantum objects. Furthermore, these objects are weakly interacting with their environments. Therefore the cell membrane and the electrolytes surrounding it provide a *heat bath* making a thermodynamical treatment necessary. From a statistical point of view, an individual channel has a probability of being open, p_k, such that the number of open channels is the expectation value

$$O_k = p_k\, N_k\,. \tag{1.14}$$

Inserting (1.14) into (1.12) yields the conductance

$$g_k = p_k\, N_k\, \gamma_k = p_k\, \bar{g}_k\,. \tag{1.15}$$

The problem of determining the probability p_k is usually tackled by modeling Markov chains [24, 25]. The simplest approach is a two-state Markov process shown in Fig. 1.11, where C and O denote the closed and the open state, respectively, while α, β are transition rates.

The state probabilities of the Markov chain in Fig. 1.11 obey a *master equation* [30, 31]

$$\frac{\mathrm{d}p_k}{\mathrm{d}t} = \alpha_k\,(1 - p_k(t)) - \beta_k\, p_k(t)\,, \tag{1.16}$$

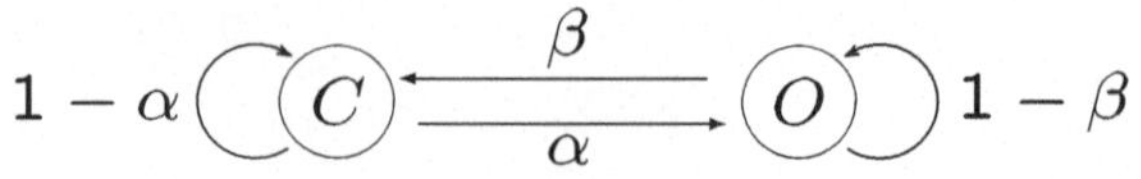

$$1 - \alpha \; \overset{\beta}{\underset{\alpha}{C \rightleftharpoons O}} \; 1 - \beta$$

Fig. 1.11. Two-state Markov model of a voltage-gated ion channel

whose transition rates are given by the thermodynamic Boltzmann weights

$$\alpha_k = e^{\frac{W(C \to O)}{k_B T}} , \tag{1.17}$$

where $W(C \to O)$ is the necessary amount of energy that has to be supplied by the heat bath to open the channel pore.

Channel proteins consist of amino acids that are to some extent electrically polarized [19–21]. The gate blocking the pore is assumed to be a subunit with charge Q. Call $W_0(C \to O)$ the work that is necessary to move Q through the electric field generated by the other amino acids to open the channel pore. Superimposing this field with the membrane potential U yields the total transition energy

$$W(C \to O) = W_0(C \to O) + QU . \tag{1.18}$$

If $QU < 0$, $W(C \to O)$ is diminished and the transition $C \to O$ is facilitated [12], thereby increasing the rate α_k according to

$$\alpha_k(U) = e^{\frac{W_0(C \to O) + QU}{k_B T}} . \tag{1.19}$$

The equations (1.15, 1.16, 1.19) describe the functioning of voltage-gated ion channels [12,13,15,23–25]. Yet, voltage-gated resistors are also well-known in electric engineering: *transistors* are *transi*ent resi*stors*. Though not usual in the literature, I would like to use the transistor symbol to denote voltage-gated ion channels here (Fig. 1.12). In contrast to batteries, resistors and capacitors, which are *passive* building blocks of electronic engineering, transistors are *active* components thus justifying our choice for active membranes.

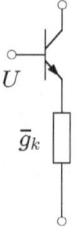

Fig. 1.12. Equivalent circuit for a population of voltage-gated ion channels. The maximal conductance $\bar{g}_k$ is reached when the transistor is in saturation

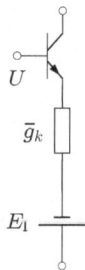

Fig. 1.13. Equivalent circuit for a population of voltage-gated ion channels of kind k selective for the ion sort I embedded in a membrane with resting potential E_I

Corresponding to Fig. 1.7, the equivalent circuit for a population of voltage-gated channels of kind k permeable for ions I supplied by their respective resting potential E_I is provided in Fig. 1.13.

The Hodgkin-Huxley Equations

Now we are prepared to derive the Nobel-prize-winning Hodgkin-Huxley equations for the action potential [32] (see also [12–15, 23–25]). Looking again at Fig. 1.8, one easily recognizes that an increase of the sodium conductance leads to a more positive membrane potential, or, to a *depolarization*, while an increasing conductance either of potassium or of chloride entails a further negativity, or *hyperpolarization* of the membrane potential. These effects are in fact achieved by voltage-gated sodium and potassium channels which we refer here to as AN and AK, respectively. Embedding these into the cell membrane yields the equivalent circuit shown in Fig. 1.14.[3]

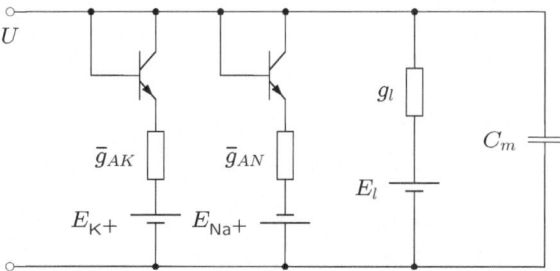

Fig. 1.14. Equivalent circuit for the Hodgkin-Huxley equations (1.25, 1.27–1.29)

---

[3] I apologize to all electrical engineers for taking their notation rather symbolically. Certainly, this circuit has neither protection resistors nor voltage stabilizers and should not be reproduced. Sorry for that!

The first and second branches represent the voltage-gated potassium and sodium channels, respectively. The third is taken from the stationary descriptions of the leakage potential (Sect. 1.2.2) while the capacitance is now necessary to account for the dynamics of the membrane potential. According to Kirchhoff's First Law, the total current through the circuit adds up to an injected current I_m,

$$I_m = I_{AK} + I_{AN} + I_l + I_C \,. \tag{1.20}$$

The partial currents are

$$I_{AK} = p_{AK}\,\bar{g}_{AK}(U - E_{K^+}) \tag{1.21}$$

$$I_{AN} = p_{AN}\,\bar{g}_{AN}(U - E_{Na^+}) \tag{1.22}$$

$$I_l = g_l(U - E_l) \tag{1.23}$$

$$I_C = C_m \frac{dU}{dt}\,, \tag{1.24}$$

where (1.21, 1.22) are produced from (1.15) and (1.8), (1.23) is actually (1.11) and (1.24) is the temporal derivative of (1.1). Taken together, the membrane potential $U(t)$ obeys the differential equation

$$C_m \frac{dU}{dt} + p_{AK}\,\bar{g}_{AK}(U - E_{K^+}) + p_{AN}\,\bar{g}_{AN}(U - E_{Na^+}) + g_l(U - E_l) = I_m \,. \tag{1.25}$$

Equation (1.25) has to be supplemented by two master equations: (1.16) for the open probabilities p_{AK}, p_{AN} and the rate equations (1.19) for α_{AK}, α_{AN}.

Unfortunately, this approach is inconsistent with the experimental findings of Hodgkin and Huxley [32]. They reported two other relations

$$p_{AK} = n^4; \qquad p_{AN} = m^3 h\,, \tag{1.26}$$

where n, m and h now obey three master equations

$$\frac{dn}{dt} = \alpha_n (1 - n) - \beta_n\, n \tag{1.27}$$

$$\frac{dm}{dt} = \alpha_m (1 - m) - \beta_m\, m \tag{1.28}$$

$$\frac{dh}{dt} = \alpha_h (1 - h) - \beta_h\, h\,. \tag{1.29}$$

The equations (1.25, 1.27–1.29) are called Hodgkin-Huxley equations [12–15, 23–25, 32]. They constitute a four-dimensional nonlinear dynamical system controlled by the parameter I_m. Figure 1.15 displays numerical solutions for three different values of I_m.

Figure 1.15 illustrates only two of a multitude of dynamical patters of the Hodgkin-Huxley system. Firstly, it exhibits a threshold behavior that is due to a Hopf bifurcation [18]. For subthreshold currents (solid line: $I_m = 7.09\,\mu A$), one observes a damped oscillation corresponding to a stable fixed point in

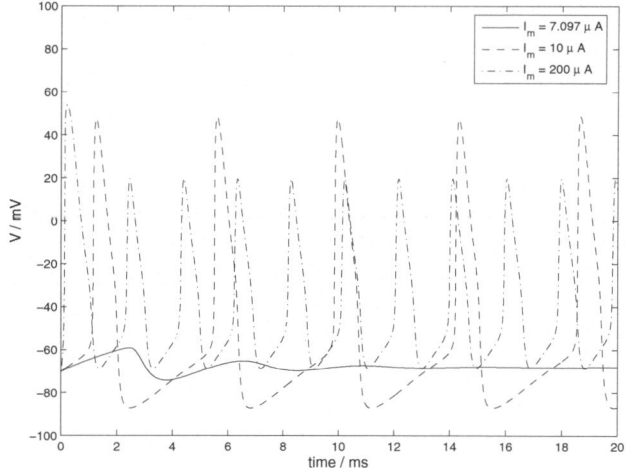

Fig. 1.15. Numeric solutions of the Hodgkin-Huxley equations (1.25, 1.27–1.29) according to the Rinzel-Wilson model (Sect. 1.4.3) for three different values of the control parameter I_m. *Solid: subthreshold current $I_m = 7.09\,\mu A$; dashed: super-threshold current $I_m = 10\,\mu A$; dashed-dotted: even higher current $I_m = 200\,\mu A$*

the phase space. If the control parameter I_m exceeds a certain threshold θ, this fixed point destabilizes and a limit cycle emerges (dashed line: $I_m = 10\,\mu A$). Secondly, further heightening of I_m leads to limit cycles of increased frequencies (dashed-dotted line: $I_m = 200\,\mu A$). This *regular spiking* dynamics explains the law of all-or-nothing as well as the encoding principle by frequency modulation in the nervous system [19–21].

In order to interpret the Hodgkin-Huxley equations (1.25, 1.27–1.29) biologically, we have to consider (1.26) first. It tells that our simple two-state Markov chain (Fig. 1.11) is not appropriate. Instead, the description of the active potassium channel requires a four-state Markov chain comprising three distinct closed and one open state [24, 25]. However, (1.26) allows for another instructive interpretation: According to a fundamental theorem of probability theory, the joint probability of disjunct events equals the product of the individual probabilities upon their stochastic independence. Since $p_{AK} = n^4$, we can assume the existence of four independently moving *gating charges* within the channel molecule. Correspondingly, for the sodium channel we expect three independent gating charges and one inhibiting subunit since $p_{AN} = m^3 h$. This is supported by patch clamp measurements where the channel's pores were blocked by the Fugu's fish tetradotoxin [19–21]. Although the blocked channel could not pass any ions, about three brief currents were observed. We can imagine these charges as key cylinders that have to be brought into the right positions to unlock a cylinder lock (thus opening the channel).

The emergence of an action potential results from different kinetics of the ion channels. If the cell membrane is slightly depolarized by the current I_m, the opening rate α_n for the sodium channels increases, thus entailing a further depolarization of the membrane. The positive feed-back loop started in this way leads to a torrent of inflowing sodium until the peak of the action potential is reached. Then, the membrane potential is positive in comparison to the extracellular space, causing voltage-gated potassium channels to open. Due to its negative equilibrium potential, potassium leaves the cell thereby hyperpolarizing the interior. Contrastingly, the hyperpolarization of the membrane reduces the open probability of the sodium channels, which become increasingly closed. Another positive feed-back loop enhances the hyperpolarization thereby overshooting the resting potential. While the potassium channels change very slowly back to their closed state, the sodium channels become additionally inactivated by a stopper subunit of the channel molecule whose kinetics is governed by the h term. This inhibition process is responsible for the refractory time prohibiting the occurrence of another action potential within this period.

1.3.2 Presynaptic Potentials

A spike train, generated in the way described by the Hodgkin-Huxley equations, travels along the axon and, after several branches, reaches the presynaptic terminals. Here, the composition of the membrane changes again. Voltage-gated calcium channels are present in addition to the voltage-gated potassium and sodium channels, and can be described by another branch in Fig. 1.14. The class of voltage-gated calcium channels is quite extensive and they operate generally far from the linear (ohmic) domain of the Goldman-Hodgkin-Katz equation [13, 15, 24, 25]. However, according to Johnston & Wu [24], an ohmic treatment of presynaptic Ca^{2+} channels is feasible such that their current is given by

$$I_{AC} = l^5 \, \bar{g}_{AC} \left(U - E_{Ca^{2+}} \right), \tag{1.30}$$

where l obeys another master equation

$$\frac{dl}{dt} = \alpha_l \left(1 - l \right) - \beta_l \, l. \tag{1.31}$$

In the absence of an injected current ($I_m = 0$), the presynaptic potential $U(t)$ is then governed by the differential equation

$$C_m \frac{dU}{dt} + I_{AK} + I_{AN} + I_{AC} + I_l = 0. \tag{1.32}$$

Neglecting calcium leakage, the current (1.30) leads to an enhancement of the intracellular concentration $[Ca^{2+}]_{int}$ that is described by a continuity equation [12]

$$\frac{d[Ca^{2+}]_{int}}{dt} = -\frac{I_{AC}}{q N_A V}. \tag{1.33}$$

Here, $q = 2e$ is the charge of the calcium ion (e denoting the elementary charge). Avogadro's constant N_A scales the ion concentration to moles contained in the volume V. The accumulation of calcium in the cell plasma gives rise to a cascade of metabolic reactions. Calcium does not only serve as an electric signal; it also acts as an important messenger and chemical reagent, enabling or disenabling the functioning of enzymes.

The movement of neurotransmitter into the synaptic cleft comprises two sub-processes taking place in the presynaptic terminal: Firstly, transmitter must be allocated, and secondly, it must be released. The allocation of transmitter depends on the intracellular calcium concentration (1.33), while it is stochastically released by increased calcium currents (1.30) as a consequence of an arriving action potential with a probability p.

In the resting state, transmitter vesicles are anchored at the cytoskeleton by proteins called *synapsin*, which act like a wheel clamp. The probability to loosen these joints increases with the concentration $[\text{Ca}^{2+}]_{\text{int}}$. Liberated vesicles wander to one of a finite number Z of *active zones* where vesicles can fuse with the terminal membrane thereby releasing their content into the synaptic cleft by the process of *exocytosis* [19–21]. Allocation means that $Y \le Z$ active zones are provided with vesicles, where

$$Y = \kappa([\text{Ca}^{2+}]_{\text{int}}) \, Z \tag{1.34}$$

is the average number of occupied active zones, and $\kappa([\text{Ca}^{2+}]_{\text{int}})$ is a monotonic function of the calcium concentration that must be determined from the reaction kinetics between calcium and synapsin mediated by kinases. The release of transmitter is then described by a *Bernoulli process* started by an arriving action potential. The probability that k of the Y occupied active zones release a vesicle is given by the *binomial distribution*

$$p(k, Y) = \binom{Y}{k} p^k (1 - p)^{Y-k} \, . \tag{1.35}$$

For the sake of mathematical convenience, we shall replace the binomial distribution by a normal distribution

$$\rho(k, Y) = \frac{1}{\sqrt{2\pi y(1 - p)}} \exp\left[-\frac{(k - y)^2}{2y(1 - p)}\right], \tag{1.36}$$

where $y = Yp$ is the average number of transmitter releasing active zones. Assuming that a vesicle contains on average $n_T = 5000$ transmitter molecules [19, 20], we can estimate the mean number of transmitter molecules that are released by an action potential as

$$T = n_T Y p = n_T Z p \, \kappa([\text{Ca}^{2+}]_{\text{int}}) \, . \tag{1.37}$$

Correspondingly, the expected number of transmitter molecules released by k vesicles is given by

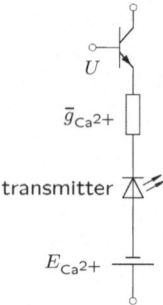

Fig. 1.16. Equivalent circuit for the calcium-controlled transmitter release (indicated by the arrows of the LED)

$$T(k) = \frac{n_T Y}{\sqrt{2\pi y(1-p)}} \exp\left[-\frac{(k-y)^2}{2y(1-p)}\right].$$ (1.38)

Finally, we need an equivalent circuit symbol for the transmitter release. Electronics suggests the use of the LED symbol (light-emitting diode). Connected all together, the calcium controlled transmitter release might be represented by the branch shown in Fig. 1.16.

1.3.3 Postsynaptic Potentials

After being poured out into the synaptic cleft of a chemical synapse, transmitter molecules diffuse to the opposite postsynaptic membrane, unless they have not been decomposed by enzymic reactions. There, they dock onto *receptor molecules*, which fall into two classes: *ionotropic receptors* are actually transmitter-gated ion channels, whereas *metabotropic receptors* are proteins that, once activated by transmitter molecules, start metabolic processes from second messenger release up to gene expression. At particular pathways, they control the opening of other ion channels gated by intracellular reaction products. The directly transmitter-gated channels are fast and effective, while the intracellularly gated channels react very slowly [19–21, 33]. In this section, I shall treat two distinct examples from each receptor class.

Excitatory Postsynaptic Potentials

One important transmitter-gated ion channel is (among others, such as the AMPA, GABA$_A$, and NMDA receptors) the nACh receptor that has nicotine as an antagonist. It becomes open if three or four molecules of the neurotransmitter acetylcholine (ACh) dock at its surface rising into the synaptic cleft. These molecules cause shifts of the electric polarization within the molecule which opens the gate in the pore. This process can be modeled by a Markov chain similarly to the exposition in Sect. 1.3.1. However, another treatment is also feasible, using *chemical reaction networks* [30, 33].

The open nACh channel is conductive for sodium as well as for potassium ions, such that its reversal (resting) potential is provided by the Goldman equation (1.10). Yet the sodium conductance is slightly larger than that for potassium yielding a net current of inflowing sodium ions. Since this current is depolarizing, the nACh channels constitute *excitatory synapses*. Therefore, they generate *excitatory postsynaptic potentials* (EPSP). On the other hand, hyperpolarizing channels, such as the $GABA_A$ channel, constitute *inhibitory synapses* generating *inhibitory postsynaptic potentials* (IPSP).

Let us once more consider a membrane patch of area A containing N_{nACh} receptors. Again, let O_{nACh} be the number of momentarily opened and C_{nACh} the number of closed channels. According to (1.12), the conductance of all open channels connected in parallel is then $g_{nACh} = O_{nACh} \gamma_{nACh}$. Opening of the channels can now be described by the *chemical reaction equation*

$$C + 3T \rightleftarrows O \,, \tag{1.39}$$

where C denotes the closed and O the opened molecules. T stands for the transmitter ACh. Because in each single reaction, three molecules T react with one molecule C to produce one molecule O, the corresponding *kinetic equation* [30, 31, 33] comprises a cubic nonlinearity,

$$\frac{dO}{dt} = \nu_1 CT^3 - \nu_2 O \,, \tag{1.40}$$

where ν_1 denotes the production and ν_2 the decomposition rate of open channels in (1.39). These reaction rates depend on the temperature of the heat bath and probably on metabolic circumstances such as phosphorylation. This equation has to be supplemented by a *reaction-diffusion equation* for the neurotransmitter reservoir in the synaptic cleft

$$\frac{dT}{dt} = \nu_2 O - \nu_3 TE - \sigma T \,, \tag{1.41}$$

where $\nu_2 O$ is the intake of transmitter due to decaying receptor-transmitter complexes, which is the same as the loss of open channels in (1.40), $\nu_3 TE$ is the decline due to reactions between the transmitter with enzyme E, and σT denotes the diffusion out of the synaptic cleft. Its initial condition $T(t = 0)$ is supplied by (1.38). Taken together, the equations (1.40, 1.41) describe the *reaction-diffusion kinetics* of the ligand-gated ion channel nACh.

Expressing the electric conductivity (1.12) through the maximal conductivity $\bar{g}_{nACh}$,

$$g_k = \frac{O_{nACh}}{N_{nACh}} \bar{g}_{nACh} \,, \tag{1.42}$$

suggests a new equivalent circuit symbol for ligand-gated channels. The conductance is controlled by the number of transmitter molecules, i.e. the number of particular particles in the environment. This corresponds to the phototransistor in electronic engineering which is controlled by the number of photons

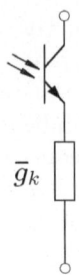

Fig. 1.17. Equivalent circuit for a population of ligand-gated ion channels of kind k

collected by its base. Hence, I would like to suggest the circuit shown in Fig. 1.17 as an equivalent to the nACh receptor.

In order to compute the postsynaptic potential, the circuit in Fig. 1.17 has to be connected in parallel with the leakage conductance and the membrane capacitance as in Fig. 1.18.

The EPSP for the nACh receptor then obeys the equations

$$C_m \frac{dU}{dt} + \frac{O_{\mathrm{nACh}}}{N_{\mathrm{nACh}}} \bar{g}_{\mathrm{nACh}}(U - E_{\mathrm{nACh}}) + g_l(U - E_l) = 0 \qquad (1.43)$$

together with (1.40, 1.41), and initial condition (1.38).

However, instead of solving these differential equations, most postsynaptic potentials can be easily described by alpha functions as synaptic gain functions [12–16, 18]

$$U^{\mathrm{PSP}}(t) = E_{\mathrm{PSP}} \, \alpha^2 t \, \mathrm{e}^{-\alpha t} \, \Theta(t) \,, \qquad (1.44)$$

where α is the characteristic time constant of the postsynaptic potential (PSP) and

$$\Theta(t) = \begin{cases} 0 & \text{for} \quad t \le 0 \\ 1 & \text{for} \quad t > 0 \end{cases} \qquad (1.45)$$

is Heaviside's jump function.

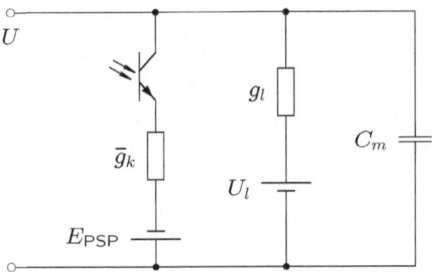

Fig. 1.18. Equivalent circuit for the postsynaptic potential generated by ligand-gated channels of kind k with reversal potential PSP

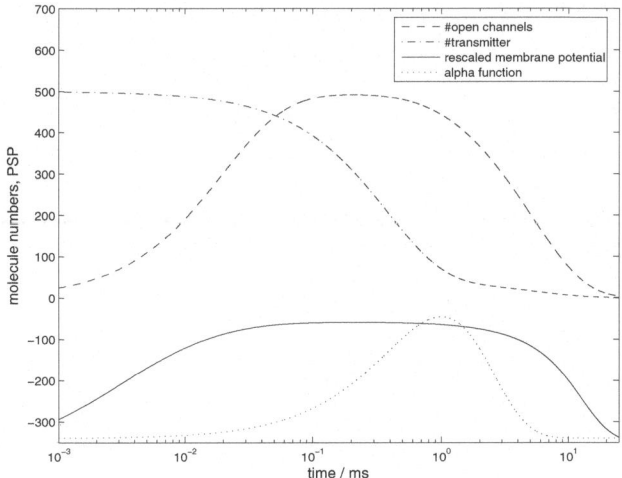

Fig. 1.19. Numeric solutions of the kinetic equations (1.40, 1.41, 1.43) for the nACh receptor. *Dashed-dotted*: the number of ACh molecules (max = 500); *dashed*: the number of open nACh channels (max = 500); *solid*: the EPSP $U(t)$; *dotted*: an alpha function (1.44). The time axis is logarithmically scaled; the functions are rescaled and (if necessary) shifted for better visibility

Figure 1.19 displays the numerical solution of equations (1.40, 1.41, 1.43) for arbitrarily chosen parameters together with a "fitted" alpha function for comparison. Obviously, the correspondence is not that large.

Inhibitory Postsynaptic Potentials

Synapses are excitatory if they open sodium or calcium channels with more positive reversal potentials compared to the resting state. Their neurotransmitters are generally acetylcholine (ACh) or the amino acid glutamate. Contrastingly, most inhibitory synapses employ the amino acids glycine or GABA (gamma-amino-butyric-acid) to open potassium or chloride channels with more negative reversal potentials. While the $GABA_A$ receptor is transmitter-gated such as the nACh receptor discussed in the previous section, the $GABA_B$- and mACh receptors (having the toadstool toxin *muscarine* as an antagonist) activate intracellular *G proteins* which subsequently open G protein-gated potassium channels [19–21]. The activation of G protein-gated potassium channels comprises the following chemical reactions [12]:

$$R_0 + T \rightleftarrows R^* \rightleftarrows D \qquad (1.46)$$
$$R^* + G_0 \rightleftarrows RG^* \rightarrow R^* + G^*$$
$$G^* \rightarrow G_0$$
$$C + n\,G^* \rightleftarrows O,$$

where R_0 is the metabotropic GABA$_B$ receptor in its resting state, T the transmitter GABA, R^* the transmitter-activated receptor on the one hand, and D the same transmitter-receptor complex in its inactivated state on the other hand; furthermore, G_0 is the G protein in its resting state, $(RG)^*$ a short-lived activated receptor-G protein complex and G^* the activated G protein; finally, C is the G protein-gated potassium channel in its closed state and O in the open state. The channel possesses n docking sites for G protein molecules. Translating (1.46) into kinetic equations and adding (1.41) yields

$$\frac{dR_0}{dt} = -\nu_1\,R_0 T + \nu_2\,R^* \tag{1.47}$$

$$\frac{dT}{dt} = -\nu_1\,R_0 T + \nu_2\,R^* - \nu_{11}\,TE - \sigma T \tag{1.48}$$

$$\frac{dR^*}{dt} = \nu_1\,R_0 T - \nu_2\,R^* + \nu_3\,D - \nu_4 R^* \tag{1.49}$$

$$-\nu_5\,R^* G_0 + \nu_6\,(RG)^* + \nu_8\,(RG)^*$$

$$\frac{dD}{dt} = -\nu_3\,D + \nu_4 R^* \tag{1.50}$$

$$\frac{dG_0}{dt} = -\nu_5\,R^* G_0 + \nu_6\,(RG)^* + \nu_7\,G^* \tag{1.51}$$

$$\frac{d(RG)^*}{dt} = \nu_5\,R^* G_0 - \nu_6\,(RG)^* - \nu_8\,(RG)^* \tag{1.52}$$

$$\frac{dG^*}{dt} = -\nu_7\,G^* + \nu_8\,(RG)^* + \nu_{10}\,O \tag{1.53}$$

$$\frac{dO}{dt} = \nu_9\,C G^{*^n} - \nu_{10}\,O \tag{1.54}$$

for the metabolic dynamics. Equations (1.47–1.54) together with (1.43) describe the inhibitory GABA-ergic potential

$$C_m \frac{dU}{dt} + \frac{O}{O + C}\,\bar{g}_{GP}(U - E_{K^+}) + g_l(U - E_l) = 0\,, \tag{1.55}$$

where $\bar{g}_{GP}$ denotes the maximal conductance and E_{K^+} the reversal potential of the G protein-gated potassium channels.

Temporal Integration

Each action potential arriving at the presynaptic terminal causes the (binomially distributed) release of one or more vesicles that pour their total amount of transmitter molecules into the synaptic cleft. Here, transmitter molecules react either with ionotropic or with metabotropic receptors which open — more or less directly — ion channels such that a postsynaptic current

$$I^{PSC}(t) = \frac{O_k(t)}{N_k}\,\bar{g}_k(U - E_k)\,, \tag{1.56}$$

either excitatory or inhibitory, flows through the "phototransistor" branch of Fig. 1.17. This current gives rise to the EPSP or IPSP according to (1.43, 1.55). Since these potentials were determined for the transmitter released by one action potential, we can consider them as *impulse response functions*. Inserting I^{PSC} into (1.43, 1.55) and shifting the resting potential to $E_l = 0$ yields the inhomogeneous linear differential equation

$$\tau \frac{dU}{dt} + U = -\frac{I^{\mathrm{PSC}}}{C_m}, \tag{1.57}$$

with $\tau = C_m/g_l$ as the characteristic *time constant* of the membrane patch. If we describe the current I^{PSC} by a pulse of height I_0 at time t_0,

$$I^{\mathrm{PSC}}(t) = I_0 \delta(t - t_0), \tag{1.58}$$

the solution of (1.57) is given by a Green's function $U^{\mathrm{PSP}} = G(t, t')$ [13,18,22]. By virtue of this Green's function, we can easily compute the postsynaptic potential evoked by an arbitrary spike train

$$I^{\mathrm{PSC}}(t) = I_0 \sum_k \delta(t - t_k) \tag{1.59}$$

as the *convolution product*

$$U^{\mathrm{PSP}}(t) = \int G(t, t') \, I^{\mathrm{PSC}}(t) \, dt' = G * I^{\mathrm{PSC}}. \tag{1.60}$$

Inserting (1.59) into (1.60) gives

$$U^{\mathrm{PSP}}(t) = I_0 \sum_k G(t, t_k). \tag{1.61}$$

If the action potentials are generated by the presynaptic neuron in the regular spiking mode with frequency f (see Sect. 1.3.1), the event times are given by

$$t_k = \frac{k}{f}. \tag{1.62}$$

Eventually, (1.61, 1.62) lead to

$$U^{\mathrm{PSP}}(t) = I_0 \sum_k G\left(t, \frac{k}{f}\right). \tag{1.63}$$

Figure 1.20 displays two postsynaptic potentials obtained by the convolution of the Green's function

$$G(t, t') = \Theta(t - t') \cdot \frac{I_0}{C_m} \exp\left(-\frac{t - t'}{\tau}\right)$$

with regular spike trains.

By means of the convolution mechanism, an analogue continuously varying membrane potential is regained from a frequency-modulated spike train. This process is called *temporal integration*.

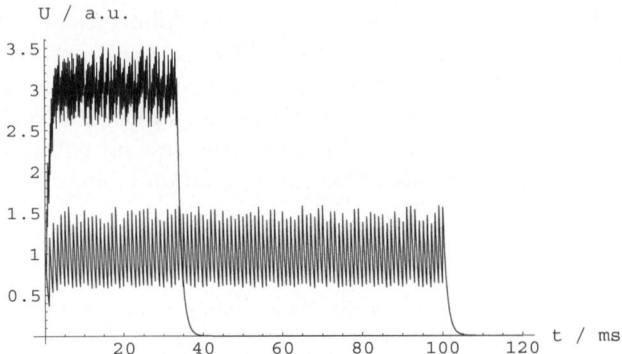

Fig. 1.20. Temporal integration of postsynaptic pulse responses for a lower (*lower curve*) and a higher regular spiking frequency (*upper curve*)

1.4 Neuron Models

In the preceding sections, we completed our construction kit for neurophysical engineering. In the remaining ones, we are going to apply these building blocks. There are three main threads of neural modeling. The first one builds point models, where all kinds of ion channels are connected in parallel. Secondly, compartment models additionally take into account the cable properties of cell membranes that are responsible for *spatial integration* processes. However, all these models are computationally very expensive. Therefore several simplifications and abstractions have been proposed to cope with these problems especially for the modeling of *neural networks*. Nevertheless, for the simulation of relatively small networks of point or compartment models, powerful software tools such as GENESIS [34], or NEURON [35] have been developed.

1.4.1 Point Models

In the point model account, all membrane patches of a nerve cell are assumed to be equipotential, disregarding its spacial extension [12, 15, 36, 37]. In our equivalent circuits, this assumption is reflected by connecting all different ion channels in parallel. Figure 1.21 shows such a point model.

In order to simulate the neuron in Fig. 1.21, all discussed differential equations are to be solved simultaneously. Neural networks then consist of many circuits of this form that are "optically" coupled, i.e. by the transmitter releasing and receiving devices at both ends of one circuit. The efficacy of the coupling between two neurons i and j is expressed by the *synaptic weight* w_{ij}. Physiologically, these weights depend on the maximal synaptic conductances $\bar{g}_{PSP}$.

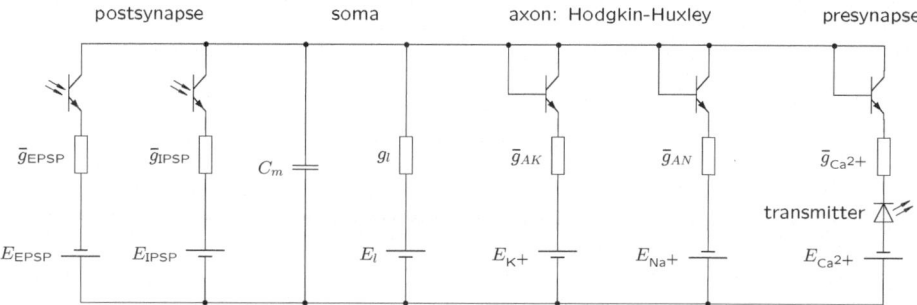

Fig. 1.21. Equivalent circuit for a neural point model

1.4.2 Compartment Models

Point models have one serious disadvantage: They completely disregard the spatial extension of the cell and the fact that different membrane patches, e.g. at the soma, the axon, the terminals, or the dendrites, exhibit different potentials (cf. Chap. 2). The gradients between these sites then lead to ion currents through the cell plasma thus contributing to the mechanisms of *spatial integration*. Moreover, currents moving back through the extracellular space give rise to the *somato-dendritic field potentials* (DFP). These fields sum to the local field potentials (LFP) at a mesoscopic and to electrocorticogram (ECoG) and electroencephalogram (EEG) at macroscopic scales [38–42] (cf. Chaps. 8 and 7 in this volume).

Correctly, the spatiotemporal dynamics of neuronal membranes must be treated by the *cable equation* [12, 13, 15, 18, 22, 24, 25]. This is a second-order partial differential equation for the membrane potential $U(r, t)$. For the sake of numerical simulations, its discretized form leads to compartment models where individual membrane patches are described by the equivalent circuits discussed in the previous sections [12, 13, 15, 16, 18, 40, 41]. As an example, I shall create an equivalent circuit for a three-compartment model for the cortical pyramidal cells that is able to describe somato-dendritic field potentials.

Pyramidal cells are characterized by their axial symmetry. They consist of an *apical dendritic tree* comprising only excitatory synapses and a *basal dendritic tree* where mainly inhibitory synapses are situated. Both types of synapses are significantly separated in space thus forming a dipole of current sources (the inhibitory synapses) and sinks (the excitatory synapses) [38, 39, 41, 42]. The extracellular current flows from the sources to the sinks through a non-negligible resistance R_{out} which entails the somato-dendritic field. Therefore, we divide the pyramidal cell into three compartments: the first represents the apical dendrites, the second the basal dendrites, and the third takes firing into account. Figure 1.22 depicts its equivalent circuit. The internal resistance R_{int} accounts for the *length constant* of the neuron and contributes also to the synaptic weights (see Chap. 7).

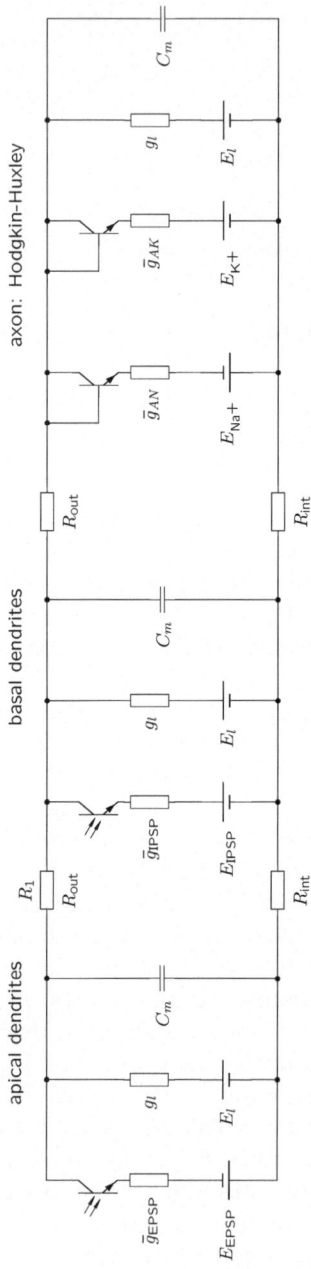

Fig. 1.22. Equivalent circuit for a three-compartment model of a pyramidal cell

In this model, the extracellular current I_{out} flowing from the inhibitory to the excitatory compartment through R_1 entails the dendritic field potential

$$U^{\text{DFP}} = \frac{I_{\text{out}}}{R_1}. \tag{1.64}$$

Note that I_{out} is large for a large difference between the EPSP and the IPSP, i.e. when both types of synapses are synchronously activated. In this case, however, the remaining current $I - I_{\text{out}}$ flowing through the axonal compartment can be too small to elicit action potentials. Therefore, DFP and spiking are inversely related with each other [41].

1.4.3 Reduced Models

The computational complexity of conductance models prevents numerical simulations of large neural networks. Therefore, simplifications and approximations have been devised and employed by several authors [10–18, 43–50].

In the following, we shall consider networks composed from n model neurons. Their membrane potentials U_i ($1 \leq i \leq n$) span the *observable state space*, such that $\boldsymbol{U} \in \mathbb{R}^n$; note that the proper phase space of the neural network might be of higher dimension. The observables U_i depend on the total postsynaptic current

$$I_i^{\text{PSC}} = -\sum_{j=1}^{n} w_{ij} I_j - I_i^{\text{ext}}, \tag{1.65}$$

where w_{ij} is the *synaptic weight* of the connection from unit j to unit i, dependent on the synaptic gain $\bar{g}_{ij}$ that evolves during learning, thus reflecting *synaptic plasticity* (see Chap. 2), and the intracellular resistances (see Chap. 7). The capacitance in (1.57) has been deliberately neglected and I_i^{ext} denotes the externally controlled input to neuron i.

The McCulloch-Pitts Model

The coarsest simplification by McCulloch and Pitts [16, 18, 51, 52] replaces the involved Hodgkin-Huxley system (1.25, 1.27–1.29) by a threshold device with only two states: $X_i \in \{0, 1\}$ where 0 denotes the inactivated, silent, and 1 denotes the activated, firing state. The dynamics of a network of n McCulloch-Pitts units is governed by the equations

$$X_i(t + 1) = \Theta(I_i^{\text{PSC}} - \theta_i), \tag{1.66}$$

where t is the discretized time, θ_i the activation threshold for unit i, and $I_i = X_i$ have been identified.

Integrate-and-fire Models

The next step to make threshold units biologically more plausible is by taking the passive membrane properties as described by (1.57) into account. This leads to the class of (leaky) integrate-and-fire models [12, 13, 15, 16, 18, 46, 50]:

$$\tau_i \frac{dU_i}{dt} + U_i = I_i^{\mathrm{PSC}} \tag{1.67}$$
$$X_i(t_k) = \Theta(U_i(t_k) - \theta_i)$$
$$U_i(t_k) \leftarrow E.$$

Here, $U_i(t)$ describes the membrane potential of unit i, X_i and θ_i model the action potentials and the firing thresholds as in (1.66), and t_k are the firing times where the membrane potential is reset to its resting value E (indicated by the arrow).

Rate Models

In Sects. 1.3.1 and 1.3.3, we saw that the principles of frequency modulation are exploited for neural en- and decoding — at least for regular spiking dynamics. Therefore, it seems to be appropriate to replace the exact time-course of a spike train by its frequency, *firing rate*, or firing probability [53, 54]. The latter approach leads to the problem of determining the value

$$R_i(t) = \mathrm{Prob}(U_i(t) \geq \theta_i) = \int d^{n-1}u \int_{\theta_i}^{\infty} \rho(\boldsymbol{u}, t)\, du_i, \tag{1.68}$$

where we have to regard the membrane potentials $\boldsymbol{U}(t)$ as a multivariate stochastic variable in the observable space with expectation values $\bar{U}_i(t)$ and probability density function $\rho(\boldsymbol{u}, t)$. The first integral in (1.68) provides the marginal distribution in the ith observable subspace. The stochasticity assumption is justified by our treatment of the presynaptic potential in Sect. 1.3.2. Because every action potential starts a Bernoulli process which describes how many vesicles are to be released, this element of stochasticity propagates along the synapse. As we have characterized the distribution of the number of released vesicles by (1.36), the postsynaptic currents are normally distributed about their voltage-dependent means $\bar{I}_i(\bar{\boldsymbol{U}})$,

$$I_i^{\mathrm{PSC}} = -\sum_j w_{ij}\left(\bar{I}_j(\bar{\boldsymbol{U}}(t)) + \eta_j(t)\right), \tag{1.69}$$

where $\eta_j(t)$ are independent normally distributed stochastic processes with

$$\langle \eta_j(t) \rangle = 0 \tag{1.70}$$
$$\langle \eta_j(t), \eta_k(t') \rangle = Q_{jk}\, \delta(t - t').$$

Therefore, (1.57) has to be read as a stochastic differential (Langevin) [30,31] equation

$$\frac{dU_i}{dt} = K_i(\boldsymbol{U}) - \sum_j \alpha_i w_{ij} \eta_j(t)) \,, \tag{1.71}$$

where

$$K_i(\boldsymbol{U}) = -\alpha_i U_i - \sum_j \alpha_i w_{ij} \bar{I}_j(\bar{\boldsymbol{U}}) \tag{1.72}$$

are the deterministic drifting forces and $\sum_j \alpha_i w_{ij} \eta_j$ are stochastic fluctuations, obeying

$$\left\langle \sum_j \alpha_i w_{ij} \eta_j(t), \sum_l \alpha_k w_{kl} \eta_l(t') \right\rangle = R_{ik}\delta(t - t') \,, \tag{1.73}$$

with

$$R_{ik} = \sum_{jl} \alpha_i w_{ij} \, \alpha_k w_{kl} \, Q_{jl} \,; \tag{1.74}$$

here, we substituted $\alpha_i = \tau_i^{-1}$.

The probability distribution density $\rho(\boldsymbol{u}, t)$ is then obtained by solving the Fokker-Planck equation [30,31] associated to (1.71),

$$\frac{\partial \rho}{\partial t} = \sum_i \frac{\partial}{\partial u_i}[K_i(\boldsymbol{u})\rho] + \frac{1}{2} \sum_{ik} R_{ik} \frac{\partial^2 \rho}{\partial u_i \partial u_k} \,. \tag{1.75}$$

In order to solve (1.75), we assume that the currents $\bar{I}_j$ do not explicitly depend on the mean membrane potential, and that they change rather slowly in comparison to the density ρ (the "adiabatic ansatz"). Then, (1.75) is linear and hence solved by the Gaussians

$$\rho(u, t) = \frac{1}{\sqrt{2\pi\sigma_U^2(t)}} \exp\left[-\frac{(u - \bar{U}(t))^2}{2\sigma_U^2(t)} \right] \tag{1.76}$$

as its stationary marginal distributions, where $\bar{U}(t)$ and $\sigma_U^2(t)$ have to be determined from $\bar{I}(t)$ and R_{ik}. Integrating (1.68) with respect to (1.76) yields the spike rate

$$R_i = f(\bar{U}_i) = \frac{1}{2} \operatorname{erfc}\left(\frac{\theta_i - \bar{U}_i}{\sqrt{2}\sigma_U} \right) \,, \tag{1.77}$$

with "erfc" denoting the complementary error function. In such a way, the stochastic threshold dynamics are translated into the typical *sigmoidal activation functions* $f(x)$ employed in computational neuroscience [7–9,11–13,15, 16,18].

Gathering (1.67, 1.77), a leaky integrator model [46] is obtained as

$$\tau_i \frac{dU_i}{dt} + U_i = \sum_j w_{ij} \, f(U_j) \,. \tag{1.78}$$

An alternative derivation of (1.78) can be found in Chap. 7 by disregarding the postsynaptic impulse response functions $G(t, t')$. If these are taken into account, instead an integro-differential equation

$$\tau_i \frac{\mathrm{d}U_i}{\mathrm{d}t} + U_i = \sum_j w_{ij} \int_{-\infty}^{t} G(t - t') f(U_j(t')) \, \mathrm{d}t' \qquad (1.79)$$

applies.

The Rinzel-Wilson Model

The models to be discussed next are approximations for the full Hodgkin-Huxley equations (1.25, 1.27–1.29). Following Rinzel, Wilson and Trappenberg [14,16], the Hodgkin-Huxley equations exhibit two separated time-scales: at the fast scale, the opening of the sodium channels characterized by $m(t)$ happens nearly instantaneously such that $m(t)$ can be replaced by its stationary value m_∞. On the other hand, the opening rate for the potassium channels n and the inactivation rate h for the sodium channels exhibit an almost linear relationship $h = 1 - n$. The corresponding substitutions then lead to a two-dimensional system for each neuron i.

$$I_i = C_m \frac{\mathrm{d}U_i}{\mathrm{d}t} + n_i^4 \, \bar{g}_{AK}(U_i - E_{K^+}) + \qquad (1.80)$$
$$+ m_\infty^3 (1 - n_i) \, \bar{g}_{AN}(U_i - E_{Na^+}) + g_l(U_i - E_l)$$
$$\frac{\mathrm{d}n_i}{\mathrm{d}t} = \alpha_n (1 - n_i) - \beta_n \, n_i$$

which was applied for the plot in Fig. 1.15.

The FitzHugh-Nagumo Model

The same observation as in above led FitzHugh and Nagumo to their approximation of the Hodgkin-Huxley equations [13, 14, 18, 43, 50]. Here, a general linear relation $h = a - bn$ is used in combination with a coordinate transformation and rescaling to arrive at the *Bonhoeffer-Van-der-Pol-*, or likewise, *FitzHugh-Nagumo equations*,

$$\frac{\mathrm{d}U_i}{\mathrm{d}t} = U_i - \frac{1}{3}U_i^3 - W_i + I_i \qquad (1.81)$$
$$\frac{\mathrm{d}W_i}{\mathrm{d}t} = \phi(U_i + a - bW_i),$$

with parameters ϕ, a, b.

The Morris-Lecar Model

Originally, the *Morris-Lecar model* was devised to describe the spiking dynamics of potassium- and calcium-controlled muscle fibers [12–14,18,50]. After introducing dimensionless variables and rescaled parameters, they read

$$\frac{dU_i}{dt} = -m_\infty \bar{g}_{AC}(U_i - 1) - W_i \, \bar{g}_{AK}(U_i - E_{K^+}) - g_l(U_i - E_l) + I_i \quad (1.82)$$

$$\frac{dW_i}{dt} = \alpha_W (1 - W_i) - \beta_W \, W_i \,.$$

The Morris-Lecar model has been extensively employed during the Summer School, see Chaps. 9, 11, 12, 14.

The Hindmarsh-Rose Model

The FitzHugh-Nagumo and Morris-Lecar models have the disadvantage that they do not have a bursting regime in their parameter space [50]. In order to overcome this obstacle, a third dimension for the phase space is necessary. The *Hindmarsh-Rose equations*, which exhibit this third dimension, are [14,47]

$$\frac{dU_i}{dt} = V_i - U_i^3 + 3U_i^2 + I_i - W_i \quad (1.83)$$

$$\frac{dV_i}{dt} = 1 - 5U_i^2 - V_i$$

$$\frac{dW_i}{dt} = r[s(U_i - U_0) - W_i] \,,$$

with parameters r, s, U_0.

For applications of the Hindmarsh-Rose model in this book, see Chap. 6.

The Izhikevich Model

Making use of arguments from bifurcation theory, Izhikevich [49] approximated the Hodgkin-Huxley equations by the two-dimensional flow

$$\frac{dU_i}{dt} = 0.04U_i^2 + 5U_i + 140 - U_i + I_i \quad (1.84)$$

$$\frac{dV_i}{dt} = a(bV_i - U_i) \,, \quad (1.85)$$

disrupted by an auxiliary after-spike resetting

$$\text{if} \quad U_i \geq 30 \, \text{mV}, \quad \text{then} \begin{cases} U_i & \leftarrow & E \\ V_i & \leftarrow & V_i + c \end{cases}$$

with parameters a, b, c, E, where E denotes the resting potential.

A comprehensive comparison of different spiking neuron models with respect to their biological plausibility and computational complexity can be found in [50].

Also the Izhikevich model has been used during the Summer School. These results are presented in Chap. 13.

1.4.4 Neural Field Theories

For very large neural networks, a continuum approximation by spatial coarse-graining suggests itself [55–70]. Starting from the rate equation (1.79), the sum over the nodes connected with unit i has to be replaced by an integral transformation of a neural field quantity $U(x,t)$, where the continuous parameter x now indicates the position i in the network. Correspondingly, the synaptic weights w_{ij} turn into a kernel function $w(x,y)$. In addition, for large networks, the propagation velocity c of neural activation has to be taken into account. Therefore, (1.79) assumes the retarded form

$$\tau(x)\,\frac{\partial U(x,t)}{\partial t}+U(x,t)=\int\limits_{-\infty}^{t}\mathrm{d}t'\int\mathrm{d}x'w(x,x')G(t-t')f\left[U\left(x',t'-\frac{|x-x'|}{c}\right)\right],$$

(1.86)

which can be transformed into a wave equation under additional assumptions. For further details, consult Chap. 8 and the references above.

1.5 Mass Potentials

Neural field theories [55–70] as well as population models of cortical modules [39, 71–81] (see also Chap. 7) describe mass potentials such as LFP or EEG as spatial sums of the EPSPs and IPSPs of cortical pyramidal cells. In these accounts, the somato-dendritic field potential (DFP) of an infinitesimally small volume element of cortical tissue, or of a single neuron, respectively, is described [79] by

$$U^{\mathrm{DFP}} = U^{\mathrm{EPSP}} + U^{\mathrm{IPSP}}$$

(1.87)

when $U^{\mathrm{EPSP}} > 0, U^{\mathrm{IPSP}} < 0$.[4] Unfortunately, this description is at variance with the physiological origin of the DFP. Looking at the equivalent circuit of the three-compartment model in Fig. 1.22, one easily recognizes that simultaneously active excitatory and inhibitory synapses give rise to a large voltage drop along the resistor R_1 separating both kind of synapses in space. Therefore, a large extracellular current yields a large DFP according to (1.64). On the other hand, the sum in (1.87) becomes comparatively small since EPSP and IPSP almost compensate each other. Therefore, the geometry and anatomy of pyramidal cells and cortex have to be taken into account. To this end, I shall mainly review the presentation of Nunez and Srinivasan [42, 82] in the following.

---

[4] The signs in (1.87) are physiologically plausible (cf. (1.43), (1.55)), whereas Jansen et al. [75, 76], Wendling et al. [77, 78], and David and Friston [79] assume that EPSP and IPSP both have positive signs such that their estimate for the DFP reads $U^{\mathrm{DFP}} = U^{\mathrm{EPSP}} - U^{\mathrm{IPSP}}$ (cf. Chap. 5).

1.5.1 Dendritic Field Potentials

If the reader takes a look at Fig. 8.1 in Chap. 8, she or he sees three triangular knobs that are the cell bodies of three pyramidal cells. Starting from
their bases, axons proceed downwards like the roots of plants. In the other direction, they send strongly branched trees of dendrites towards the surface of
the cortex. Pyramidal cells exhibit roughly an axonal symmetry and they are
very densely packed in parallel, forming a fibrous tissue. Excitatory and inhibitory synapses are spatially separated along the dendritic tree: Excitatory
synapses are mainly situated at the apical (i.e. the top-most) dendrites, while
inhibitory synapses are arranged at the soma and the basal dendrites of the
cells. This arrangement is functionally significant as the inhibitory synapses
very effectively suppress the generation of action potentials by establishing
short-cuts [41].

From the viewpoint of the extracellular space, inhibitory synapses act as
current sources while excitatory synapses are current sinks. The extracellular
space itself can be regarded as an electrolyte with (volume-) conductance $\sigma(\boldsymbol{x})$,
where $\boldsymbol{x}$ indicates the dependence on the spatial position. From Maxwell's
equations for the electromagnetic field, a continuity equation

$$-\nabla \cdot (\sigma \nabla \phi) + \frac{\partial \rho}{\partial t} = 0 \qquad (1.88)$$

can be derived for the "wet-ware" [42, 82]. Here, $\phi(\boldsymbol{x})$ denotes the electric
potential and $\rho(\boldsymbol{x}, t)$ the charge density, and $\boldsymbol{j} = -\sigma \nabla \phi$ is the current density
according to Ohm's Law (1.5). Assuming that the conductivity $\sigma(\boldsymbol{x})$ is piece-
wise constant in the vicinity of a pyramidal cell, σ can be removed from the
scope of the first gradient, yielding

$$-\sigma \Delta \phi + \frac{\partial \rho}{\partial t} = 0 \,. \qquad (1.89)$$

Next, we have to describe the change of the current density. Setting

$$\frac{\partial \rho(\boldsymbol{x})}{\partial t} = \sum_i I_i \delta(\boldsymbol{x} - \boldsymbol{x}_i) \qquad (1.90)$$

describes the postsynaptic transmembrane currents in the desired way as point
sources and sinks located at $\boldsymbol{x}_i$. When we insert (1.90) into (1.89), we finally
arrive at a Poisson equation

$$\sigma \Delta \phi = \sum_i I_i \delta(\boldsymbol{x} - \boldsymbol{x}_i) \qquad (1.91)$$

in complete analogy to electrostatics.

Equation (1.91) can be easily solved by choosing appropriate boundary
conditions that exclude the interiors and the membranes of the cells from the

domain of integration.[5] Integrating (1.91) over the extracellular space gives

$$\phi(\boldsymbol{x}) = \frac{1}{4\pi\sigma} \sum_i \frac{I_i}{r_i}, \qquad (1.92)$$

where $\boldsymbol{x}$ denotes the observation site and $r_i = |\boldsymbol{x}-\boldsymbol{x}_i|$ abbreviates the distance between the point sources and sinks I_i and $\boldsymbol{x}$.

If the distance of the observation site $\boldsymbol{x}$ is large in comparison to the respective distances of the current sources and sinks from each other, the potential $\phi(\boldsymbol{x})$ can be approximated by the first few terms of a multipole expansion,

$$\phi(\boldsymbol{x}) = \frac{1}{4\pi\sigma} \left(\frac{1}{x} \sum_i I_i + \frac{1}{x^3} \sum_i I_i \, \boldsymbol{x}_i \cdot \boldsymbol{x} + \dots \right). \qquad (1.93)$$

Now, x denotes the distance of the observation point from the center of mass of the current cloud I_i. Due to the conservation of charge, the monopole term vanishes, whereas the higher order multipoles strongly decline with $x \to \infty$. Therefore, only the dipole term accounts for the DFP,

$$\phi^{\mathrm{DFP}}(\boldsymbol{x}) = \frac{1}{4\pi\sigma} \frac{1}{x^3} \sum_i I_i \, \boldsymbol{x}_i \cdot \boldsymbol{x} = \frac{1}{4\pi\sigma} \frac{\boldsymbol{p} \cdot \boldsymbol{x}}{x^3}, \qquad (1.94)$$

where the dipole moment of the currents

$$\boldsymbol{p} = I(\boldsymbol{x}_1 - \boldsymbol{x}_2) = I\boldsymbol{d} \qquad (1.95)$$

can be introduced when a current source $+I$ and a sink $-I$ are separated by the distance d. The unit vector $\boldsymbol{d}/d$ points from the source to the sink.

Equation (1.95) now suggests a solution for the problem with (1.87). The DFP is proportional to the dipole moment which depends monotonically on the absolute value I. Assuming that the excitatory and the inhibitory branch in the equivalent circuit in Fig. 1.22 are symmetric, the dipole moment, which is determined by I_{out}, depends on the difference between the (positive) EPSP and the (negative) IPSP,

$$U^{\mathrm{DFP}} = U^{\mathrm{EPSP}} - U^{\mathrm{IPSP}}, \qquad (1.96)$$

such that a simple change of the sign corrects (1.87).

---

[5] My presentation here deviates from that given by Nunez and Srinivasan [42] who assume quasi-stationary currents $[\nabla \cdot \boldsymbol{j} = 0]$. As a consequence of (1.88), the change of the charge density would also vanish leading to a trivial solution. In order to circumvent this obstacle, Nunez and Srinivasan [42, p. 166] decompose the current into an "ohmic" part $-\sigma \nabla \phi$ and peculiar "impressed currents" $\boldsymbol{J}_S$ corresponding to EPSC and IPSC that cross the cell membranes. However, they concede that "the introduction of this pseudo-current may, at first, appear artificial and mysterious" aiming at the representation of the boundary conditions. This distinction is actually unnecessary when boundary conditions are appropriately chosen [82]. Nevertheless, Nunez' and Srinivasan's argument became very popular in the literature, e.g. in [4, 67].

1.5.2 Local Field potentials

Since pyramidal cells are aligned in parallel, they form a dipole layer of thickness d when they are synchronized within a cortical module. Subsequently, we will identify such modules with the anatomical columns (cf. Chap. 8) in order to compute the collective DFP, i.e. the local field potential (LFP) generated by a mass of approximately 10,000 pyramidal cells.

The current differential dI is then proportional to the infinitesimal area in cylindrical coordinates

$$dI = j dA, \qquad (1.97)$$

where the current density j is assumed to be a constant scalar within one column. Hence, the differential of the potential $d\phi$ at a distance z perpendicular to a cortical column of radius R that is contributed by the current dI is given by

$$d\phi(\boldsymbol{x}) = \frac{1}{4\pi\sigma} \frac{j\,\boldsymbol{d}\cdot(\boldsymbol{x}-\boldsymbol{x}')}{|\boldsymbol{x}-\boldsymbol{x}'|^3} dA, \qquad (1.98)$$

where $\boldsymbol{x}'$ varies across the area of the module. Making use of the geometry depicted in Fig. 1.23 yields

$$d\phi(z) = \frac{1}{4\pi\sigma} \frac{jd\sqrt{r^2+z^2}\cos\vartheta}{(r^2+z^2)^{3/2}} r\,dr\,d\varphi$$

$$= \frac{jd}{4\pi\sigma} \frac{z\sqrt{r^2+z^2}}{\sqrt{r^2+z^2}(r^2+z^2)^{3/2}} r\,dr\,d\varphi$$

$$= \frac{jd}{4\pi\sigma} \frac{rz}{(r^2+z^2)^{3/2}} dr d\varphi$$

$$\phi(z) = \frac{jd}{4\pi\sigma} \int_0^{2\pi} d\varphi \int_0^R dr\, \frac{rz}{(r^2+z^2)^{3/2}}.$$

Performing the integration then gives the LFP perpendicular to the dipole layer

$$\phi^{\mathrm{LFP}}(z) = \frac{jd}{2\sigma}\left(1 - \frac{z}{\sqrt{R^2+z^2}}\right). \qquad (1.99)$$

1.5.3 Electroencephalograms

Equation (1.99) describes the summed potential resulting from the synchronized synaptic activity of all pyramidal neurons in a column in a distance z above the cortical gray matter. By integrating over a larger domain of cortical tissue, e.g. over a macrocolumn, one obtains an estimator of the electrocorticogram (ECoG) [83]. In order to compute the electroencephalogram (EEG), one has to take the different conductances of skull and scalp into account. Nunez and Srinivasan [42] discuss different scenarios with different

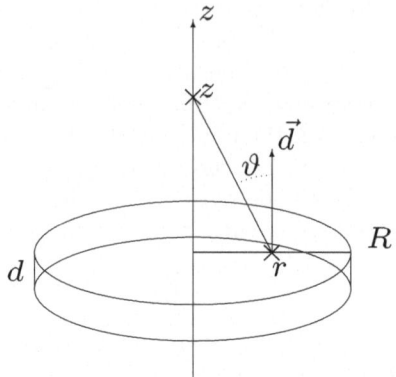

Fig. 1.23. Geometry of a cortical column

geometries. In the simplest case, only one interface between a conductor (G_1) with conductance σ_1 and an isolator G_2 (conductance $\sigma_2 = 0$) is considered. According to the respective interpretation, either the skull, or the air above the subject's head is regarded to be the isolator.[6]

In one case, one has to consider the potential generated by a point source (or sink) I at a distance $-h$ from the interface in the semi-space G_1 by attaching a mirror source (or sink) I' at a distance h from the interface in order to solve Dirichlet's boundary problem if $\boldsymbol{x} \in G_2, z > 0$ [42]. In the other case, one has to replace the source (or sink) I in the semi-space G_1 by another source (or sink) $I + I''$ if $\boldsymbol{x} \in G_1, z < 0$. The geometrical situation is shown in Fig. 1.24.

In the semi-space G_2 relevant for the EEG measurement, (1.91) is then solved by

$$\phi^{\mathrm{DFP}}(\boldsymbol{x}) = \frac{1}{2\pi(\sigma_1 + \sigma_2)} \frac{I}{\sqrt{r^2 + (z+h)^2}} . \tag{1.100}$$

When G_2 is assumed to be an isolator, we set $\sigma_2 = 0$, $\sigma_1 \equiv \sigma$. Hence the potential in the semi-space G_2 is simply twice the potential in a homogeneous medium. Provided that all current sources and sinks are distributed in G_1, the superposition principle entails

$$\phi^{\mathrm{DFP}}(\boldsymbol{x}) = \frac{1}{2\pi\sigma} \sum_i \frac{I_i}{\sqrt{r_i^2 + (z+h_i)^2}} . \tag{1.101}$$

From (1.99) follows

---

[6] Occasionally a misunderstanding occurs in the literature where ionic currents and dielectric displacement, i.e. polarization, are confused [84, 85]. Do we really measure sodium or potassium ions that have traversed the skull during an EEG measurement, or is the skull merely a polarizable medium?

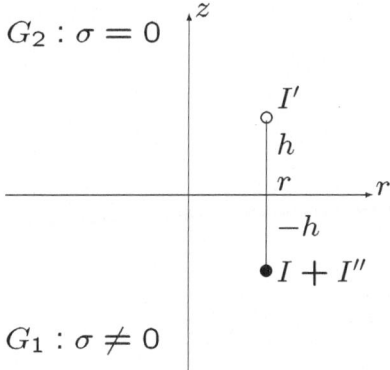

Fig. 1.24. Geometry of Dirichlet's boundary problem for the EEG

$$\Phi(z) \equiv \phi^{\mathrm{EEG}} = \frac{jd}{\sigma} \left(1 - \frac{z}{\sqrt{R^2 + z^2}} \right) \tag{1.102}$$

for the EEG generated by a cortical column measurable at the scalp.

1.5.4 Mean Fields

In this last subsection, I shall discuss the question of whether mass potentials such as LFP or EEG are mere epiphenomena [42, 86], or whether they play a functional role in the organization of brain functioning. If the latter were the case, they would be described as *order parameters* which couple as *mean fields* onto the microscopic neurodynamics with the ability to enslave its behavior [30, 87].

In order to encounter this problem, one has to estimate the average field strengths and voltage differences that are generated by synchronized activity of all pyramidal cells of a cortical module. These quantities then have to be compared with experimental findings on the susceptibility of nerve cells through electromagnetic fields. As mentioned above, the spiking threshold is around $\theta = -50\,\mathrm{mV}$, i.e. the membrane must be polarized by $\Delta U = 10\,\mathrm{mV} - 20\,\mathrm{mV}$ from its resting value given by the Nernst equation (1.7). This corresponds to an electric field strength $E = 10^6\,\mathrm{V/m}$ for a thickness of 5 nm of the cell membrane [88].

On the other hand, neurophysiological experiments have revealed that much smaller field strengths of about 1 V/m entail significant changes of neural excitability [89–93]. Event-related potentials can be modulated by values around 4 V/m. In the hippocampus, where pyramidal cells are very densely packed, effective field strengths are in the range of 5–7 V/m, whereas 10–15 V/m are needed in the cerebellum [91].

To estimate the field strength generated by a cortical column, we have to solve the Dirichlet boundary problem within the electrolyte G_1 as shown in

Fig. 1.24. The potential of a point source (or sink) I at $\boldsymbol{x} \cong (0, 0, z)$ is given in the semi-space G_1 as

$$\phi(\boldsymbol{x}) = \frac{I}{4\pi\sigma_1} \left(\frac{1}{\sqrt{r^2 + (z+h)^2}} + \frac{\sigma_1 - \sigma_2}{\sigma_1 + \sigma_2} \frac{1}{\sqrt{r^2 + (z-h)^2}} \right) . \qquad (1.103)$$

Since we assume G_2 to be an isolator again ($\sigma_2 = 0$), the superposition principle yields

$$\phi(\boldsymbol{x}) = \frac{1}{4\pi\sigma} \sum_i I_i \left(\frac{1}{\sqrt{r_i^2 + (z+h_i)^2}} + \frac{1}{\sqrt{r_i^2 + (z-h_i)^2}} \right) . \qquad (1.104)$$

Next, we apply (1.104) to a current dipole made up by a source (or sink) I at $(r, \varphi, l - d/2)$ and a sink (or source) $-I$ at $(r, \varphi, l + d/2)$, where the dipole's center is situated in a distance l below the interface:

$$\phi(\boldsymbol{x}) = \frac{I}{4\pi\sigma} \left(\frac{1}{\sqrt{r^2 + (z+l-d/2)^2}} + \frac{1}{\sqrt{r^2 + (z-l+d/2)^2}} - \frac{1}{\sqrt{r^2 + (z+l+d/2)^2}} + \frac{1}{\sqrt{r^2 + (z-l-d/2)^2}} \right) .$$

Approximating the quotients by $(1+x)^{-1/2} \approx 1 - x/2$ gives

$$\phi(\boldsymbol{x}) = \frac{Ild}{2\pi\sigma r^3} , \qquad (1.105)$$

i.e. the potential depends only on the radial direction in the conductor. Therefore, the field strength is given by the r-component of the gradient

$$E = -\frac{\partial}{\partial r} \phi(\boldsymbol{x}) = \frac{3Ild}{2\pi\sigma r^4} . \qquad (1.106)$$

In order to replace the column by an equivalent current dipole moment generating the same field, we have to compute the current density through the surface of the column according to (1.102) from the measured scalp EEG. Rearrangement of (1.102) yields

$$j = \frac{\sigma}{d} \frac{\Phi(z)}{1 - \frac{z}{\sqrt{R^2 + z^2}}} . \qquad (1.107)$$

Then, the current through the column would be

$$I = j\pi R^2 \qquad (1.108)$$

if the tissue were a continuum as presupposed in Sect. 1.5.2. Here, $R \approx 150\,\mu m$ is the radius of a column. By contrast, one has to take into account that a

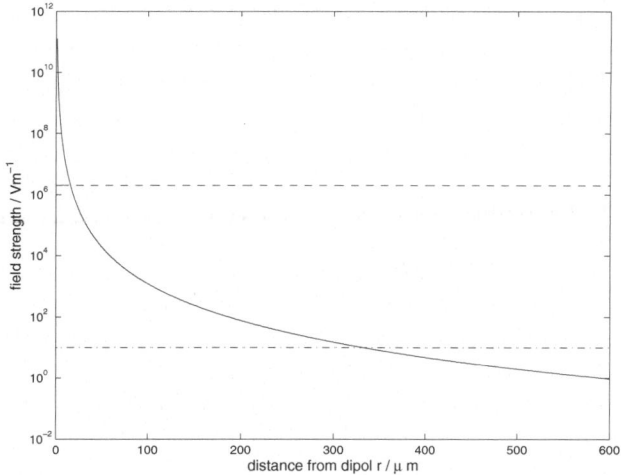

Fig. 1.25. Electric field strength estimated from the EEG depending on the distance of an equivalent dipole. The horizontal lines indicate critical field strengths for evoking action potentials (*upper line*), and for detectable physiological impact (*bottom line*)

column contains about 10,000 pyramidal cells. Thus, the current along a single pyramidal cell is

$$I_{\mathrm{Pyr}} = \frac{j\pi R^2}{N} \qquad (1.109)$$

with $N = 10,000$. Inserting (1.109) and (1.107) into (1.106) gives with $z = l$

$$E(r) = \frac{3lR^2}{2Nr^4} \frac{\Phi(l)}{1 - \frac{l}{\sqrt{R^2+l^2}}} . \qquad (1.110)$$

Figure 1.25 presents a plot of $E(r)$ for the parameter values $R = 150\,\mu\mathrm{m}$, $N = 10,000$, where a distance between the cortex surface and the skull $l = 8\,\mathrm{mm}$ and a peak EEG amplitude of $\Phi(l) = 100\,\mu\mathrm{V}$ have been assumed.

Additionally, Fig. 1.25 displays two lines: the upper line reflects the spiking threshold of a single neuron, $E = 10^6\,\mathrm{V/m}$; the bottom one indicates the limit of physiological efficacy, $E = 10\,\mathrm{V/m}$ [91]. These thresholds correspond to the distances $r_1 = 16.39\,\mu\mathrm{m}$, and $r_2 = 346.77\,\mu\mathrm{m}$ from the equivalent dipole. Because we took $R = 150\,\mu\mathrm{m}$ as the radius of a cortical module, r_2 reaches far into the neighboring column. With 10,000 pyramidal cells per module, their average distance amounts to $3\,\mu\mathrm{m}$, such that approximately 120 closely packed pyramidal cells can be excited by the mass potential. Interestingly, the radius r_2 coincides nicely with the size of the column. Hence, this rough estimate suggests that cortical columns are functional modules controlled by their own electric mean fields that are very likely not mere epiphenomena.

This is consistent with empirical findings. Adey [89] reported a change of the calcium conductivity of neuron membranes under the impact of sustained high-frequency fields. In the hippocampus of rats, field effects are particularly pronounced due to the extreme packing density of the pyramidal cells and the resulting low conductivity of the extracellular liquid. Suppressing synaptic transmission experimentally by reducing the amount of available extracellular calcium, leads to the emergence of spontaneous *bursting* that can be synchronized by mass potentials [91]. Bracci et al. [90] demonstrated that the synchronization of hippocampal neurons is facilitated by the application of external electric fields. They showed also that the conductance of the extracellular electrolyte is a control parameter which can be tuned in such a way that spontaneous synchronization takes place if the conductance is lowered below a critical value. In this case, the fall of the dendritic field potentials along the extracellular resistors contribute to larger LFP and EEG that in turn enslave the whole population. Most recently, Richardson, Schiff and Gluckman [92,93] studied the propagation of traveling waves through an excitable neural medium under the influence of external fields. They reported a dependence of the group velocity on the polarity of the applied field, and modeled these effects through a neural field theory analogue of (1.86).

Coming back to our neuron models, the impact of mass potentials can be formally taken into account by introducing a *mean field* coupling. Concerning, for example, the Hodgkin-Huxley equations (1.25, 1.27–1.29), one has to replace the membrane potential U_i of neuron i by a shifted value

$$U_i' = U_i - \sum_j U_j^{\mathrm{DFP}} , \tag{1.111}$$

where either the dendritic field potential is given by (1.64), or, in a continuum account, the whole sum is provided by (1.99). This idea has been expressed by Braitenberg and Schüz [94, p. 198], in that a cortical module controls its own activation thresholds.

1.6 Discussion

In this chapter, I have sketched the basic principles of neurophysics, i.e. the biophysics of membranes, neurons, neural networks and neural masses. Let me finally make some concluding remarks on neural modeling and descriptions. I hope the reader has recognized that there is no unique *physical model* of *the* neuron, or of *the* neural network of the brain. Even a single neuron can be described by models from different complexity classes. It can be regarded as a continuous system governed by a nonlinear partial differential equation which describes its cable properties. Decomposing the cable into compartments, one obtains either compartment models comprised of lots of coupled ordinary differential equations, or point models that are still described by many coupled ordinary differential equations, one for the kinetics of each population of ion channels. Further simplifying these models, one eventually arrives at the coarsest McCulloch-Pitts neuron [51].

On the other hand, each neuron model dictates the dimensionality of its phase space and, as its projection, its observable space. Observables provide the interface to experimental neuroscience in that they should be *observable*. The best theoretical neuron model is not much good if it contains quantities that are not observable in real experiments. In most cases, observables are membrane potentials either collected from the axons, i.e. action potentials, or measured from the dendro-somatic membranes such as EPSP and IPSP. However, a few electrode tips put into a couple of neurons will not provide sufficiently many observables for describing the behavior of a neural network. At the network level, the abovementioned problems greatly worsen with increasing size and complexity of the network, ending in an unmanageable number of degrees of freedom for continuum models.

In this case, spatial coarse-graining [62] is the method of choice. By averaging activity across regions of appropriate size, one obtains mass potentials such as LFP or EEG. LFP is experimentally observable through the use of multi-electrode arrays placed into the extracellular space. Each sensor collects averaged dendritic field potentials from several thousands of neurons, as well as some spiking activity in its vicinity. On the other hand, ECoG (intracranial EEG) and EEG are gathered by electrodes placed at the cerebral membrane or at the scalp, respectively. Each electrode registers the mean activity of billions of neurons. Using conventional 32, 64, or 128 channel amplifiers thereby collapses the huge microscopic observable space of single neurons and the large mesoscopic observable space of LFP to a macroscopic observable space of 32, 64, or 128 dimensions.

As we have seen in Sect. 1.5.3, such mass potentials are not in the least irrelevant because they serve as order parameters [30,87], both indicating and causing macroscopic ordering of the system. Yet there is another important aspect of mass potentials. They do not only comprise a spatial coarse-graining by definition, but also provide a coarse-graining of the high-dimensional microscopic phase space. Consider a mean field observable

$$F(x) = \sum_i f_i(\boldsymbol{x}), \tag{1.112}$$

where the sum extends over a population of n neurons and f_i denotes a projection of the microscopic state $\boldsymbol{x} \in X$ onto the i-th coordinate axis measuring the activation $f_i(\boldsymbol{x}) = x_i$ of the ith neuron. Obviously, the outcomes of F may have multiple realizations as the terms in the sum in (1.111) can be arbitrarily arranged. Therefore, two neural activation vectors $\boldsymbol{x}, \boldsymbol{y}$ can lead to the same value $F(\boldsymbol{x}) = F(\boldsymbol{y})$ (e.g. when $f_i(\boldsymbol{x}) - \epsilon = f_j(\boldsymbol{x}) + \epsilon$, $i \neq j$), so that they are indistinguishable by means of F. Beim Graben and Atmanspacher [95] call such microstates *epistemically equivalent*. All microstates that are epistemically equivalent to each other form an equivalence class, and, as it is well known from set theory, all equivalence classes partition the phase space X. If the equivalence classes of F in X form a finite partition $\mathcal{Q} = \{A_1, \ldots A_I\}$ of X, one can assign symbols a_i from an alphabet $\mathbf{A}$ to the cells A_i and obtain a *symbolic dynamics* [96–98]. In this way, experimentally well-defined meso- and macroscopic brain observables, LFP and EEG, form a coarse-grained description of the underlying microscopic neurodynamics.

Atmanspacher and beim Graben [99,100] discuss this coarse-graining with respect to its stability properties. The microscopic dynamics $x \mapsto \Phi^t(x)$ where the *flow* Φ solves the microscopic differential equations, is captured by transitions from one symbol to another one $a_i \mapsto a_j$. If these transitions can be described by an ergodic Markov chain, the symbolic dynamics exhibits particular stability properties. If the Markov chain is aperiodic, a distinguished thermal equilibrium state can be constructed for the symbolic description. If, contrarily, the Markov chain is periodic, the system possesses stable fixed point or limit cycle attractors. Atmanspacher and beim Graben argue that in both cases, the concept of *contextual emergence* applies where higher-level descriptions emerge from contingently supplied contexts that are not merely reducible to lower-level descriptions. As an application, Atmanspacher and beim Graben [99,100] demonstrated the contextual emergence of neural correlates of consciousness [101] from neurodynamics where arbitrary contexts are given by phenomenal families partitioning the space of phenomenal experiences [102]. Other examples were discussed by beim Graben [103] and Dale and Spivey [104] where symbolic cognition emerges from partitioned dynamical systems.

The problem of finding reasonable macroscopic observables for neural networks has been addressed by Amari [52]. He considered random networks of McCulloch-Pitts neurons (cf. Chaps. 3, 5, 7), and defined a proper macrostate as a macroscopic observable such as (1.112) if two conditions hold: Firstly, the temporal evolution of the observable should be compatible with the coarse-graining, and, secondly, the observable should be structurally stable against topological deformations of the network. The second requirement is closely related to ergodicity of the resulting symbolic dynamics [99]. Accordingly, the first demand entails that all initial conditions that are mapped onto the same value of a macrostate are epistemically equivalent. As an example for a good macrostate, at least for his toy-model, Amari [52] provided the mass potential (1.112). Hence, macroscopic descriptions of neural masses provide important insights into the functional organization of neurodynamical systems. They are by far more then mere epiphenomena.

Acknowledgements

This chapter is based on notes from a lecture course "Introduction to Neurophysics", I taught together with J. Kurths, D. Saddy, and T. Liebscher in the years 2000 and 2002 at the University of Potsdam for the DFG Research Group "Conflicting Rules in Cognitive Systems". Section 1.5 contains parts of my thesis [82] that were previously available only in German. Section 1.6 presents some results from a cooperation with H. Atmanspacher. S. J. Nasuto,

J. J. Wright and C. Zhou helped me improving this chapter. I greatly acknowledge their respective contributions and inspiring discussions.

References

1. J. von Neumann. *The Computer and the Brain*. Yale University Press, New Haven (CT), 1958. Partly reprinted in J. A. Anderson and E. Rosenfeld (1988), p. 83ff.
2. Z. W. Pylyshyn. *Computation and Cognition: Toward a Foundation for Cognitive Science*. MIT Press, Cambrigde (MA), 1986.
3. J. R. Anderson. *Cognitive Psychology and its Implications*. W. H. Freeman and Company, New York (NY), 4th edition, 1995.
4. M. Kutas and A. Dale. Electrical and magnetic readings of mental functions. In M. Rugg, editor, *Cognitive Neuroscience*, pp. 197–242. Psychology Press, Hove East Sussex, 1997.
5. R. C. O'Reilly and Y. Munakata. *Computational Explorations in Cognitive Neuroscience. Understanding the Mind by Simulating the Brain*. MIT Press, Cambridge (MA), 2000.
6. M. S. Gazzaniga, R. B. Ivry, and G. R. Mangun, editors. *Cognitive Neuroscience. The Biology of the Mind*. W. W. Norton, New York (NY), 2nd edition, 2002.
7. J. A. Anderson and E. Rosenfeld, editors. *Neurocomputing. Foundations of Research*, Vol. 1. MIT Press, Cambridge (MA), 1988.
8. J. A. Anderson, A. Pellionisz, and E. Rosenfeld, editors. *Neurocomputing. Directions for Research*, Vol. 2. MIT Press, Cambridge (MA), 1990.
9. P. S. Churchland and T. J. Sejnowski. *The Computational Brain*. MIT Press, Cambridge (MA), 1994.
10. F. Riecke, D. Warland, R. de Ruyter van Steveninck, and W. Bialek. *Spikes: Exploring the Neural Code*. Computational Neurosciences. MIT Press, Cambridge (MA), 1997.
11. M. A. Arbib, editor. *The Handbook of Brain Theory and Neural Networks*. MIT Press, Cambridge (MA), 1998.
12. C. Koch and I. Segev, editors. *Methods in Neuronal Modelling. From Ions to Networks*. Computational Neuroscience. MIT Press, Cambridge (MA), 1998.
13. C. Koch. *Biophysics of Computation. Information Processing in Single Neurons*. Computational Neuroscience. Oxford University Press, New York (NY), 1999.
14. H. R. Wilson. *Spikes, Decisions and Actions. Dynamical Foundations of Neuroscience*. Oxford University Press, New York (NY), 1999.
15. P. Dayan and L. F. Abbott. *Theoretical Neuroscience*. Computational Neuroscience. MIT Press, Cambridge (MA), 2001.
16. T. P. Trappenberg. *Fundamentals of Computational Neuroscience*. Oxford University Press, Oxford (GB), 2002.
17. R. P. N. Rao, B. A. Olshausen, and M. S. Lewicky, editors. *Probabilistic Models of the Brain: Perception and Neural Function*. MIT Press, Cambridge (MA), 2002.
18. W. Gerstner and W. Kistler. *Spiking Neuron Models. Single Neurons, Populations, Plasticity*. Cambridge University Press, Cambridge (UK), 2002.

19. E. R. Kandel, J. H. Schwartz, and T. M. Jessel, editors. *Principles of Neural Science*. Appleton & Lange, East Norwalk, Connecticut, 1991.

20. E. R. Kandel, J. H. Schwartz, and T. M. Jessel, editors. *Essentials of Neural Science and Behavior*. Appleton & Lange, East Norwalk, Connecticut, 1995.

21. J. G. Nicholls, A. R-Martin, B. G. Wallace, and P. A. Fuchs. *From Neuron to Brain*. Sinauer, Sunderland (MA), 2001.

22. H. C. Tuckwell. *Introduction to Theoretical Neurobiology*, Vol. 1. Cambridge University Press, Cambridge (UK), 1988.

23. H. C. Tuckwell. *Introduction to Theoretical Neurobiology*, Vol. 2. Cambridge University Press, Cambridge (UK), 1988.

24. D. Johnston and S. M.-S. Wu. *Foundations of Cellular Neurophysiology*. MIT Press, Cambridge (MA), 1997.

25. B. Hille. *Ion Channels of Excitable Membranes*. Sinauer, Sunderland, 2001.

26. A. Einstein. Eine neue Bestimmung der Moleküldimensionen. *Annalen der Physik*, 19:289–306, 1906.

27. S. B. Laughlin, R. R. de Ruyter van Steveninck, and J. C. Anderson. The metabolic cost of neural information. *Nature Neuroscience*, 1(1): 36–41, 1998.

28. W. W. Orrison Jr., J. D. Lewine, J. A. Sanders, and M. F. Hartshorne. *Functional Brain Imaging*. Mosby, St. Louis, 1995.

29. N. K. Logothetis, J. Pauls, M. Augath, T. Trinath, and A. Oeltermann. Neurophysiological investigation of the basis of the fMRI signal. *Nature*, 412: 150–157, 2001.

30. H. Haken. *Synergetics. An Introduction*, Vol. 1 of *Springer Series in Synergetics*. Springer, Berlin, 1983.

31. N. G. van Kampen. *Stochastic Processes in Physics and Chemistry*. Elsevier, Amsterdam, 1992.

32. A. L. Hodgkin and A. F. Huxley. A quantitative description of membrane current and its application to conduction and excitation in nerve. *J. Physiol.*, 117: 500–544, 1952.

33. I. Swameye, T. G. Müller, J. Timmer, O. Sandra, and U. Klingmüller. Identification of nucleocytoplasmatic cycling as a remote sensor in cellular signaling by databased modeling. *Proceedings of the National Academy of Sciences of the U.S.A.*, 100(3): 1028–1033, 2003.

34. J. M. Bower and D. Beeman. *The Book of GENESIS. Exploring Realistic Neural Models with the GEneral NEural SImulation System*. Springer, New York (NY), 1998.

35. J. W. Moore and M. L Hines. *Simulations with NEURON*. Duke and Yale University, 1994.

36. A. Destexhe, D. Contreras, and M. Steriade. Cortically-induced coherence of a thalamic-generated oscillation. *Neuroscience*, 92(2): 427–443, 1999.

37. C. Bédard, H. Kröger, and A. Destexhe. Modeling extracellular field potentials and the frequency-filtering properties of extracellular space. *Biophys. J.*, 86(3): 1829–1842, 2004.

38. O. Creutzfeld and J. Houchin. Neuronal basis of EEG-waves. In *Handbook of Electroencephalography and Clinical Neurophysiology*, Vol. 2, Part C, pp. 2C-5–2C-55. Elsevier, Amsterdam, 1974.

39. W. J. Freeman. *Mass Action in the Nervous System*. Academic Press, New York (NY), 1975.

40. D. T. J. Liley, D. M. Alexander, J. J. Wright, and M. D. Aldous. Alpha rhythm emerges from large-scale networks of realistically coupled multicompartmental model cortical neurons. *Network: Comput. Neural Syst.*, 10: 79–92, 1999.

41. A. J. Trevelyan and O. Watkinson. Does inhibition balance excitation in neocortex? *Prog. Biophys. Mol. Biol.*,, 87: 109–143, 2005.

42. P. L. Nunez and R. Srinivasan. *Electric Fields of the Brain: The Neurophysics of EEG.* Oxford University Press, New York, 2006.

43. R. FitzHugh. Impulses and physiological states in theoretical models of nerve membrane. *Biophys. J.*, 1: 445–466, 1961.

44. T. Pavlidis. A new model for simple neural nets and its application in the design of a neural oscillator. *Bull. Math. Biol.*, 27: 215–229, 1965.

45. R. B. Stein, K. V. Leung, M. N. Oğuztöreli, and D. W. Williams. Properties of small neural networks. *Kybernetik*, 14:223–230, 1974.

46. R. B. Stein, K. V. Leung, D. Mangeron, and M. N. Oğuztöreli. Improved neuronal models for studying neural networks. *Kybernetik*, 15: 1–9, 1974.

47. J. L. Hindmarsh and R. M. Rose. A model of neuronal bursting using three coupled first-order differential equations. *Proceedings of the Royal Society London*, B221:87–102, 1984.

48. N. F. Rulkov. Modeling of spiking-bursting neural behavior using two-dimensional map. *Phys. Rev. E*, 65: 041922, 2002.

49. E. M. Izhikevich. Simple model of spiking neurons. *IEEE Trans. Neural Networks*, 14(6): 1569–1572, 2003.

50. E. M. Izhikevich. Which model to use for cortical spiking neurons? *IEEE Trans. Neural Networks*, 15(5): 1063–1070, 2004.

51. W. S. McCulloch and W. Pitts. A logical calculus of ideas immanent in nervous activity. *Bull. Math. Biophys.*, 5:115–133, 1943. Reprinted in J. A. Anderson and E. Rosenfeld (1988) [7], p. 83ff.

52. S. Amari. A method of statistical neurodynamics. *Kybernetik*, 14: 201–215, 1974.

53. D. J. Amit. *Modeling Brain Function. The World of Attractor Neural Networks.* Cambridge University Press, Cambridge (MA), 1989.

54. A. Kuhn, A. Aertsen, and S. Rotter. Neuronal integration of synaptic input in the fluctuation-driven regime. *J. Neurosci.*, 24(10): 2345–2356, 2004.

55. J. S. Griffith. A field theory of neural nets: I. derivation of field equations. *Bull. Math. Biophys.*, 25:111–120, 1963.

56. J. S. Griffith. A field theory of neural nets: II. properties of the field equations. *Bull. Math. Biophys.*, 27: 187–195, 1965.

57. H. R. Wilson and J. D. Cowan. A mathematical theory of the functional dynamics of cortical and thalamic nervous tissue. *Kybernetik*, 13: 55–80, 1973.

58. P. L. Nunez, editor. *Neocortical Dynamics and Human EEG Rhythms.* Oxford University Press, New York (NY), 1995.

59. V. K. Jirsa and H. Haken. Field theory of electromagnetic brain activity. *Phys. Rev. Lett.*, 77(5): 960–963, 1996.

60. V. K. Jirsa and H. Haken. A derivation of a macroscopic field theory of the brain from the quasi-microscopic neural dynamics. *Physica D*, 99: 503–526, 1997.

61. J. J. Wright and D. T. J. Liley. Dynamics of the brain at global and microscopic scales: Neural networks and the EEG. *Behavioral and Brain Sciences*, 19: 285–320, 1996.

62. D. T. J. Liley, P. J. Cadusch, and J. J. Wright. A continuum theory of electro-cortical activity. *Neurocomputing*, 26–27: 795–800, 1999.

63. P. A. Robinson, C. J. Rennie, J. J. Wright, H. Bahramali, E. Gordon, and D. L. Rowe. Prediction of electroencephalic spectra from neurophysiology. *Phys. Rev. E*, 63, 2001. 021903.

64. C. J. Rennie, P. A. Robinson, and J. J. Wright. Effects of local feedback on dispersion of electrical waves in the cerebral cortex. *Phys. Rev. E.*, 59(3): 3320–3329, 1999.

65. P. A. Robinson, C. J. Rennie, J. J. Wright, and P. D. Bourke. Steady states and global dynamics of electrical activity in the cerebral cortex. *Phys. Rev. E.*, 58(3): 3557–3571, 1998.

66. J. J. Wright, C. J. Rennie, G. J. Lees, P. A. Robinson, P. D. Bourke, C. L. Chapman, E. Gordon, and D. L. Rowe. Simulated electrocortical activity at microscopic, mesoscopic, and global scales. *Neuropsychopharmacology*, 28: S80–S93, 2003.

67. V. K. Jirsa. Information processing in brain and behavior displayed in large-scale scalp topographies such as EEG and MEG. *Int. J. Bifurcation and Chaos*, 14(2): 679–692, 2004.

68. J. J. Wright, C. J. Rennie, G. J. Lees, P. A. Robinson, P. D. Bourke, C. L. Chapman, E. Gordon, and D. L. Rowe. Simulated electrocortical activity at microscopic, mesoscopic and global scales. *Int. J. Bifurcation and Chaos*, 14(2): 853–872, 2004.

69. J. J. Wright, P. A. Robinson, C. J. Rennie, E. Gordon, P. D. Burke, C. L. Chapman, N. Hawthorn, G. J. Lees, and D. Alexander. Toward an integrated continuum model of cerebral dynamics: the cerebral rhythms, synchronous oscillation and cortical stability. *Biosystems*, 63: 71–88, 2001.

70. V. K. Jirsa and J. A. S. Kelso. Spatiotemporal pattern formation in neural systems with heterogeneous connection toplogies. *Phys. Rev. E.*, 62(6): 8462–8465, 2000.

71. H. R. Wilson and J. D. Cowan. Excitatory and inhibitory interactions in localized populations of model neurons. *Biophys. J.*, 12: 1–24, 1972.

72. F. H. Lopes da Silva, A. Hoecks, H. Smits, and L. H. Zetterberg. Model of brain rhythmic activity: The alpha-rhythm of the thalamus. *Kybernetik*, 15: 27–37, 1974.

73. F. H. Lopes da Silva, A. van Rotterdam, P. Bartels, E. van Heusden, and W. Burr. Models of neuronal populations: The basic mechanisms of rhythmicity. In M. A. Corner and D. F. Swaab, editors, *Perspectives of Brain Research*, Vol. 45 of *Prog. Brain Res.*, pp. 281–308. 1976.

74. W. J. Freeman. Simulation of chaotic EEG patterns with a dynamic model of the olfactory system. *Biol. Cybern.*, 56: 139–150, 1987.

75. B. H. Jansen, G. Zouridakis, and M. E. Brandt. A neurophysiologically-based mathematical model of flash visual evoked potentials. *Biol. Cybern.*, 68: 275–283, 1993.

76. B. H. Jansen and V. G. Rit. Electroencephalogram and visual evoked potential generation in a mathematical model of coupled cortical columns. *Biol. Cybern.*, 73: 357–366, 1995.

77. F. Wendling, J. J. Bellanger, F. Bartolomei, and P. Chauvel. Relevance of nonlinear lumped-parameter models in the analysis of depth-EEG epileptic signals. *Biol. Cybern.*, 83: 367–378, 2000.

78. F. Wendling, F. Bartolomei, J. J. Bellanger, and P. Chauvel. Epileptic fast activity can be explained by a model of impaired GABAergic dendritic inhibition. *Eur. J. Neurosci.*, 15: 1499–1508, 2002.

79. O. David and K. J. Friston. A neural mass model for MEG/EEG: coupling and neuronal dynamics. *Neuroimage*, 20: 1743–1755, 2003.

80. O. David, D. Cosmelli, and K. J. Friston. Evaluation of different measures of functional connectivity using a neural mass model. *Neuroimage*, 21: 659–673, 2004.

81. O. David, L. Harrison, and K. J. Friston. Modelling event-related respones in the brain. *Neuroimage*, 25: 756–770, 2005.

82. P. beim Graben. *Symbolische Dynamik Ereigniskorrelierter Potentiale in der Sprachverarbeitung*. Berichte aus der Biophysik. Shaker Verlag, Aachen, 2001.

83. C. Baumgartner. Clinical applications of source localisation techniques — the human somatosensory cortex. In F. Angelieri, S. Butler, S. Giaquinto, and J. Majkowski, editors, *Analysis of the Electrical Activity of the Brain*, pp. 271–308. Wiley & Sons, Chichester, 1997.

84. W. Lutzenberger, T. Elbert, B. Rockstroh, and N. Birbaumer. *Das EEG*. Springer, Berlin, 1985.

85. N. Birbaumer and R. F. Schmidt. *Biologische Psychologie*. Springer, Berlin, 1996.

86. S. Zschocke. *Klinische Elektroenzephalographie*. Springer, Berlin, 1995.

87. A. Wunderlin. On the slaving principle. In R. Graham and A. Wunderlin, editors, *Lasers and Synergetics*, pp. 140–147, Springer, Berlin, 1987.

88. J. Dudel, R. Menzel, and R. F. Schmidt, editors. *Neurowissenschaft. Vom Molekül zur Kognition*. Springer, Berlin, 1996.

89. W. R. Adey. Molecular aspects of cell membranes as substrates for interaction with electromagnetic fields. In E. Başar, H. Flohr, H. Haken, and A. J. Mandell, editors, *Synergetics of the Brain*, pp. 201–211, Springer, Berlin, 1983.

90. E. Bracci, M. Vreugdenhil, S. P. Hack, and J. G. R. Jefferys. On the synchronizing mechanism of tetanically induced hippocampal oscillations. *J. Neurosci.*, 19(18): 8104–8113, 1999.

91. J. G. R. Jefferys. Nonsynaptic modulation of neuronal activity in the brain: Electric currents and extracellular ions. *Physiol. Rev.*, 75: 689–723, 1995.

92. K. A. Richardson, S. J. Schiff, and B. J. Gluckman. Electric field control of seizure propagation: From theory to experiment. In S. Boccaletti, B. Gluckman, J. Kurths, L. M. Pecora, R. Meucci, and O. Yordanov, editors, *Proceeding of the 8th Experimental Chaos Conference 2004*, pp. 185–196, American Institute of Physics, Melville (NY), 2004.

93. K. A. Richardson, S. J. Schiff, and B. J. Gluckman. Control of traveling waves in the mammalian cortex. *Phys. Rev. Lett.*, 94: 028103, 2005.

94. V. Braitenberg and A. Schüz. *Cortex: Statistics and Geometry of Neuronal Connectivity*. Springer, Berlin, 1998.

95. P. beim Graben and H. Atmanspacher. Complementarity in classical dynamical systems. *Found. Phys.*, 36(2): 291–306, 2006.

96. D. Lind and B. Marcus. *An Introduction to Symbolic Dynamics and Coding*. Cambridge University Press, Cambridge (UK), 1995.

97. P. beim Graben, J. D. Saddy, M. Schlesewsky, and J. Kurths. Symbolic dynamics of event–related brain potentials. *Phys. Rev. E.*, 62(4): 5518–5541, 2000.

98. P. beim Graben and J. Kurths. Detecting subthreshold events in noisy data by symbolic dynamics. *Phys. Rev. Let.*, 90(10): 100602, 2003.

99. H. Atmanspacher and P. beim Graben. Contextual emergence of mental states from neurodynamics. *Chaos and Complexity Letters*, 2(2/3), 151–168, 2007.

100. H. Atmanspacher. Contextual emergence from physics to cognitive neuro-science. *J. of Consciousness Stud.*, 14(1–2): 18–36, 2007.

101. T. Metzinger, editor. *Neural Correlates of Consciousness.* MIT Press, Cam-bridge (MA), 2000.

102. D. J. Chalmers. What is a neural correlate of consciousness? In Metzinger [101], Chap. 2, pp. 17–39, 2000.

103. P. beim Graben. Incompatible implementations of physical symbol systems. *Mind and Matter*, 2(2): 29–51, 2004.

104. R. Dale and M. J. Spivey. From apples and oranges to symbolic dynamics: A framework for conciliating notions of cognitive representation. *J. Exp. & Theor. Artific. Intell.*, 17(4): 317–342, 2005.

Synapses and Neurons:
Basic Properties and Their Use in Recognizing Environmental Signals

Henry D. I. Abarbanel[1,2,3], Julie S. Haas[1] and Sachin S. Talathi[1,2]

[1] Institution for Nonlinear Science
  haas@ucsd.edu
[2] Department of Physics, University of California San Diego
  talathi@physics.ucsd.edu
[3] Marine Physical Laboratory (Scripps Institute for Oceanography)
  hdia@jacobi.ucsd.edu

2.1 Introduction

This chapter incorporates two lectures presented at the 2005 Helmholtz School in Potsdam, Germany in September, 2005. The goal of this chapter is to cover two topics, both briefly:

- some basics of the biophysics of neurons and synapses. Many students had studied this material, and for them it was a lightning fast review. Some students had not studied this material, and for them it was a lightning fast introduction. See the book by Johnston and Wu [1] to make the introduction slower.
- the use of these basics to develop a neural time delay circuit and explore its use in the fundamental nervous system task of recognizing signals sent from the sensory system to a central nervous system as spike trains [2].

In a broad sense, modeling or computation in neuroscience takes place at two levels:

- Top-down: Start with the analysis of those macroscopic aspects of an animal's behavior that are robust, reproducible and important for survival. Try to represent the nervous system as a collection of circuits, perhaps based on biological physics, but perhaps not, needed to perform these function. The top-down approach is a speculative "big picture" view.
- Bottom-up: Start from a description of individual neurons and their synaptic connections; use facts about the details of their dynamical behavior from observed anatomical and electrophysiological data. Using these data, the pattern of connectivity in a circuit is reconstructed. Using the patterns of connectivity (the "wiring diagram") along with the dynamical features

of the neurons and synapses, bottom-up models have been able to predict functional properties of neural circuits and their role in animal behavior.

The point of view in these lectures is distinctly "bottom-up." This is a much harder task than "top-down" approaches as it relies on observations to dictate the road ahead and the constraints in navigating that road. In our opinion, biological phenomena, neurobiological as well as others, are complex because the networks are designed by the necessity of performing functions for living things. They are not designed by optimality principles so far as we know.

The point of view we take here is that if one can understand in a well formulated predictive and quantitative mathematical model how biological processes are constructed in nature, then general principles for the design and construction of these can be analyzed. This requires working closely with experiments as guides, making predictions with models that are never totally correct, and refining those models within the framework of the outcome of experiments suggested by the models or formulated on other grounds.

Some will find the effort required rather formidable, but looking back on the historical interplay between experiment and theory that, say, led to the 20th century uncovering of the structure of quantum theory from the application of the Planck heat radiation law to wave equations, one should be ready for the engagement.

The dynamical point of view emphasized in this chapter is expanded in a review article appeared in *Reviews of Modern Physics* [3].

2.2 Lightning Fast (Review, Introduction) to Neural and Synaptic Dynamics

2.2.1 Neurons-points

The basic biophysical phenomena involved in the operation of neurons and a phenomenological set of equations describing them were identified by Hodgkin, Katz, Huxley, and many others in the mid part of the 20th century. Neurons are cells producing electrical signals through protein channels penetrating their plasma membrane allowing ions to flow into or out of the cells with permeabilities and rates often controlled by the voltage $V(t)$ across the membrane. There are two competing sources of current leading to the voltage difference across the membrane: ions flowing from higher concentrations $C(x,t)$ to lower concentrations giving rise to a current

$$J_{\text{concentration}}(x,t) = -D\frac{\partial C(x,t)}{\partial x},\qquad(2.1)$$

with D the diffusion coefficient; and ions flowing because of the electric field associated with the electrostatic potential difference between the inside and the outside of the cell, giving rise to a competing current

$$J_{\text{electric}}(x,t) = -D\frac{zF}{RT}C(x,t)\frac{\partial V(x,t)}{\partial x}. \tag{2.2}$$

z is the magnitude of the ion charge in units of $|e|$, F is the Faraday constant 96485.34 C/mol, T is the temperature, and R is the universal gas constant 8.3144 J/mol K.

Approximating the gradient of the voltage across the thickness l of the cell membrane by $V(t)/l$, we have for the total current

$$J(t) = -D\left\{\frac{\partial C(x,t)}{\partial x} + \frac{zF}{RT}C(x,t)\frac{V(t)}{l}\right\}. \tag{2.3}$$

Denoting the intracellular concentration of the ion in question as C_{in} and the extracellular concentration as C_{out}, one may solve for the current $J(t)$, resulting in the Goldman-Hodgkin-Katz (GHK) equation

$$J(t) = -\frac{V(t)zFD}{RTl}\frac{C_{\text{in}} - C_{\text{out}}e^{-\frac{zFV(t)}{RT}}}{1 - e^{-\frac{zFV(t)}{RT}}}. \tag{2.4}$$

The GHK current has a zero at the *reversal potential or Nernst potential* V_{rev} for each ionic species

$$\begin{aligned}V_{\text{rev}} &= \frac{RT}{zF}\ln\frac{C_{\text{out}}}{C_{\text{in}}}\\ &= \frac{61.5\,\text{mV}}{z}\ln\frac{C_{\text{out}}}{C_{\text{in}}}.\end{aligned} \tag{2.5}$$

Using the values of intracellular and extracellular concentrations for various common ions present in nerve cells, we have the resting potentials listed in Fig. 2.1.

In the study of many neurons and their processes (axons carrying signals to other receiver cells and dendrites receiving signals from other transmitter cells), Hodgkin-Huxley (HH), along with many others, formulated a quite general, phenomenological form for ion currents that are gated, or controlled, by the membrane voltage. This form is

$$I_{\text{voltage-gated}}(t) = g_{\text{ion}}m(t)^p h(t)^q(V(t) - V_{\text{rev}}), \tag{2.6}$$

where g_{ion} is the maximal conductance of the ion channel, p and q are integers, and $Z(t) = \{m(t), h(t)\}$ are activation and inactivation variables depending on the voltage $V(t)$ through phenomenological first order kinetic equations (master equations)

$$\begin{aligned}\frac{dZ(t)}{dt} &= \alpha_Z(V(t))(1 - Z(t)) - \beta_Z(V(t))Z(t)\\ &= \frac{Z_0(V(t)) - Z(t)}{\tau(V(t))},\end{aligned} \tag{2.7}$$

Ion Name	Intracellular Concentration nM	Extracellular Concentration nM	Resting Potential
Na^+	18	145	56 mV
K^+	135	3	-101 mV
Cl^-	120	7	-76 mV
Ca^{2+}	0.1	3000	136 mV

Fig. 2.1. Resting potentials for various common ions in nerve cells

where $\alpha_Z(V)$ and $\beta_Z(V)$ are empirically determined functions of voltage whose functional form is dictated by considerations of activations of gating variables. A good discussion of this formalism is found in Chap. 2 by Christopher Fall and Joel Keizer of the Joel Keizer memorial volume *Computational Cell Biology* [4]. The activation variable $m(t)$ normally resides near zero for voltages near V_{rev} and rises to order unity when the voltage rises toward positive values. The inactivation variables $n(t)$ and $h(t)$ normally reside near unity and decrease towards zero as the voltage rises.

In the original HH model, we have three channels, one for Na^+ ions, one for K^+ ions, and one for a generalized lossy effect called a "leak" channel. With an added DC current, the HH equations are

$$C\frac{dV(t)}{dt} = -(g_{Na}m(t)^3 h(t)(V(t) - V_{Na}) \tag{2.8}$$
$$+ g_K n(t)^4(V(t) - V_K) + g_L(V(t) - V_L)) + I_{DC},$$

along with kinetic equations for $m(t), h(t)$, and $n(t)$ (2.7). Using $g_{Na} = 120\,\text{mS cm}^{-2}, g_K = 36\,\text{mS cm}^{-2}, g_L = 0.3\,\text{mS cm}^{-2}, V_{Na} = 55\,\text{mV}, V_K = -72\,\text{mV}, V_L = -49\,\text{mV}$ and $C = 1\,\mu\text{F cm}^{-2}$, we can numerically solve these HH equations for various values of I_{DC}. The standard HH equations for the $\alpha_Z(V)$ and the $\beta_Z(V)$ are given in Fig. 2.2.

For low values of I_{DC}, the neuron is at rest at various voltages. Above a threshold, action potentials are observed as shown in Fig. 2.3 for $I_{DC} = 0.55\,\mu\text{A}$. The behavior of the K^+ activation variable $n(t)$ is shown in Fig. 2.4. The action potential comes from Na^+ flowing rapidly into the cell from higher concentrations outside the cell. The K^+ activation variable shows the K^+

$$\alpha_m(V) = -0.1 \frac{35 + V}{e^{-(35 + V)/10} - 1} \qquad \beta_m(V) = e^{-(60 + V)/18}$$

$$\alpha_n(V) = 0.07 \, e^{-(60 + V)/20} \qquad \beta_n(V) = \frac{1}{e^{-(30 + V)/10} + 1}$$

$$\alpha_h(V) = -0.01 \frac{50 + V}{e^{-(50 + V)} - 1} \qquad \beta_h(V) = 0.125 \, e^{-(60 + V)/80}$$

Fig. 2.2. Empirical forms for the voltage-gating functions in the original HH equations

channel opening subsequently and allowing K^+ to flow out of the cell, thus assisting in terminating the action potential (cf. Chap. 1).

The HH model has four degrees of freedom, and with the parameters we have used exhibits a limit cycle or periodic behavior. To represent the phase space variation of a limit cycle solution of a differential equation generically takes three coordinates [5], however, we are fortunate here in that we can see the full limit cycle in two dimensions as shown in Fig. 2.5. In Fig. 2.6 we show the functions $m_0(V), h_0(V)$, and $n_0(V)$ along with the associated voltage dependent times, $\tau(V)$.

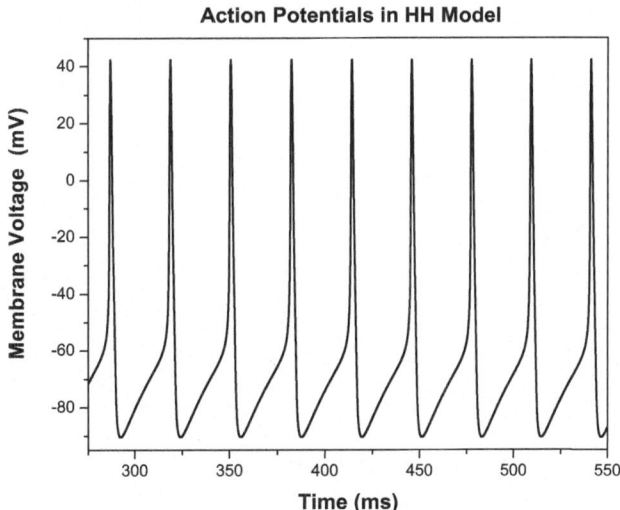

Fig. 2.3. Action potentials in the HH model

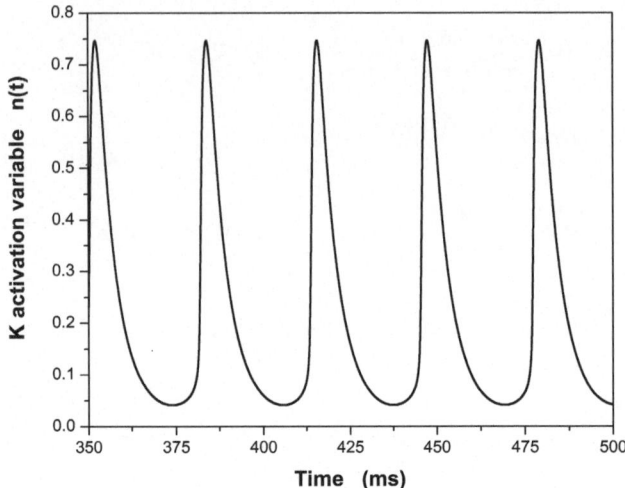

Fig. 2.4. The potassium activation variable $n(t)$ in the standard HH model

The HH model is phenomenological in origin, yet has the distinct scientific advantage of identifying measurable quantities in each expression. It comes from a "bottom-up" analysis of cellular dynamics. Reduced models of neural behavior often lose the latter property, and this makes the connection with how biology solves problems less satisfactory. In the "bottom-up" approach we are following, one of the goals is to use "biological parts" to construct neural circuits, not so much for simplicity at times but for the ability to connect

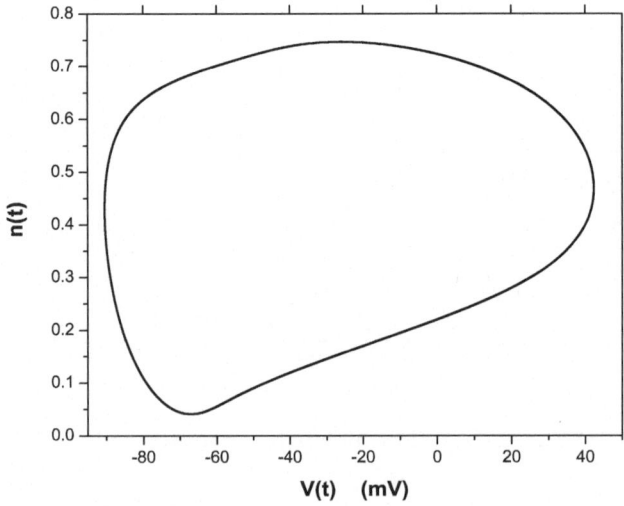

Fig. 2.5. The $(n(t), V(t))$-plane during action potential generation by the HH model

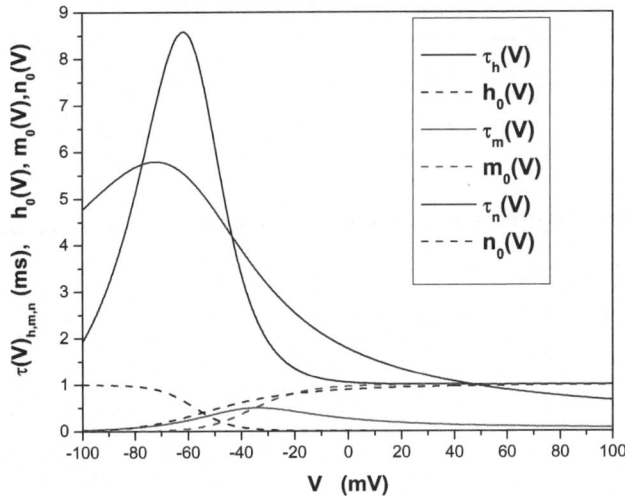

Fig. 2.6. Activation $\{m_0(V), n_0(V)\}$ and inactivation $h_0(V)$ variables and the time constants associated with them for the standard HH model. Note that the time scale of Na^+ activation $\tau_m(V)$ is much smaller than the times for the others, indicating the rapidity with which the Na^+ channels open

with biological networks and biophysical realizations of those networks. If one invents a lovely "top-down" network but has no way to establish whether it is implemented by biology, that may prove to be amusing applied mathematics but it tells us little about how biology, evolution and environment, achieved its functional goals.

Not all real neurons are periodic action potential generators as we have seen of the HH model neurons. Indeed, even the HH model has complex behavior for parameter values outside the range quoted. It is interesting, and perhaps sobering, to have a look at some real neuron data which is quite different, lest one settle into thinking of neurons as periodic oscillators.

In Fig. 2.7, we show a long time series of (scaled) membrane voltage measured from a neuron in a rhythm generator circuit in the digestive system of a California spiny lobster (not the tasty kind, alas). The circuit is comprised of fourteen neurons, see Fig. 2.8. The data is from an isolated LP neuron; it was physically isolated from the rest. The data were taken every 0.2 ms for several minutes. Only a portion of the data is shown.

The membrane voltage shows a nearly periodic oscillation at about 1 Hz, and on top of the peaks of this oscillation occur bursts of spikes with frequencies in the range of 30–50 Hz. While inspection of a time series by eye is not a recommended diagnostic, look carefully at the graphic and you will notice the nonperiodic nature of the signal. The tools for analyzing such a signal are given in [3, 5]. We can achieve some, again visual, insight if we reconstruct a proxy version of the full phase space of the system in a manner which imitates

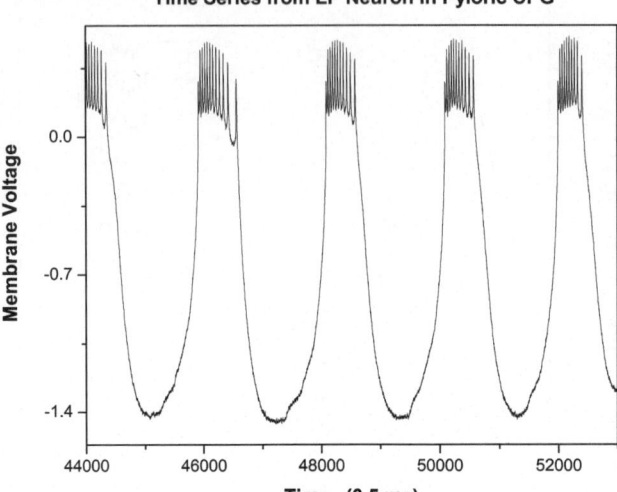

Fig. 2.7. Experimental membrane voltage data from an isolated LP neuron in the Pyloric Central Pattern Generator

the plot of $n(t)$ versus $V(t)$ we presented for the HH model. In Fig. 2.9, we plot the membrane voltage against itself, but time delayed. This is the start of an unfolding of the full phase space of the neuron [5] from the voltage measurements which are a projection of the data. In this figure, we see the region of nearly regular behavior, as well as a structured picture of the spiking activity.

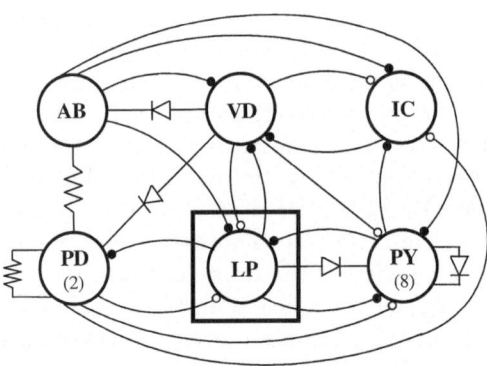

Fig. 2.8. Circuit diagram of the Pyloric Central Pattern Generator. The data in Fig. 2.7 were taken from the LP neuron, in the box, after it was physically isolated from the remainder of the circuit

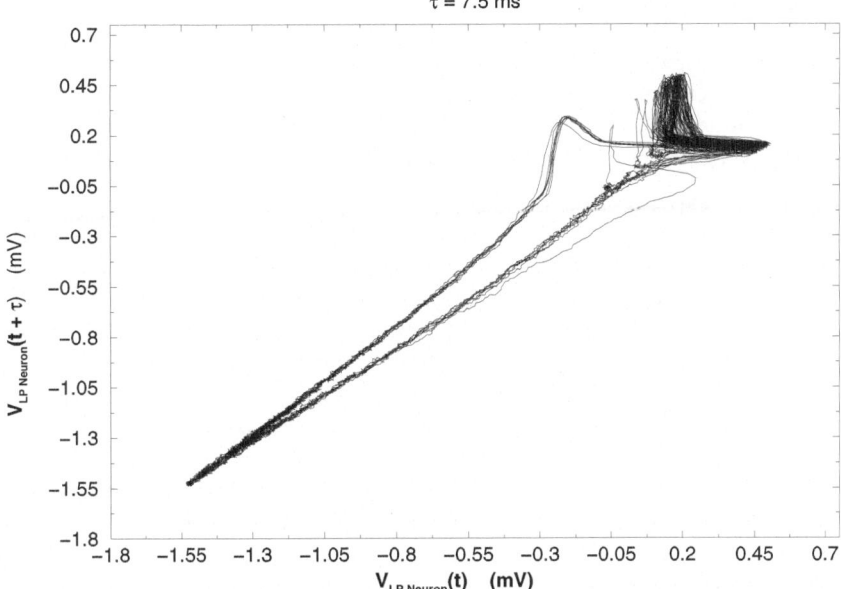

Fig. 2.9. Two dimensional reconstructed phase space plot [5] for the LP membrane voltage. This indicates that two dimensions is not sufficient to unfold the projection of the attractor onto the voltage axis; that takes four dimensions

2.2.2 Neurons-non-points

Of course, neurons are not points; they have spatial structure, and this can be very important. The point neuron which we have constructed so far is an abstraction which represents the idea that the membrane voltage is constant over the spatial extent of a neuron at any given time. To address spatially varying voltages across a nerve cell, it is common practice to make a many-compartment model comprised of the various sections of a neuron (see Chap. 1). Each section is assumed to be an equipotential, and one connects them together by ohmic couplings.

One might construct an HH model determining the voltage in the soma of a neuron $V_S(t)$ and another HH model for a dendrite compartment with potential $V_d(t)$. In the soma equation, one would then include a coupling term $g_{SD}(V_d(t) - V_S(t))$; similarly, in the dendrite equation. This allows one to more realistically represent the distribution of ion channels across a spatially extended neuron and to introduce interesting time delays as potential variations propagate from one part of the neuron to another. We do not explore this here as there is an extensive literature on compartment neuron models [6].

2.2.3 Synapses

The isolated neuron of HH type we have constructed is an interesting dynamical system having so far no functional role in describing biological systems. We must connect these nonlinear oscillators to create a network. There are two major types of connections among neurons:

- Ohmic or gap junction or electrotonic connections are implemented by means of a protein which penetrates the membrane of two adjacent cells and allows the flow of the various intracellular ion species. The current entering an HH neuron when the two connected cells have voltage $V_1(t)$ and $V_2(t)$ is (for the equation for $V_1(t)$) $I_{\text{gapjunction}}(t) = g_{GJ}(V_2(t) - V_1(t))$.
- Chemical synaptic connections, which activate when an action potential signal from a transmitting (presynaptic) cell arrives at the termination of an axon process. This starts processes which release neurotransmitters of various compositions to diffuse across a gap (the synaptic cleft) and dock on receptors penetrating the membrane of the receiving (or postsynaptic) cell. The docking of the neurotransmitter changes the permeability of the channel associated with the receptor and allows ions to flow. When those neurotransmitter molecules undock, the receptor closes down the ion flow. The times for the docking and undocking of the neurotransmitter on receptors vary substantially with the receptor type.

  Each synaptic connection can be represented by a maximal conductance g, a function $S(t)$ taking values between zero and unity representing the percentage of the maximum allowed docked neurotransmitter presently on postsynaptic receptors, and an "ohmic term" $(V_{\text{postsynaptic}}(t) - V_{\text{reversal}})$. This makes for a net current $I_{\text{synaptic}}(t) = gS(t)(V_{\text{postsynaptic}}(t) - V_{\text{reversal}})$.

  - Excitatory synaptic connections have reversal potentials near 0 mV and allow mixtures of Na^+, Ca^{2+} and K^+ ions to flow. They tend to cause the postsynaptic voltage to rise from its resting value upon receipt of a presynaptic action potential, thus exciting the postsynaptic cell toward action potential generation. The rise, often small, in the postsynaptic potential is called an "excitatory postsynaptic potential" or EPSP.

    There are two quite important excitatory synapses we need to know about for our discussion of synaptic plasticity:

    - AMPA receptors, which are the main excitatory connection in many nervous systems and where synaptic strength change (plasticity) occurs. AMPA connections are fast, opening and closing in a few ms.
    - NMDA receptors, which have a very high permeability to Ca^{2+} ions, are blocked by Mg^{2+} ions at low membrane voltages, and allow the entry of Ca^{2+} for order of 100 ms after being opened. In addition to the form of the synaptic current noted above, one must add a multiplicative term

$$B(V) = \frac{1}{1 + 0.288\,e^{-0.062V}}$$

to the NMDA synaptic current to represent the Mg^{2+} block:
$I_{NMDA}(t) = g_{NMDA} S_{NMDA}(t) B(V_{post})(V_{post}(t) - V_{reversal})$.

- Inhibitory synaptic connections have reversal potentials near $-80\,mV$ and allow Cl^- ions to flow. They tend to cause the postsynaptic voltage to fall from its resting value upon receipt of a presynaptic action potential, thus inhibiting the postsynaptic cell from action potential generation. The fall, often small, in the postsynaptic potential is called an "inhibitory postsynaptic potential" or IPSP.

A useful equation for $S(t)$ associates two time constants with its temporal evolution: one for docking the neurotransmitter and one for undocking it. The following describes this:

$$\frac{dS(t)}{dt} = \frac{S_0(V_{pre}(t)) - S(t)}{\tau(S_1 - S_0(V_{pre}(t)))}, \tag{2.9}$$

where $S_0(x)$ is zero for negative arguments and rises rapidly to unity for $x \geq 0$. A convenient form for this is

$$S_0(x) = 0.5\,(1 + \tanh(120(x - 0.1))). \tag{2.10}$$

This equation tells us that when the presynaptic voltage is below $0\,mV$, so no action potential is present, $S(t)$ decays to zero with a time constant τS_1. This is the undocking time. When an action potential arrives at the presynaptic terminal, S_0 rises rapidly to unity, driving $S(t)$ toward unity with a time constant $\tau(S_1 - 1)$. This is the docking time. This formulation applies for both inhibition and excitation when the rise time of the synapse is comparable to the width of the action potential, as in the case of fast excitatory AMPA synapses and fast inhibitory $GABA_A$ synapses. In order to model synapses

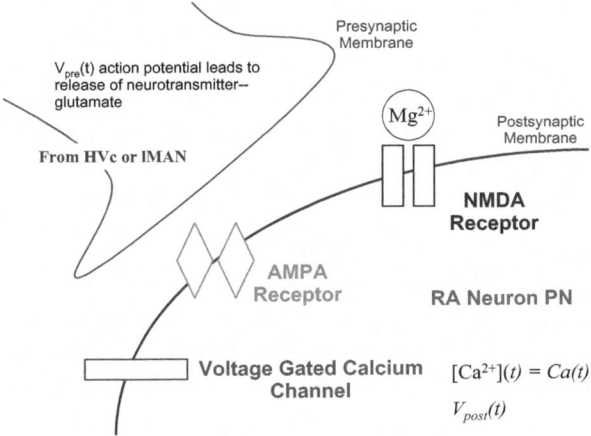

Fig. 2.10. Graphic depicting the arrival of an action potential at a synaptic terminal

with slower rise times such as NMDA, we need coupled first order kinetic equations of the form given in (2.10).

Figure 2.10 is a "cartoon" indicating the action at a synapse and postsynaptic neuron.

The ion channels discussed in the context of the HH model are called voltage-gated channels. The synaptic connections just described are called ligand-gated. There are other synaptic connections which do not allow ion flow after receipt of a presynaptic potential but through changes in the properties of the receptor-induced postsynaptic biochemical processes. These metabotropic receptors are discussed in Chap. 1.

2.3 Synaptic Plasticity

An important dynamical process in nervous systems is the activity dependent change in synaptic strength associated with both inhibitory and excitatory synaptic connections. Though far from "proven", it is widely believed that changes in these connection strengths among neurons produce the rewiring occurring when the nervous systems learns.

As early as 1973, Lomø and Bliss [7] showed that if one presents a series of spikes (a tetanus) with interspike intervals (ISIs) as small as 10 ms, namely a spiking frequency of 100 Hz, to certain hippocampal cells through an excitatory AMPA synapse, the baseline EPSP before and after the presentation of the tetanus showed increased amplitude that persisted for hours after the presentation. This was called "long term potentiation" or LTP. Experiments by Malinow and Miller [8] that presented lower frequency tetani and controlled the postsynaptic voltage at the same time showed that one could induce long lasting decreases in EPSPs. This is called long term depression or LTD.

Experiments in the 1990s showed that both LTP and LTD could be induced at excitatory AMPA synapses by pairing isolated single presynaptic and postsynaptic spikes. An evoked presynaptic spike arrives at the synaptic terminal at t_{pre} and a postsynaptic spike is induced by a short current injection at time t_{post}. As a function of the time difference $\tau = t_{\text{post}} - t_{\text{pre}}$, one observes both LTP and LTD. This is nicely summarized in the data of Bi and Poo [9] shown in Fig. 2.11.

To provide an explanation of these phenomena, several groups [10–12] have made biophysically based plasticity models founded on the observation [13] that postsynaptic Ca^{2+} concentration is critical to inducing the competing biochemical pathways in the postsynaptic cell.

In our formulation [11], we attribute a dynamical degree of freedom $P(t)$ to kinases which lead to potentiation and another $D(t)$ to phosphatases which lead to depression. We hypothesize first order kinematics for each of these

$$\frac{dP(t)}{dt} = f_P(\Delta[Ca^{2+}](t))(1 - P(t)) - \beta_P P(t)$$

$$\frac{dD(t)}{dt} = f_D(\Delta[Ca^{2+}](t))(1 - D(t)) - \beta_D D(t), \qquad (2.11)$$

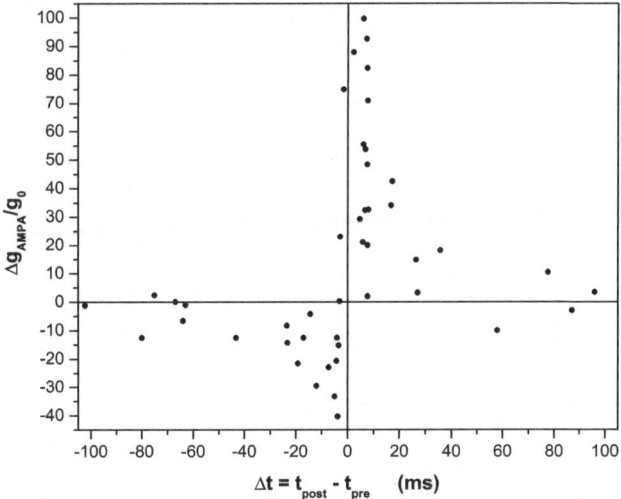

Fig. 2.11. Data from Bi and Poo [9] on LTP and LTD induction by presentation of a presynaptic spike at t_pre and stimulation of a postsynaptic spike at t_post as a function of $\tau = t_\mathrm{post} - t_\mathrm{pre}$

with $f_P(\Delta[\mathrm{Ca}^{2+}](t))$ and $f_D(\Delta[\mathrm{Ca}^{2+}](t))$ functions of the time dependent elevation of intracellular Ca^{2+} over its equilibrium value C_0, and β_P and β_D being rates for the return of each process to zero. The change in excitatory synaptic strength is hypothesized to be proportional to the nonlinear competition between these processes

$$\frac{\mathrm{d}g_E(t)}{\mathrm{d}t} = g_0(P(t)D(t)^\eta - D(t)P(t)^\eta), \qquad (2.12)$$

with g_0 being a baseline conductance.

The competition of these two processes is supported by the data from O'Connor, et al. [14] and presented in Fig. 2.12. To develop these data, the potentiation processes mediated by kinases were first blocked by the application of K252a, and then the depression processes mediated by phosphatases were blocked by okadaic acid. One can clearly see from their very nice data the presence of competing dynamical postsynaptic mechanisms.

The time dependence of intracellular calcium $[\mathrm{Ca}^{2+}](t)$ is determined by a rate equation of the form

$$\frac{\mathrm{d}[\mathrm{Ca}^{2+}](t)}{\mathrm{d}t} = \frac{C_0 - [\mathrm{Ca}^{2+}](t)}{\tau_C} + \mathrm{Sources}(t) \qquad (2.13)$$

where $\tau_C \approx 20\,\mathrm{ms}$ is the relaxation rate of $[\mathrm{Ca}^{2+}]$ back to C_0. The sources include AMPA currents, NMDA currents, and voltage-gated Ca^{2+} channels [11]. Coupling this to a **HH** model of the postsynaptic cell allows us to simulate

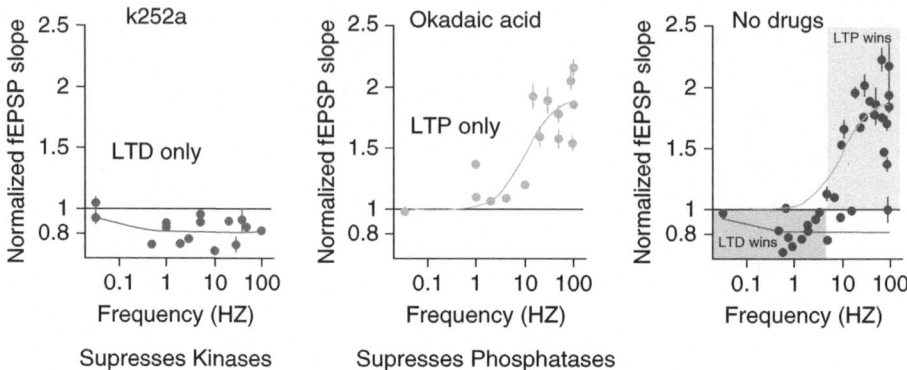

LTD and LTP are separable

Supresses Kinases   Supresses Phosphatases

Fig. 2.12. Separation of the biochemical postsynaptic processes associated with LTP (kinases) and LTD (phosphatases) (adapted from [14])

the effect of an experimental electrophysiological stimulation protocol — for example, one presynaptic spike at t_{pre} and one postsynaptic spike at t_{post} — to determine the postsynaptic membrane voltage $V_{post}(t)$, the intracellular Ca^{2+} time course $[Ca^{2+}](t)$ and from those quantities deduce the change in $g_E(t)$ from its value g_0 before the electrophysiological induction, to the value $g_0(1 + \int_{-\infty}^{\infty} dt((P(t)D(t)^\eta - D(t)P(t)^\eta)))$ after the induction.

In Fig. 2.13, we show the result of a calculation of $\frac{\Delta g(\tau)}{g_0}$ for such a model.

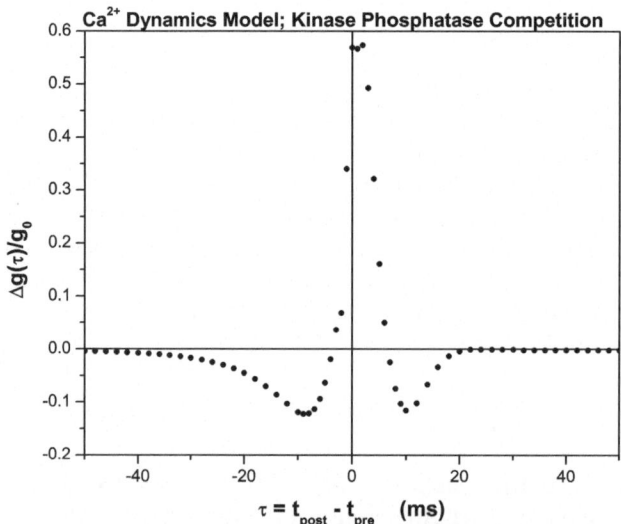

Fig. 2.13. $\frac{\Delta g(\tau)}{g_0}$ for Ca^{2+} dynamics model

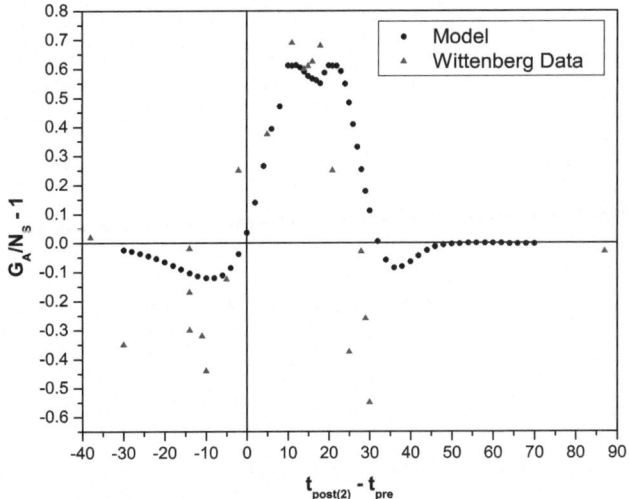

Fig. 2.14. Data from Gayle Wittenberg as a function of $\tau = t_{\mathrm{post}(2)} - t_{\mathrm{pre}}$ compared to our model calculation based on Ca^{2+} postsynaptic dynamics

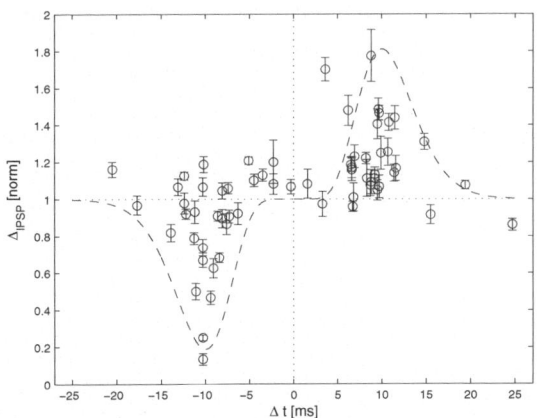

Fig. 2.15. Spike timing plasticity at an inhibitory synapse: Summary results of change in postsynaptic IPSP initial slope, expressed as a function of $\Delta t = t_{\mathrm{post}} - t_{\mathrm{pre}}$. No change is represented by normalized IPSP equal to unity (equal to the baseline value before pairing). Each point represents data from one cell. Change is evaluated as the mean IPSP slope over the 20 min. following pairings, normalized to the mean of the slopes for 15 minutes preceding pairings. Cells for which the change in IPSP slope was significant ($p < .01$, ANOVA) are plotted in blue. Empirical fit to the observed data is given by $\Delta_{\mathrm{IPSP}} = 1 + \frac{g_0}{\beta \mathrm{e}^{-\beta}} \alpha^{\beta} \Delta t |\Delta t|^{\beta-1} \mathrm{e}^{-\alpha|\Delta t|}$, with $g_0 = 0.8$, $\alpha = 1$, and $\beta = 10$

Gayle Wittenberg [15] has performed experiments in which she presents cells in the hippocampus with a single presynaptic spike at t_{pre} and two postsynaptic spikes separated by Δt. She used the time of the second postsynaptic spike and t_{pre} to determine $\tau = t_{\text{post}(2)} - t_{\text{pre}}$, and her results are shown in Fig. 2.14. Along with her data is the result of a calculation by our laboratory for the same process based on a Ca^{2+} dynamics model.

Finally, we note, and we will use below, the experimental evidence for spike timing dependent plasticity at an inhibitory synapse. This is shown in Fig. 2.15 and comes from data by [16].

That ends the lightning fast review (or introduction). As suggested, the introductory book by Johnston and Wu [1], or Chap. 1 expands on these topics in rather more detail.

2.4 Synchronization and Plasticity

It is widely thought, though hardly proven, that synchronization among populations of neurons can play an important role in their performing important functional activity in biological neural networks. Here we look at the microcircuit of one periodically oscillating HH neuron driving another periodically oscillating HH neuron. The question we ask is whether synaptic plasticity has an interesting effect on the ability of these two neurons to synchronize. The answer is yes, and though interesting, does not answer a biological question yet. Let's look at what we can show, then pose some harder questions for further investigation.

The setup we examine is that of a periodically oscillating HH neuron with a period T_1. We control T_1 by injection of a selected level of DC current. This is the "transmitter". The receiver, or postsynaptic neuron, is another HH neuron oscillating with a period T_2^0 before receiving synaptic input from neuron 1. When the synaptic current has begun, the receiver neuron changes its period to T_2. A schematic of the setup is in Fig. 2.16. The coupling is through an excitatory synapse with current $I_{\text{synapse}}(t) = g_E(t)S_E(t, V_{\text{pre}}(t))(V_{\text{post}}(t) - V_{\text{reversal}})$.

We have selected the postsynaptic neuron to be our two-compartment model as described in [17] and set it to produce autonomous oscillations with a period T_2^0. This period is a function of the injected DC current into the somatic compartment. We hold this fixed while we inject a synaptic AMPA current

$$I_{\text{synapse}}(t) = g_{\text{AMPA}}(t)S_A(t)(V_{\text{post}}(t) - V_{\text{rev}}), \qquad (2.14)$$

into the postsynaptic somatic compartment. $V_{\text{post}}(t)$ is the membrane voltage of this postsynaptic compartment. $g_{\text{AMPA}}(t)$ is our time-dependent maximal AMPA conductance, and $S_A(t)$ satisfies

$$\frac{dS_A(t)}{dt} = \frac{1}{\tau_A} \frac{S_0(V_{\text{pre}}(t)) - S_A(t)}{S_{1A} - S_0(V_{\text{pre}}(t))} \qquad (2.15)$$

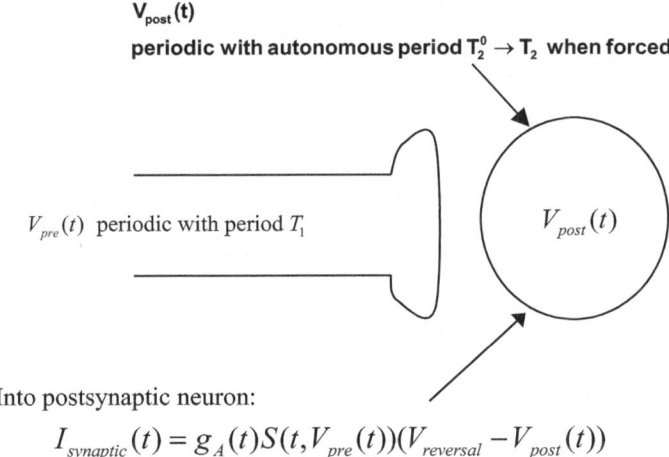

$V_{post}(t)$

periodic with autonomous period $T_2^0 \rightarrow T_2$ when forced

$V_{pre}(t)$ periodic with period T_1

$V_{post}(t)$

Into postsynaptic neuron:

$$I_{synaptic}(t) = g_A(t)S(t, V_{pre}(t))(V_{reversal} - V_{post}(t))$$

Fig. 2.16. Setup for exploring effect of a dynamical synapse on synchronization of two periodic neural HH neurons

as described above. $V_{pre}(t)$ is the periodic presynaptic voltage which we adjust by selecting the injected DC current into the presynaptic HH neuron. We call the period of this oscillation T_1.

When $g_{AMPA} = 0$, the neurons are disconnected and oscillate autonomously. When $g_{AMPA}(t) \neq 0$, the synaptic current into the postsynaptic neuron changes its period of oscillation from the autonomous T_2^0 to the driven value of T_2, which we evaluate for various choices of T_1. We expect from general arguments [18] that there will be regimes of synchronization where $\frac{T_1}{T_2}$ equal ratios of integers over the range of frequencies $\frac{1}{T_1}$ presented presynaptically. This will be true both for fixed g_{AMPA} and when $g_{AMPA}(t)$ varies as determined by our model.

In Fig. 2.17 we present $\frac{T_1}{T_2}$ as function of the frequency $\frac{1000}{T_1}$ (T_1 is given in milliseconds, so this is in units of Hz) for fixed $g_{AMPA} = 0.1 \, \text{mS cm}^{-2}$ and for $g_{AMPA}(t)$ determined from our model. This amounts to a choice for the baseline value of the AMPA conductance. The fixed g_{AMPA} results are in filled upright triangles and, as expected, show a regime of one-to-one synchronization over a range of frequencies. One also sees regions of two-to-one and hints of five-to-two and three-to-one synchronization. These are expected from general arguments on the parametric driving of a nonlinear oscillator by periodic forces.

When we allow g_{AMPA} to change in time according to the model we have discussed, we see (unfilled inverted triangles) a substantial increase in the regime of one-to-one synchronization, the appearance of some instances of

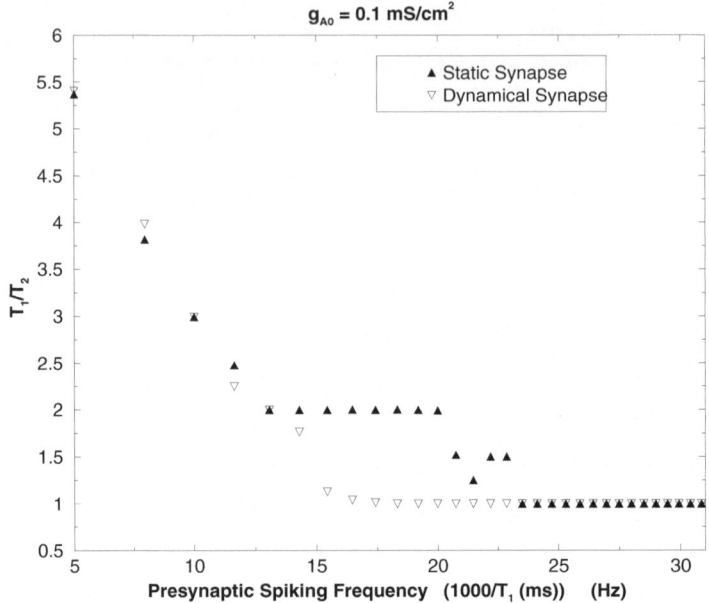

Fig. 2.17. T_1/T_2, the ratio of the interspike interval T_1 of the presynaptic neuron to the interspike interval T_2 of the postsynaptic neuron, is plotted as a function of the presynaptic input frequency, $1000/T_1$ Hz, for a synapse starting at a base AMPA conductance of $g_{\mathrm{AMPA}}(t=0) = 0.1\,\mathrm{mS\,cm}^{-2}$. We see that the one-to-one synchronization window is broadened when the static synapse is replaced by a plastic synapse

three-to-two synchronization, and a much smaller regime with two-to-one synchronization. This suggests that the one-to-one synchronization of oscillating neurons, which is what one usually means by neural synchrony, is substantially enhanced when the synaptic coupling between neurons is allowed to vary by the rules we have described.

2.5 Marking Time Biologically

Biological systems have to mark time to keep pace with events and transmit information. There are three distinct ways known to us by which this is accomplished. One is to use the same principle of a delay line in physics: a signal has a propagation velocity v along some cable, and then a signal traversing a length of this cable L takes a time $\frac{L}{v}$. This is manifest in the interaural time differences used by the barn owl, for example, in actively locating prey or passively detecting sources of sound. With a velocity $v \approx 5\,\frac{m}{s}$ and axon lengths of a few mm, time delays as short as tens of microseconds are used [19–22].

Time delays of the order of hours or days are connected with circadian rhythms and are marked using limit cycle oscillators. A detailed model of the biochemical processes thought to underly the $\approx$ 24 h circadian rhythm is found in recent work by Forger and Peskin [23, 24], where a limit cycle oscillator with a period slightly more than 24 h is identified and analyzed.

Environmental signals are passed on in an animal from its sensory systems to central nervous system neurons as sequences of spikes with interspike intervals (ISIs) of a few to hundreds of milliseconds. Since these spikes are essentially identical in waveform, all information contained in these sequences are in the ISIs. Our focus is on these signals.

There are many examples of sensitive stimulus-response properties characterizing how neurons respond to specific stimuli. These include whisker-selective neural response in barrel cortex [25, 26] of rats and motion sensitive cells in the visual cortical areas of primates [27, 28].

One striking example is the selective auditory response of neurons in the songbird telencephalic nucleus HVC [29–32]. Projection neurons within HVC fire sparse bursts of spikes when presented with auditory playback of the bird's own song (BOS) and are quite unresponsive to other auditory inputs. Nucleus NIf, through which auditory signals reach HVC [32–36], also strongly responds to BOS in addition to responding to a broad range of other auditory stimuli. NIf projects to HVC, and the similarity of NIf responses to the auditory input and the subthreshold activity in HVC neurons suggests that NIf could be acting as a nonlinear filter for BOS, preferentially passing that important signal on to HVC. It was these examples from birdsong that led us to address the ISI reading problem we consider here.

How can these neural circuits be sensitive to specific sequences of ISIs? One way is that they act as a nonlinear filter for such sequences. The circuit is trained on a particular sequence which the animal has found important to recognize with great sensitivity. We identify the ISI sequence we wish to recognize as a set of times $S_{\text{ISI}} = \{T_0, T_1, T_2, \ldots, T_N\}$ coming from a set of spike times $S_{\text{spikes}} = \{t_0, t_1, \ldots, t_{N+1}\}$ with $T_j = t_{j+1} - t_j$. If we have a set of time delay units which is trained to create a signal at τ_j after it receives a spike at time t_j, then we can use the output sequence $\tau_0, \tau_1, \ldots, \tau_N$ coming from the input S_{spikes} as a comparison to the original ISI sequence S_{ISI}. The comparison can be achieved by introducing both into a detection circuit which fires only when two spikes occur within a few ms of each other. If the detection unit fires, the correct ISI T_j has matched the time delay τ_j.

How are we to build a time delay circuit with tunable synapses? For this, we turn to the observations of Kimpo et al. [37] in which they identify a pathway in the birdsong system which quite reliably produces a time delay of 50 ± 10 ms when a short burst of spikes is introduced into its entry point. The importance of this timing in the birdsong systems has been examined in [38]; not surprisingly, given the tone of these lectures, it is connected with a specific timing required by spike timing plasticity in the stabilization of

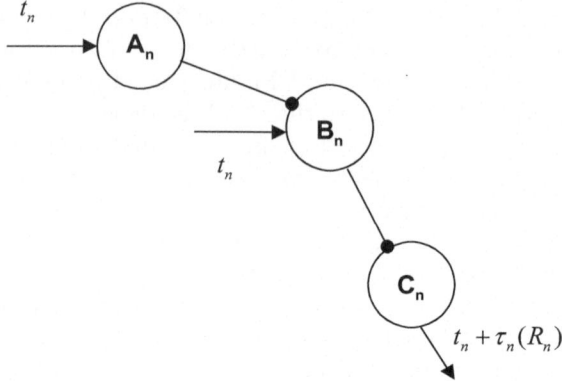

Fig. 2.18. Time delay circuit adapted from the birdsong system

the bird's song. How are we to "tune" τ_j? For this, we use the spike time dependent inhibitory plasticity rules observed by Haas [16].

The time delay circuit is displayed in Fig. 2.18. There are three neurons in the circuit, and they are connected by inhibitory synapses. Neuron A receives an excitatory input signal from some source at time t_n. It is at rest when the source is quiet, and when activated, it inhibits neuron B. Neuron B receives an excitatory input from the same source at the same time t_n. Neuron B oscillates periodically when there is no input from the source. Neuron B inhibits neuron C. Neuron C produces periodic spiking in the absence of inhibition from neuron B. The tunable synaptic strength is that connecting neuron B to neuron C. The dimensionless number R_n is the magnitude of the B→C maximal conductance relative to some baseline conductance.

When the inhibition from neuron B to neuron C is released by the inhibitory signal from neuron A to neuron B, neuron C rebounds and produces an action potential some time later. This is due to the intrinsic stable spiking of neuron C in the absence of any inhibition from neuron B.

This time delay is dependent on the strength of the B→C inhibition, as the stronger that is set the further below threshold neuron C is driven and the further it must rise in membrane voltage to reach the action potential threshold. This means the larger the B→C inhibition, the longer the time delay produced by the circuit. Other parameters in the circuit, such as the cellular membrane time constants, set the scale of the overall time delay.

The direct excitation of neuron B by the signal source is critical. It serves to reset the phase of the neuron B oscillation, as a result of which the spike from neuron C is measured with respect to the input signal and thus makes the timing of the circuit precise relative to the arrival of the initiating spike.

Without this excitation to neuron B, the phase of its oscillation is uncorrelated with the arrival time of a signal from the source, and the time delay of the circuit varies over the period of oscillation of neuron B. This is not a desirable outcome.

We have constructed this circuit using HH conductance based neurons and realistic synaptic connections. The dynamical equations for the three HH neurons shown in Fig. 2.18 are these:

$$C_M \frac{dV_i(t)}{dt} = g_{\text{Na}} m(t, V_i(t))^3 h(t, V_i(t))(V_i(t) - V_{\text{Na}})$$
$$+ g_{\text{K}} n(t, V_i(t))^4 (V_i(t) - V_{\text{K}}) + g_L(V_i(t) - V_L)$$
$$+ g_{ij}^I S_I(t)(V_i(t) - V_{\text{revI}}) + I_i^{syn}(t) + I_i^{DC}, \qquad (2.16)$$

where $(i, j) = [\text{A, B, C}]$. The membrane capacitance is C_M, and $V_{\text{Na}}, V_{\text{K}}, V_L$, and V_{revI} are reversal potentials for the sodium, potassium, leak, and inhibitory synaptic connections, respectively. $m(t), h(t)$, and $n(t)$ are the usual activation and inactivation dynamical variables. g_{Na}, g_K and g_L are the maximal conductances of sodium, potassium and leak channels respectively. I_i^{DC} is the DC current into the A, B or C neuron. These are selected such that neuron A is resting at $-63.74\,\text{mV}$ in the absence of any synaptic input, neuron B is spiking at around 20 Hz, and neuron C would also be spiking at around 20 Hz in the absence of any synaptic inputs.

$I_i^{syn} = g_i^E S_E(t)(V_i(t) - V_{\text{revE}})$ is the synaptic input to the delay circuit at neuron A and B. It receives a spike from the signal source at time t_0; $g_i^E = (g_A, g_B, 0)$. The nonzero inhibitory synaptic strengths g_{ij}^I in the delay circuit are $g_{BA} = R^0 g_I$ and $g_{CB} = R g_I$. The dimensionless factors, R and R^0, set the strength of A→B and B→C inhibitory connections respectively, relative to baseline strength g_I, which is set to $1\,\text{mS cm}^{-2}$ in all the calculations presented here.

$g_A = g_B = 0.5\,\text{mS cm}^{-2}$. $g_I = 1\,\text{mS cm}^{-2}$, $R^0 = 50.0$, and R varies as given in text. $V_{\text{revE}} = 0\,\text{mV}$, and $V_{\text{revI}} = -80\,\text{mV}$. $\tau_E = 1.0\,\text{ms}$, $S_{1E} = 1.5$, $\tau_I = 1.2\,\text{ms}$, $S_{1I} = 4.6$. The DC currents in the neurons are taken as $I_A^{DC} = 0.0\,\mu\text{A cm}^{-2}$, $I_B^{DC} = 1.97\,\mu\text{A cm}^{-2}$ and $I_C^{DC} = 1.96\,\mu\text{A cm}^{-2}$.

$S_E(t)$ represents the fraction of excitatory neurotransmitter docked on the postsynaptic cell receptors as a function of time. It varies between 0 and 1 and has two time constants: one for the docking time of the neurotransmitter and one for its release time. It satisfies the dynamical equation:

$$\frac{dS_E(t)}{dt} = \frac{S_0(V_{\text{pre}}(t)) - S_E(t)}{\tau_E(S_{1E} - S_0(V_{\text{pre}}(t)))}. \qquad (2.17)$$

The docking time constant for the neurotransmitter is $\tau_E(S_{1E} - 1)$, while the undocking time is $\tau_E S_{1E}$. For neurons A and B, the presynaptic voltage is given by the incoming spike or burst of spikes arriving from some source at time t_0. For our excitatory synapses, we take $\tau_E = 1\,\text{ms}$ and $S_{1E} = 1.5$, for

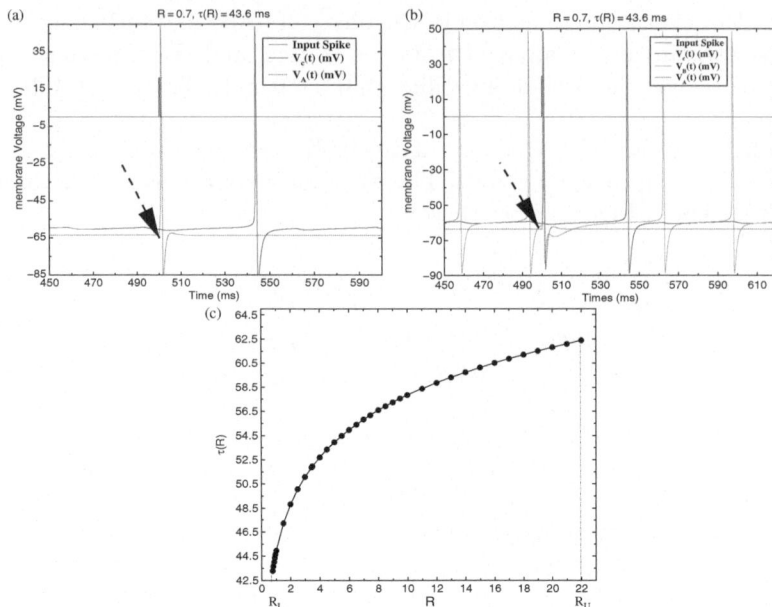

Fig. 2.19. (a) For $R = 0.7$, we show the membrane voltages of neuron A (*blue*) and neuron C (*red*) in response to single spike input (*black*) arriving at neuron A and neuron B at time $t_0 = 500$ ms. We see the output spike from neuron C occurring at $t = 543.68$ ms, corresponding to $\tau(R) = 43.6$ ms; (b) For $R = 0.7$ we again show the membrane voltages of neuron A (*blue*) and neuron C (*red*), and in addition now display the membrane voltage of neuron B (*green*). A single spike input (*black*) arrives at time $t = 500$ ms. We see that the periodic action potential generation by neuron B is reset by the incoming signal; (c) The delay $\tau(R)$ produced by the three neuron time delay unit as a function of R, the strength of the inhibitory synaptic connection B→C. All other parameters of the time delay circuit are fixed to values given in the text. For $R < R_L$, the inhibition is too weak to prevent spiking of neuron C. For $R > R_U$, the inhibitory synapse is so strong that neuron C does not produce any action potential, so effectively the delay is infinity. In Figs. 2.19(a) and 2.19(b), the arrows indicate the time of the spike input to units A and B of our delay unit

a docking time of 0.5 ms and an undocking time of 1.5 ms. These times are characteristic of AMPA excitatory synapses.

Similarly, $S_I(t)$ represents the percentage of inhibitory neurotransmitter docked on the postsynaptic cell as a function of time. It satisfies the following equation:

$$\frac{\mathrm{d}S_I(t)}{\mathrm{d}t} = \frac{S_0(V_{\mathrm{pre}}(t)) - S_I(t)}{\tau_I(S_{1I} - S_0(V_{\mathrm{pre}}(t)))}, \tag{2.18}$$

aa

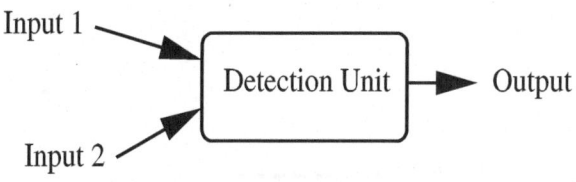

b

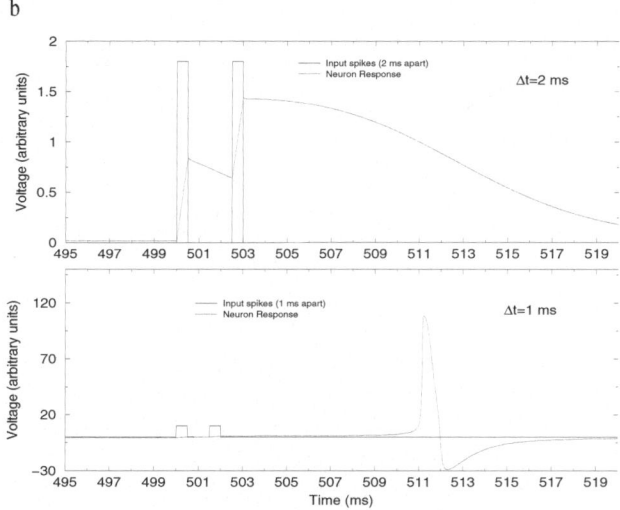

Fig. 2.20. (a) Schematic of the detection unit. It receives two input spikes with various time intervals between them. It responds with a spike if the two inputs are within 1 ms of each other; (b) *Top panel* The scaled response of the detection unit when two inputs arrive within 2 ms of each other. We see that the integrated input arriving at this delay does not result in neuron spiking. In the *bottom panel*, we show the scaled neuron response to two input spikes arriving within 1 ms of each other. The detection unit produces a spike output, indicating coincidence detection

where we select $\tau_I = 1.2$ ms and $S_{1I} = 4.6$ for a docking time of 4.32 ms and undocking time of 5.52 ms. The range of time delays produced by the three neuron delay circuit depends sensitively on the docking and undocking times of this synapse.

For these values, we find $\tau(R)$ as shown in Fig. 2.19(c). For R too small, $R < R_L$ in Fig. 2.19(c), the inhibition from B $\to$ C does not prevent the production of action potentials. For R too large, $R > R_U$ in Fig. 2.19(c), the C neuron is inhibited so strongly that it never spikes. Over the range of $R_L \le R \le R_U$, we typically find that $\tau(R)$ has a range of about 20 ms within an overall scale of about 10 ms to 100 ms.

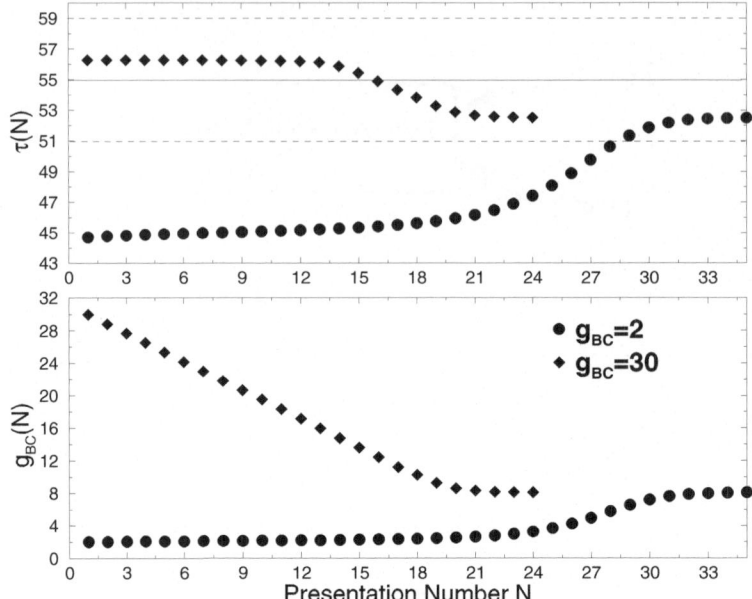

Fig. 2.21. Training an IRU to learn an ISI of $T = 55$ ms. The initial values of $g_{BC}(N = 0)$ are set to explore the two scenarios described in the text. $\tau(R)$ (*top panel*) and g_{BC} (*bottom panel*) are plotted as function of the number of presentations of the training sequence N. The resolution limit $\delta = 4$ ms is shown in dotted lines for $\tau(R)$ and $T = 55$ ms is shown as a solid line

The ISI recognition unit is now constructed as follows:

- A circuit we call the "spike separation unit" [2] separates the individual spikes at time t_0, t_1, t_{N+1} in the sequence. These are then presented to a set of time delay units in pairs $\{t_j, t_{j+1}\}$.
- After introducing an excitatory feedback from neuron C to neuron C of each time delay unit, we use the inhibitory plasticity rule in Fig. 2.15 to adjust R_j until $\tau_j \approx t_{j+1} - t_j = T_j$. When this occurs, the training is completed.
- A "detection unit" comprised of a neuron or neurons fires when two spikes within a few ms of each other are received, but not when a single spike is received. The operation of the detection unit is illustrated in Fig. 2.20. When the detection unit fires, the information that the replica ISI sequence $\tau_0(R_0), \tau_1(R_1), \ldots, \tau_N(R_N)$ has matched the desired sequence S_{ISI} is passed on to other functions.

As an example of the training of a delay unit, we show in Fig. 2.21 the training of one time delay unit to a time delay of 55 ms. The spike sequence $t_0, t_1 = t_0 + T_0$ with $T_0 = 55$ ms, is presented $N = 0, 1, 2, \ldots$ times. We present the spike sequence many times $N = 0, 1, 2, \ldots$ to the IRU to train the time delays to accurately reflect the individual ISIs in the sequence. In Fig. 2.21, we show results from training two IRU units tuned to detect an ISI of $T = 55\,\text{ms}$. The first IRU has $g_{BC}(N = 0) = 2$, corresponding to $\tau(R) \approx 43\,\text{ms}$, so $T > \tau(R)$. The second has $g_{BC}(N = 0) = 30$ leading to $\tau(R) \approx 60$ ms, so $T < \tau(R)$. Each IRU trains itself on the given ISI input presented $N = 1, 2, \ldots$ times. In the detection unit, we set the time within which the spikes must arrive at 4 ms. This is a resolution which approximates the refractory period of a typical neuron.

As a final comment, we return to the matter which led to a discussion of the sensitivity of a neural circuit, such as that in the sensory-motor junction of the birdsong system [32]. We suggest, based on the construction discussed here, that within NIf (or possibly before NIf in the auditory pathway) there will be neural structures which select for specific ISI sequences. In the first phase of song learning, called the sensory phase, we suggest that a genetically determined circuit with time delays nearly those appropriate to the tutor's song is tuned by repeated presentation of the song. This tunes the nonlinear filter for ISI sequences we just constructed. Once this has happened, the birdsong development moves into its next phase, the sensori-motor phase, where the muscles of the bird's songbox are trained by other plasticity events [38] through auditory feedback, filtered by the ISI recognition structure. Once the bird's own song matches the tutor's song, the training of the songbox is completed.

Acknowledgements

This work was partially funded by a grant from the National Science Foundation, NSF PHY0097134. HDIA and SST are partially supported by the NSF sponsored Center for Theoretical Biological Physics at UCSD.

References

1. Johnston, D. and S. Miao-Sin Wu: *Foundations of Cellular Neurophysiology*, (MIT Press, Cambridge, MA, 1994).
2. Abarbanel, H. D. I. and S.S. Talathi: Phys. Rev. Lett. **96**, 148104 (2006).
3. Rabinovich, M. I., P. Varona, A. I. Selverston, and H. D. I. Abarbanel: Rev. Mod. Phys. **78**, 1213–1265 (2006).
4. Fall, C. P., E. S. Marland, J. M. Wagner, and J. J. Tyson, Editors, *Computational Cell Biology*, (Springer-Verlag, New York, 2002).
5. Abarbanel, H. D. I: *Analysis of Observed Chaotic Data*, (Springer-Verlag, New York, 1996).
6. Traub, R. D. and R. Miles: *Neuronal Networks of the Hippocampus*, (Cambridge University Press, 1991).
7. Bliss T. V. and T. I. Lomø: J. Physiol. **232**, 331–356 (1973).
8. Malinow, R. and J. P. Miller: Nature **320**, 529–530 (1986).

9. Bi, G-Q and M-m Poo: Annu. Rev. Neurosci. **24**, 139–166 (2001).
10. Castellani, G. C., E. M. Quinlan, L. N. Cooper, and H. Z. Shouval: Proc. Natl. Acad. Sci. USA. **98**, 12772–12777 (2001).
11. Abarbanel, H. D. I., L. Gibb, R. Huerta, and M. I. Rabinovich: Biol Cybernetics. **89**, 214–226 (2003).
12. Rubin, J. E., R. C. Gerkin, G.-Q. Bi, C. C. Chow: J. Neurophysiol. **93**, 2600–2613 (2005).
13. Yang, S.N., Y. G. Tang and R. S. Zucker: J Neurophysiol **81**, 781–787, (1999).
14. O'Conner, D., G. Wittenberg and S. S. Wang: Proc. Natl. Acad. Sci. USA. **102** 9679–9684 (2005); O'Conner, D., G. Wittenberg, and S. S. Wang: J. Neurophysiol. **94**, 1565–1573 (2005).
15. Wittenberg, G. Learning and memory in the hippocampus: Events on millisecond to week times scales. PhD Dissertation, Princeton University, (2003).
16. Haas, J., T. Nowotny and H. D. I. Abarbanel: J. Neurophys. **96**, 3305–3313 (2006).
17. Abarbanel, H.D.I., S. S. Talathi, L. Gibb, M.I. Rabinovich: Phys. Rev. E, **72**, 031914 (2005).
18. Drazin, P. G.: *Nonlinear Systems*, (Cambridge University Press, Cambridge, 1992).
19. Carr, C. E. and M. Konishi: Proc. Nat. Acad. Sci. USA. **85**, 8311–8315 (1988).
20. Carr, C. E. and M. Konishi: J. Neurosci. **10**, 3227–3246 (1990).
21. Knudsen E.L. and M. Konishi: J. Neurophysiol. **41**, 870–884 (1996).
22. Köppel, C: J. Neurosci. **17**, 3312–3321 (1997).
23. Forger, D. B. and C. S. Peskin: Proc. Natl. Acad. Sci. USA **100**, 14806–14811 (2003).
24. Forger, D. B. and C. S. Peskin: J. Theor. Biol. **230**, 533–539 (2004).
25. Welker, C: J. Comp. Neurol. **166**, 173–189 (1976).
26. Aarabzadeh, E., S. Panzeri and M. E. Diamond: J. Neurosci. **24**, 6011–6020 (2004).
27. Sugase, Y., S. Yamane, S. Ueno and K. Kawano: Nature, **400**, 869–873 (1999).
28. Buracas, G.T., A. M. Zador, M. R. DeWeese, T. D. Albright: Neuron. **9**, 59–69 (1998).
29. Lewicki, M. and B. Arthur: J. Neurosci. **16**, 6897–6998 (1996).
30. Margoliash, D: J. Neurosci. **3**, 1039–1057 (1983).
31. Margoliash, D: J. Neurosci. **6**, 1643–1661 (1986).
32. Coleman, M. and R. Mooney: J. Neurosci. **24**, 7251–7265 (2004).
33. Janata, P and D. Margoliash: J. Neurosci. **19**, 5108–5118 (1999).
34. Carr, C. E.: Ann. Rev. Neurosci. **16**, 23–243 (1993).
35. Cardin, J. A., J. N. Raskin and M. F. Schmidt: J. Neurophysiol. **93**, 2157–2166 (2005).
36. Rosen, M. J. and R. Mooney: J. Neurophysiol. **95**, 1158–1175 (2006).
37. Kimpo, R. R., F. E. Theunissen and A. J. Doupe: J. Neurosci. **23**, 5750–5761 (2003).
38. Abarbanel, H. D. I., S. S. Talathi, G. B. Mindlin, M. I. Rabinovich and L. Gibb: Phys. Rev. E. **70**, 051911, (2004).

Part II

Complex Networks

3

Structural Characterization of Networks Using the Cat Cortex as an Example

Gorka Zamora-López, Changsong Zhou and Jürgen Kurths

Nonlinear Dynamics Group, University of Potsdam
gorka@agnld.uni-potsdam.de

Summary. In this chapter, *Graph Theory* will be introduced using cat corticocortical connectivity data as an example. Distinct graph measures will be summarized and examples of their usage shown, as well as hints about the kind of information one can obtain from them. Special attention will be paid to *conflicting* points in graph theory that often generate confusion and some algorithmic tips will be provided. It is not our aim to introduce graph theory to the reader in a detailed manner, nor to reproduce what other authors have written in several extensive reviews (see Sect. 3.8).

Some of the examples placed in this chapter referring to the cat cortex are unpublished material and thus, not to be regarded as established scientific results. Otherwise, references will be provided.

3.1 Introduction

A network is an abstract manner of representing a broad range of real systems in order to be mathematically tractable. Elements of that system are represented by vertices and their interactions by links, often giving rise to complex topological structures. Links could illustrate some real physical connection: in roadmaps, cities are represented by dots (vertices) and roads by lines (links); neurons (vertices) connect to each other through axons and synapses (links); the Internet is formed by computers or servers (vertices) connected by cables (links). Abstract concepts are also suitable to be translated into networks: in physics, regular solids are represented as crystal lattices; in the social sciences, vertices might represent people and a link between them could be placed when two persons are friends; the World Wide Web is also an abstract network where vertices are web pages linked by hyperlinks pointing from one web page to another; in ecological food webs, species (vertices) are linked depending on their hierarchy in the web.

In the universe of networks, three basic types of graphs are found: *simple graphs*, whose vertices are connected by edges without directional information; *digraphs* (directed graphs), whose directed links are drawn by arrows (arcs);

and *weighted graphs*, when links represent some scalar magnitude. The weights given to a link depend on the specific system under study. In a transportation network, weights could either represent the physical distance between two cities or the number of passengers, cars, trains, etc. that travel from one city to another. In order to simplify the analytical treatment, most theoretical work has been focused in the study of "graphs". In this chapter on the contrary, the general properties of both directed and weighted networks will be introduced because corticocortical connectivity data belong to this class.

3.1.1 Brief Historical Review

Historically, the study of networks has been the domain of a branch of discrete mathematics known as *graph theory*. Since its birth in 1736, when the Swiss mathematician Leonhard Euler published the solution to the Königsberg bridge problem (consisting in finding a round trip that traversed each of the bridges of the Prussian city of Königsberg exactly once), the study of networks turned out to be useful in many different contexts. In the social sciences, the practical use of graph theory started as early as the 1920s. During the 1990s thanks to the advances in computation, the handling of very large data sets became affordable for the first time, and thus the study of the interconnectivity of many real systems became possible. The field experienced a rapid growth and nowadays is mainly known under the name of *Complex Networks*; its influence can be seen in different disciplines like sociology, life sciences, technology, physics, economics, politics, etc. We will keep in mind that network theory is just a data analysis toolkit flexible enough to be applied in many different contexts.

Apart from structural characterization, special attention has been paid in recent years to dynamical processes of networks in an attempt to understand the bridge between structure and function, which is important for the study of neuroscience. On one hand, dynamical processes can happen *within* a network, like electrical current flowing through a electrical circuit or when vertices represent some dynamical system as oscillators, neurons, etc (cf. Chaps. 1, 5–14). In this case, we might raise the question of how the complex interconnectivity of elements affects both individual and collective behavior. On the other hand, the structure of the network itself might change in time (i.e. *plasticity processes* happening in neural networks), affecting its dynamical properties (cf. Chaps. 2, 5, 7). Understanding the interrelation between structure and dynamics (function) could allow, for example, the design of flexible technological networks where deliberate change of connections could optimize the "maximum service/minimum cost" problem. In the life sciences, it could provide understanding of how internal dynamical processes drive and regulate structural changes leading to *self-organization* in the system.

3.1.2 Complexity, Networks and the Nervous System

Complex systems are typically characterized by a large number of elements that interact nonlinearly. Successful mathematical methods to treat such systems are still a demanding challenge for current and near future research. Classically, nature has been abstracted and broken up into different pieces called systems and further partitioned into semi-independent subsystems susceptible to being separately studied. Once different pieces are understood, the answers are summed up to provide a more global understanding (what is known as the *principle of linear superposition.*) The small contribution of nonlinear interactions and other unknown contributions are simply classified as noise. The Fourier transformation and its applications (a method to decompose a function as a linear superposition of some other functions) is a good example of what can be mathematically achieved thanks to the assumption of linearity. Even for systems composed of very many elements, provided linear interactions dominate, statistical methods allow their description by means of "macroscopic" properties.

But, what if a system cannot be broken up into such independent pieces? Or, what if the system can be partitioned but its overall behavior cannot be described as the linear sum of its components? We are then facing a *complex system.* We are here to show another ingredient that causes complexity in real systems: *the intricate connectivity of interactions among its elements.* Solid crystals have regular structures providing symmetries that simplify their macroscopic description. In many other systems, like star clusters or clouds of charged particles, all elements interact with each other allowing a macroscopic description in terms of "mean fields". The intricate connectivity of many real systems, on the contrary, makes such simplifications impossible. It is rather true that the complexity of the nervous system arises from many different aspects: the coexisting spatio-temporal scales of its dynamics, genetic regulation, molecular organization, the mixture of electrical and molecular signals for communication, intricate connectivity among elements at different scales etc. Each of these features require the use of specific tools and scientific methodologies, making modern neuroscience a highly interdisciplinary field.

What we are trying to emphasize in this book, and especially in this chapter, is that the complex inter-connectivity among the elements comprising the brain (at different scales) is an important aspect to be understood. Therefore, we present Graph Theory as a suitable data analysis toolkit.

3.2 The Cat Cortical Connectivity Network

One of the principal enigmas of biology, and the central issue in physiology, is the intrinsic relationship between the physical geometry of biological structures and their function. The study of structures of biological systems at different scales has been enabled by advances in optical devices, imaging and spectroscopy. However, knowledge about structure at different levels of organization in the nervous system and interconnectivity is still far from being satisfactory [1]. Knowledge about mesoscopic structures and microscopic connectivity is largely missing. Even at the macroscopic scale, interconnectivity between different cortical regions is only known for few mammal species: macaque and cat (partially that of rats also), although further work is still necessary to confirm and improve the existing data. In the case of humans, no harmless tracing technique is available yet to obtain a comprehensive map of the anatomic cortical connectivity, although potential use of non-invasive techniques such as DTI (Diffusion Tensor Imaging) from Magnetic Resonance Imaging (MRI) data is already under study, as well as post-mortem tracing studies (see Chap. 4).

For the exercises and scientific tasks of the 5th *Helmholtz Summer School on Supercomputational Physics*, the corticocortical connectivity data of the cat was chosen because it is, for the moment, the most complete of its type.[1] This data summarizes the corticocortical connections, where links represent the bundles of axons projecting between distant cortical areas through the white matter. Connections are classified as weak "1", medium "2" or strong "3" depending on the diameter of the fibres. The current data is a collation of previous reports performed by Jack W. Scannell during his Ph.D. and presented in various versions [3,4] including the thalamo-cortical connections. The network has been extensively analyzed. It has been found to be clustered and hierarchically organized both in its structural [5–7] and in its functional [8] connectivity. This network also has "small-world properties" [9]. More about its network properties will be studied in this chapter and in Chap. 4.

In Fig. 3.1, the parcelation scheme used by Scannell et al. [3] for both lateral and medial views of a single cortical hemisphere are shown as

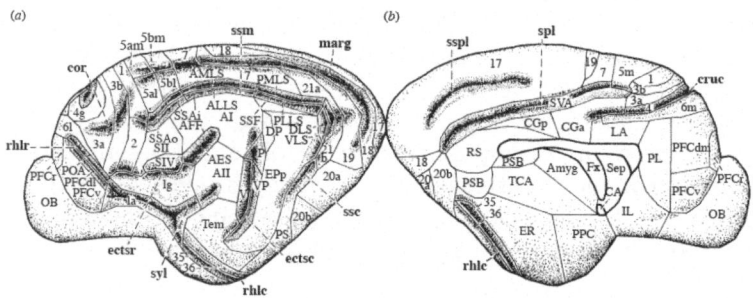

Fig. 3.1. Parcelation of a single hemisphere of the cat cortex. Reprinted from [2] with permission. Lateral view (*left*) and medial view (*right*)

---

[1] It is important to mention current efforts to improve the connectivity data of the macaque cortex, see http://www.cocomac.org, that will become the best reference in the near future.

presented in [2]. In Fig. 9.2 of Chap. 9, the corticocortical connectivity data of the cat is shown in *adjacency matrix* form. A_{ij} means that area i (row) projects a bundle of axons into area j (column). The four major clusters on the diagonal represent internal connections between areas within the visual, auditory, somatosensory-motor and frontolimbic cortex, while "off-diagonal" connections represent information exchange paths between different functional clusters. We show here the network with 53 cortical areas since this was the version used for computation during the Summer School.

3.3 Network Characterization and the Cat Cerebral Cortex

Vertices of networks are labeled with numbers from 1 to N and links between vertices i and j are represented as entries of a matrix: $A_{ij} = 1$ if i connects to j, and $A_{ij} = 0$ otherwise. This is known as the *adjacency matrix* of the network (Fig. 3.2(a)). When computing large networks, matrices become very inefficient in terms of memory allocation. A network of N vertices requires a matrix of N^2 elements, where most of the entries will be zero because real networks tend to be sparse (look at Table 1 of [10]). Thus, large amounts of memory are being wasted. Another way to represent networks is through *adjacency lists* (Fig. 3.2(b)), which consist of N lists containing only the vertices to which vertex i projects. In this case, the amount of required memory is proportional to the number of connections m in the network. All we need is a list with the N vertices and the N lists of the neighbors of each vertex. This makes a total memory requirement of $\mathcal{O}(N + \sum_i k_i) = \mathcal{O}(N + m)$, where k_i is the number of neighbors of each vertex.

For example, a network with $10{,}000$ vertices requires a matrix of 10^8 elements translating into ~ 95 MB if entries are taken as 8 bit (1 Byte) integers. Imagine our network has a density of connections $\rho = 0.1$ ($z \approx 1000$ links per vertex). The number of connections is then $m = \rho \cdot N \cdot (N - 1) = 9{,}999{,}000$.

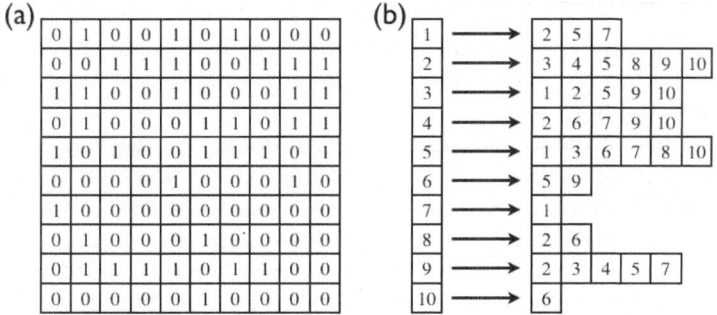

Fig. 3.2. Representation of networks: **a)** adjacency matrix and; **b)** adjacency lists

Note that we now need 16 bit (2 Byte) integers since the indices of vertices go up to 10000, thus the memory required in an *adjacency list* representation is only ~ 20 MB.

Although adjacency lists reduce memory requirements, performance may be sacrificed. Some operations run faster with adjacency lists, i.e. to calculate the degree of each node we only need to ask the system to return the size of each list while in a matrix, all elements have to be read and nonzero elements counted. However, the calculation of most graph theoretic measures requires one to enquire whether two vertices are connected or not, and this operation on a matrix is very fast because when we code something like "if net(n$_1$,n$_2$) == 1" the system only needs to visit the specific memory location for (n_1,n_2) and read the value. Using adjacency lists, the system has to look through all values in the list for node n_1 and compare each value to n_2 until it is found or the end of the list is reached. In Chap. 11, more will be discussed about suitable data structures for representing networks and performing simulations of dynamical networks.

3.3.1 Degree Distributions and Degree Correlations

The most basic property of nodes is the *vertex degree* k_i which represents the number of connections a vertex i has; this is the building block for other structural measures. In networks without *self-loops* (links going out of a node and returning to itself) and *multiple links* (more than one link connecting two vertices), the degree equals the number of neighbors of i. In *directed graphs* (digraphs), the number of connections leaving from i and the number of links that enter i are not necessarily equal, so the degree splits up into the *input degree* $(in\text{-}k_i)$ and *output degree* $(out\text{-}k_i)$; most degree-based structural measures also split accordingly. Given a network with N vertices and M connections:

$$\sum_{i=1}^{N} in\text{-}k_i = \sum_{i=1}^{N} out\text{-}k_i = M$$

In the context of weighted graphs, the natural counterpart of vertex degree is the *vertex intensity* S_i, defined as the sum of weights of i's connections. The directed versions $in\text{-}S_i$ and $out\text{-}S_i$ can equally be defined summing the weights of the input connections that i receives or the weights of the output connections leaving from i.

In order to obtain statistical information about the degrees in large networks, the *degree distribution* $P(k)$ is measured and defined as the probability that a randomly chosen vertex has degree k. Quantitatively, it is measured as the fraction of nodes in the network that have degree k, or estimated from a histogram of the degrees. However, for many real networks with scale-free or exponential distribution, a histogram provides poor statistics at high degree vertices and the *cumulative degree distribution*, $P_c(k)$, is recommended. It is defined as the fraction of nodes in the network with degree larger than k.

$P_c(k)$ requires no binning of data and is a monotonous decreasing function of k, which makes it a better measure to estimate the exponent of scale-free and exponential distributions. The only difference is that if a network has a scale-free degree distribution with exponent, $P(k) \sim k^{-\gamma}$, then the measured exponent in the cumulative distribution is $P_c(k) \sim k^{-\alpha}$ where $\alpha = \gamma - 1$.

Degree distribution alone does not tell one very much about the internal structure of a network, and other measures are required. It is interesting, for example, to look for correlations between degrees of vertices. A network will be called *assortative* when high degree vertices connect preferentially to each other, and low degree vertices to each other (correlation is thus an increasing function of k). A network is called *disassortative* when high degree vertices preferentially connect to low degree vertices (correlation is a decreasing function of k). Interestingly, as pointed out in [10], social networks are observed to show assortative behavior while many other networks (technological networks, biological networks) tend to be disassortative. Formally, degree-degree correlations are expressed by the *conditional probability* $P(k|k')$ that may be problematic to evaluate due to finite size effects.

A popular measure to evaluate degree correlations is the *average neighbors' degree*, $k_{nn}(k)$ introduced in [11]. Firstly, for each vertex i, the average degree of its neighbors is calculated ($k_{nn,i}$). Then, these values are again averaged for all nodes having degree k. In the case of directed networks, things become confusing and we may not know what to look for. Below we show only two cases:

1. is the *out-k_i* of vertex i correlated to its output neighbors' *in-k_j* degree? (Fig. 3.3(a)).
2. is the *in-k_i* of vertex i correlated to its input neighbors' *out-k_j*? (Fig. 3.3(b)).

The extension of these measurement to weighted networks consists in replacing the degrees by intensities and adjacency matrices by their weighted counterparts as presented in [12].

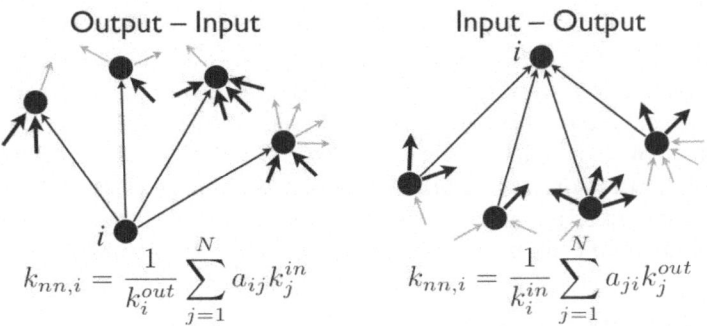

Fig. 3.3. Schematic representation of two possible combinations to calculate $k_{nn}(k)$ in directed networks

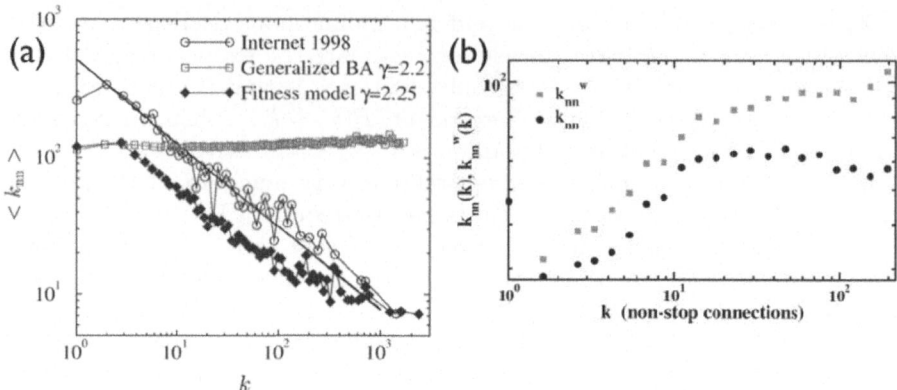

Fig. 3.4. Neighbors' degree as a measure for degree correlation: (a) $k_{nn}(k)$ of the Internet and the Barabási-Albert model. Reprinted with permission from [11]; (b) Comparison of $k_{nn}(k)$ to its weighted version for the WAN (World Airport Network) as defined in [12]. Reprinted with permission

Figure 3.4(a) (from [11]) shows the average neighbors' degree of the Internet in the year 1998 to be *disassortative*. The Internet is known to have a scale-free degree distribution with an exponent $\gamma \approx 2.2$. However, a modified Barabási-Albert model (BA) that constructs a network with similar exponent displays uncorrelated degrees, failing to catch the internal structure of the Internet. In Fig. 3.4(b) (from [12]), average neighbors' degree of the World Airport Network (WAN) is shown. In the unweighted case, connection flights between two airports are considered, while in the weighted case, the intensity of the connections represents the number of passengers flying from one airport to another in direct (non-stop) flights. The network has assortative behavior, more pronounced in the weighted case. This example illustrates the loss of information when, for simplicity, the unweighted version of a weighted network is considered. Note that in Fig. 3.4(b) the weighted $k_{nn}^w(k)$ is plotted against k only for comparative reasons, but it is more natural, for obvious reasons, to plot it against the intensity S_i of the vertex.

Figures 3.5(a) and (b) show neighbors' degree of the cat corticocortical network for the two cases previously depicted in Fig. 3.3. It is surprising to observe the asymmetry between the two cases. While *out-k_i* happens to be independent of their neighbors' *in-k_j* (Fig. 3.5(a)), the opposite case exhibits a nontrivial behavior (Fig. 3.5(b)): it looks assortative for low degree areas and then saturates to become disassortative at high degrees. The meaning of this asymmetry and its functional consequences is as yet unclear.

3.3.2 Clustering Coefficient

The concept of clustering coefficient is well illustrated by using social networks as an example: *"If person A is a friend of person B and a friend of*

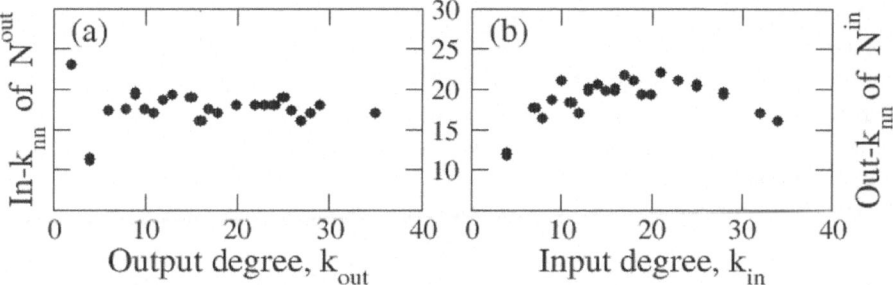

Fig. 3.5. Degree correlations of cat cortex. (a) Correlation between out-k_i of node i and its neighbors' in-k_n (b) opposite case, correlation between in-k_i of node i and its neighbors' out-k_n

person C, what is then the probability that persons B and C are also friends?" Clustering coefficient is, thus, a measure to quantify the conditional probability $P(BC \,|\, (AB \cap AC))$ (Fig. 3.6(a)). In general, given a set of vertices, the *average clustering* will quantify the cohesiveness of that set.

The most popular way of measuring C in *graphs* is to count the number of triangles. The number of paths of length 2 $\{a, b, c\}$ gives the total number of possible triangles. Taking into account that a triangle formed by vertices a, b and c contains three such paths ($\{a, b, c\}$, $\{b, c, a\}$ and $\{c, a, b\}$), the average clustering of the network is then *measured as*:

$$C = \frac{3 \times \text{number of triangles}}{\text{number of paths of length 2}}$$

Higher order versions of this method are possible by counting the number of squares and so on.

In *directed graphs*, however, it is not clear how to define a triangle. A common approach is to define first the *local clustering coefficient* C_i of a single vertex and then average over all vertices to find the *average clustering*

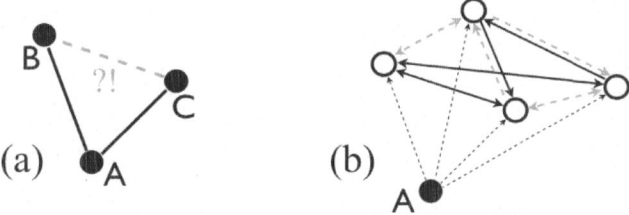

Fig. 3.6. Illustration of clustering coefficient: (a) Clustering quantifies the conditional probability of two vertices to be connected provided they share a common neighbor; (b) Local C_A measure as used in directed networks: the fraction of existing connections between neighbors of A to all the possible connections between them

C. Interestingly, C_i of a single vertex is not directly a structural measure of i itself but a property of *its neighbors* that measures "how well are the neighbors of i are connected together". Defining $\mathcal{N}_i$ as the set of neighbors of vertex i, C_i is then measured (illustrated in Fig. 3.6(b)) as the ratio between the number of *existing* links among vertices in $\mathcal{N}_i$ and the number of *possible* links in the set:

$$C_i = \frac{\sum_{j,k \in \mathcal{N}_i}(A_{jk} + A_{kj})}{k_i(k_i - 1)}$$

where k_i is the degree of vertex i. It is very important to stress that, as pointed out in [10], this approach is not exactly a measure of the conditional probability $C_i = \sum_{j,k} P(jk \,|\, (ij \cap ik))$, and it tends to overestimate the contribution of low-degree vertices due to the smaller denominator.

Again, we find different possibilities in the case of directed networks. Clustering of the "output neighbors" (using *out-k_i*, $\mathcal{N}_i^{out}$, Fig. 3.6(b)) can be measured or that of the "input neighbors" (using *in-k_i*, $\mathcal{N}_i^{in}$). Another possibility is to use both the input and output neighbors of i. However, it is erroneous to define $k_i = in\text{-}k_i + out\text{-}k_i$ because reciprocal links will be counted twice. k_i has to be now the number of *all* vertices that i is connected to (whether input or output). When reciprocal connections are highly present in the network, output and input versions should give similar results.

Once the C_i are calculated it is interesting to look for their correlations with other local properties. C_i of many real networks has been found to anticorrelate with k_i due to their modular structure [10]. This relationship has also been captured by theoretical models [13]. The cat cortex provides an illustrative example as shown in Fig. 3.7. Those areas with few neighbors have higher C_i than those with larger degree. The reason becomes clear when observing the modular structure of the adjacency matrix in Fig. 9.2 of Chap. 9. Low degree cortical areas do preferentially connect to other areas in the same functional cluster (visual, auditory, etc.) while high degree areas have connec-

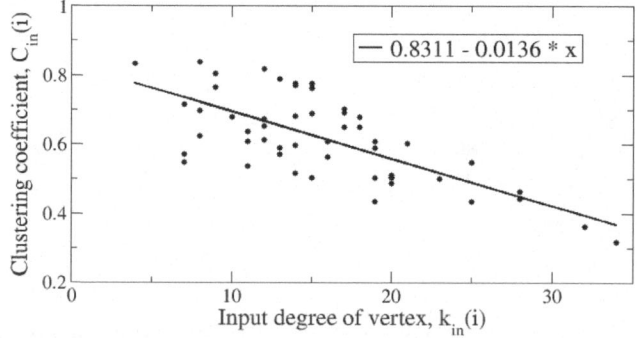

Fig. 3.7. Clustering coefficient of each cortical area C_i as function of area degree k_i. The negative slope is a signature of modular structure in the network

Table 3.1. Average clustering coefficients of the cat cortex and of its anatomical communities

Whole cortex	Visual	Auditory	Somatosensory-motor	Frontolimbic
0.61	0.64	0.87	0.79	0.72

tions all over the network. Thus, neighbors of high degree areas are gathered into semi-independent groups that rarely connect with each other.

The *average clustering coefficient C* provides a quantitative estimation of how cohesive a network is (or just a set of neighbors). Table 3.1 summarizes the internal C values of the whole cat cortical network and of its four structural clusters. As expected, the internal C of clusters is higher than that of the whole network, with auditory cortex being the most cohesive of them all. In order to measure C of a subset of vertices, the set must be first isolated from the rest of the network and then C of this sub-network measured using only internal vertices and the connections among them.

3.3.3 Distance

In networks, topological distance measures the minimum number of links crossed in order to go from a given vertex a to another vertex b. If there is a direct link from a to b, then $d(a,b) = 1$. If there is no shorter path than going from a to c and from here to b, then $d(a,b) = 2$, and so on. Usually, more than one shortest path from a to b exists. This will give rise to important measures meaningful in the context of flows in networks, whatever the flow represents: water flow in a pipeline, information flow, traffic in transportation networks, etc. In graphs, $d(a,b)$ equals $d(b,a)$, so if the *distance matrix* is defined as the matrix whose elements $d_{ab} = d(a,b)$, then it is symmetric. The *average pathlength l* of a network is defined as the average of all values in the d_{ij} matrix.

When there is no path going from a to b, then $d(a,b) = \infty$ and b is said to be *disconnected* from a. It is easier to find disconnected vertices in digraphs due to the directed nature of connections. In food-webs for example, a directed link "worms" $\rightarrow$ "birds" exists but no arrow is to be drawn like "birds" $\rightarrow$ "worms" because worms do not eat birds. Thus, $d(\text{worm}, \text{bird}) = 1$ but $d(\text{bird}, \text{worm}) = \infty$. (Note: in food webs arrows point in the direction of the energy flow.) An important matter in networks is the presence of *connected components*: isolated groups of vertices, internally connected, that cannot reach other components.

Taking connection weights into account while measuring distance is only meaningful when weights represent real metric distance between vertices, like in a roadmap. An interesting example is given by the *Via-Michelin* web page and similar services. This web page allows one to, for example, find the road trip between any two addresses within Europe. One option is to perform the search by *shortest distance* for which only metric distance information is used. Another possibility is to perform a search by *fastest trip*. In this case, other kinds of information must be used in order to give a proper "weight" to each road-segment (type of road, average traffic, etc.) that measure how quickly one can drive through them, and combined with the metric distance in order to calculate the "shortest graph distance", that in this case refers to "time".

There is an extensive literature about algorithms to find shortest paths between vertices in a network. The two most popular ones are the *Dijkstra* and the *Floyd-Warshall* algorithms. Both of them are useful for weighted digraphs since they are designed to return the *lowest cost path*, whatever the weight means in the network. The Dijkstra algorithm finds the shortest path between a source vertex s and all other vertices in time $\mathcal{O}(N^2)$, where N is the size of the network. In the case of sparse matrices ($m \ll N^2$), it can be improved up to $\mathcal{O}(N + m \cdot \log m)$. But if we want to calculate the distance between *all* pairs of vertices, then Dijkstra's algorithm has to be repeated for each vertex, $\mathcal{O}(N^3)$, the length of each path calculated and the distance matrix created. Thus, Dijkstra is the algorithm of choice when we want the shortest path (or distance) between a given pair of vertices, but when *all-to-all* distances are to be calculated, then the Floyd-Warshall algorithm is the choice. This algorithm takes the adjacency matrix of the network as an input and returns the distance matrix in time $\mathcal{O}(N^3)$, but faster than applying Dijkstra N times. It can also be implemented to return the shortest paths between each pair.

3.3.4 Centrality Measures

There is a set of measures requiring to initially calculate the distance matrix. Given a graph G and its distance matrix, the *eccentricity* e_i of a vertex i is defined by the maximum distance from i to any other vertex in the network. The *radius* of the network $\rho(G)$ is then the minimal eccentricity of all vertices and the *diameter*, $\mathrm{diam}(G)$, the maximal.

$$
\begin{aligned}
\text{eccentricity} \quad & e_i = \max(d_{ij}) : j \in V \\
\text{radius} \quad & \rho(G) = \min(d_{ij}) = \min(e_i) \\
\text{diameter} \quad & \mathrm{diam}(G) = \max(d_{ij}) = \max(e_i)
\end{aligned}
$$

The *center of the network* is composed by the set of vertices whose $e_i = \rho(G)$. The name "center" is given because these vertices are closest to any other in the network. If a signal is to be sent so that it reaches all vertices as fast as possible, then the signal should be sent from one of the vertices in the center. If an emergency center is to be placed in a city, it is desirable to strategically place it so that ambulances or firemen will arrive at any corner of the city in minimal time.

Such definitions give rise to a discrete spectrum of values, and it would be desirable to have a continuous spectrum in order to correlate them to other vertex properties. We suggest that the reader instead uses the following definition of eccentricity:

$$e_i = \frac{1}{N} \sum_{j=1}^{N} d_{ij}$$

this is nothing but the average pathlength from vertex i to any other vertex in the network.

We could also ask the opposite question: "which vertices can be most quickly reached from anywhere in the network?" In other words, where should a shopping mall be placed such that all inhabitants in a city can reach it as quickly as possible? The *status* s_i of a vertex is defined as the average of the values in the ith column of d_{ij},

$$s_i = \frac{1}{N} \sum_{j=1}^{N} d_{ji}$$

and quantifies how quickly, on average, vertex i is reached from other vertices.

Another important measure is the *betweenness centrality* that can be defined both for vertices or links. For brevity, only the case of vertices will be described here. The betweenness centrality (BC_i) of vertex i is the count of how often i is present in all the shortest paths between all pairs of vertices in the network. Note that, usually, there is more than one shortest path between two vertices, thus, *all* shortest paths between a given pair are to be accounted for. It is computationally tricky and expensive to find them all. The Dijkstra and Floyd-Warshall algorithms, for reasons of efficiency, look for only one shortest path.

The approach personally followed by the authors of this chapter was to use a modified version of the *depth-first-search algorithm* (DFS) in order to stop the tree search at a desired depth. Once the distance matrix has been calculated, the DFS algorithm is called for each pair of vertices i, j. The known distance d_{ij} is introduced into the DFS algorithm so that it will search through the whole tree but no deeper than d_{ij} steps, as depicted in Fig. 3.8(b). If $d_{ij} = \infty$, the search must be skipped. When all shortest paths in the network are found, the number of times every vertex appears as an *intermediate* vertex is counted.

BC_i is an important measure because it quantifies how much of the flow (information transmission, water flow, traffic, etc.) goes through i. Besides, BC_i does not necessarily correlate to vertex degree k_i. Imagine two independent networks that are connected by a single link. No matter how small the degree of the two vertices is at the end of that link, their BC will be very high because any path connecting the two sub-networks necessarily includes them. Thus, in robustness analysis, selective attack on vertices with high BC_i is much more relevant than attack on those with high k_i.

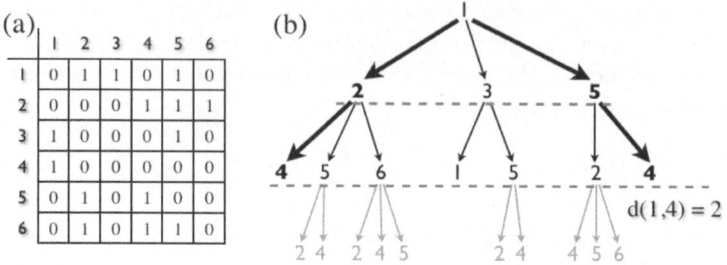

Fig. 3.8. Finding all shortest paths between a pair of vertices. (**a**) Adjacency matrix of example network; (**b**) Tree of vertex 1. Once $d(1,4) = 2$ is known, DFS algorithm is modified in order to stop the tree search at desired depth. Two distinct paths are found from 1 to 4, say $\{1 \rightarrow 2 \rightarrow 4\}$ and $\{1 \rightarrow 5 \rightarrow 4\}$, giving rise to BC(2) += 1 and BC(5) += 1

3.3.5 Matching Index

Intuitively, two vertices having the same neighbors might be performing similar functions. The matching index $MI(i,j)$ of vertices i and j is the count of common neighbors shared by both vertices, and provides an estimation of their "functional similarity". Directionality of the network brings again different possibilities since we could look for common *input* neighbors, common *output* neighbors or *both*. In order to properly normalize the measure, the number of common neighbors should be divided by the total number of distinct neighbors that i and j have. Normalizing over $(k_i + k_j)$ returns misleading results as illustrated in the following example:

$$
\begin{aligned}
1 &\rightarrow \{4,5,6\} \\
2 &\rightarrow \{4,5,7,9\} \\
3 &\rightarrow \{1,2,5,8,9,11,12\} \\
4 &\rightarrow \{1,2,5,8,9,11,12\}
\end{aligned}
$$

Vertices 1 and 2 share two $\{4,5\}$ out of five different neighbors $\{4,5,6,7,9\}$, thus $MI(1,2) = 2/5 = 0.4$. Vertices 3 and 4 share all of their seven neighbors giving rise to $MI(3,4) = 1$. In the case where we would normalize over $(k_i + k_j)$, $MI(1,2) = 2/(3+4) = 0.286$ and $MI(3,4) = 7/(7+7) = 0.5$, clearly faulty expressing the probability of common neighbors.

The matrix representation of MI is symmetric. Diagonal elements are simply ignored and receive null values. The matching index of the cat cortical network in Fig. 3.9 reflects its modular structure and includes some new information. A pair of cortical areas do not need to be connected in order to have a large overlap of neighbors, and thus similar functionality (although in the case of cat cortex, MI tends to provide higher values for connected areas). MI values within a cluster lie in general between 0.35 and 0.6 with some higher value exceptions (mainly in the sensory-motor cortex), while inter-cluster connections tend to have values below 0.35 with some exceptions dominated by the

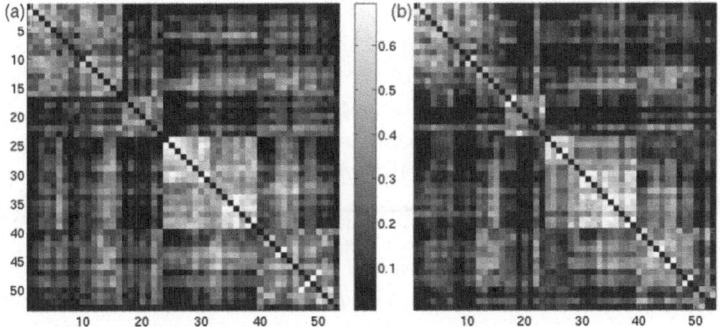

Fig. 3.9. Matching index of the cat cortical network: **a)** MI of input connections. **b)** MI of output connections

connections of high degree areas. MI versions for input neighbors in Fig. 3.9(a) and for output neighbors Fig. 3.9(b) show different internal structures of the anatomical clusters. While for the input version the four anatomical clusters (visual, auditory, somatosensory-motor and frontolimbic) are clearly observed, the output version shows subdivisions of the somatosensory-motor cortex into two clusters, and some visual areas are kicked out.

3.4 Random Graph Models

In order to understand the natural mechanisms underlying the observed structural properties of real networks, creating network models is essential. Different network classes are known to exist, e.g. technological or biological networks that share similar properties within each class and differences across them. Modeling helps us to understand both the similarities and the differences. There is a very extensive literature about random graph models and readers are strongly recommended to read some of the reviews listed in the last section of this chapter. The goal of this section is just to give a very brief but hopefully clear description of the main network models whose modification have motivated many more detailed and realistic models. We will stick to their original descriptions as *graphs* and leave directed connections and weights aside, since this simplification has permitted an extensive analytical treatment.

3.4.1 Erdős-Rényi (ER) Networks

The Erdős-Rényi (ER) model is probably the simplest random network that can be built (cf. Chaps. 5, 7). Starting from an empty set of n vertices without connections, m links are randomly added one by one so that every pair of vertices has the same probability of getting a new link. $G_{n,m}$ is the ensemble of all possible random graphs of size n and m links constructed by this procedure. Typical restrictions are 1) only one link is allowed between a pair of vertices and 2) no self-loops are allowed (self-connections leaving a vertex and re-entering itself; by contrast, see Chap. 7 for an example of an Erdős-Rényi network exhibiting self-loops). The simplicity of this model has permitted extensive analytical work of its properties, started by Erdős and Rényi themselves in the 1960s.

Its degree distribution follows a binomial distribution, becoming Poissonian in the limit of large networks. The probability of a vertex to have degree k is then:

$$P(k) = \binom{N}{k} p^k (1-p)^{N-k} \simeq \frac{z^k e^{-z}}{k!} \quad : N \to \infty$$

where p is the uniform probability of a vertex to get a link. This distribution is peaked around the average degree $z = \langle k \rangle$, which allows the description of the network in terms of z, simplifying the analytical approach.

The model drew much attention due to its percolation properties, equivalent to continuous phase transitions studied in statistical physics and thermodynamics. At the beginning of the random process, when few links are present, many independent and small connected groups of vertices appear called *connected components*, while remaining disconnected from each other (find 3 of them in Fig. 3.10). When more links are added, components grow and merge together into larger components making their *size distribution* approach a scale-free distribution — very many small components and few large ones. There is a critical number of links when the largest components merge suddenly giving rise to a *giant component*, composed of most of the nodes of the network (around 80% of them). This transition has been analytically proved and numerically corroborated to happen when $z = \langle k \rangle \approx 1$, so that $m \approx n$. Chap. 7 relates this percolation transition with another one where oscillations emerge in the networks's dynamics when super-cycles are merged from isolated ones.

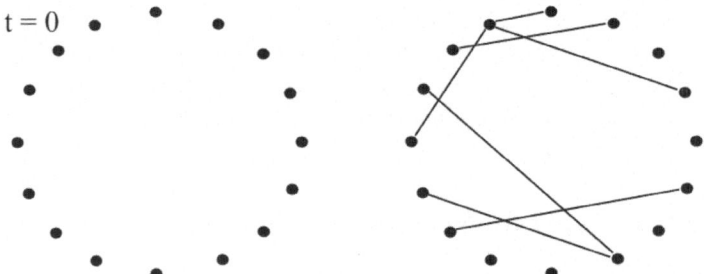

Fig. 3.10. Generation of ER random graphs. An initially empty set of n vertices is linked with uniform probability p

But, how well does the ER model resemble real networks? A universal property of real complex networks captured by this model (above the percolation threshold) is the *small-world property*. The distance between any pair of nodes, and thus the network average pathlength, is observed to be very short. However, what does "short" mean? Obviously, the larger the network, (keeping z constant) the longer the average pathlength l will become. In order to quantify what *short* means, the *small-world property* has been defined as the following upper scaling limit of the average pathlength in the network:

$$l_{upper} = \ln N$$

in the case of the ER model, average pathlength is proven to scale as:

$$l = \frac{\ln N}{\ln z} < \ln N$$

under the following conditions, $N \gg z \gg \ln N \gg 1$. This means that the network is connected (well above the percolation threshold) but connection density is not too large.

On the contrary, the ER model resembles real networks very poorly because it does not reproduce any of the other typical properties like clustering coefficient, degree correlations, etc. Real networks do not show Poissonian degree distributions. Many modifications of the ER model have been presented trying to reproduce these other properties. One to be mentioned, since it will be discussed in the next section, is the so-called *configuration model* that produces maximally random networks with a prescribed degree distribution or degree sequence [14–16]. Thus, the ER model is just an special case of the configuration-model.

3.4.2 The Watts-Strogatz (W-S) Model of "Small-world" Networks

Motivated by the high clustering observed in many real networks and by the highly unrealistic networks (lattices or random) previously used to model dynamical processes, Watts and Strogatz proposed the following model: starting from a regular lattice (a 1-d ring in this case) whose vertices all have degree z, a link is selected with uniform probability p_{rew} and randomly rewired, conserving one of its original ends (Fig. 3.11). The initial lattice had large clustering coefficient, $C = (3z - 3)/(4z - 2)$, and large average pathlength, $l \approx N/4z$ for large N. The introduction of a few shortcuts significantly decreases l while C remains nearly constant. This is true for a range of small $p_{rew} \leq 0.1$. As $p_{rew} \rightarrow 1$ (more links are rewired), the closer the network is to a random graph of the ER type. Thus, the W-S model is a p_{rew}-mediated transition between regular and ER random networks.

The following restrictions are applied to the rewiring: 1) only one of the end nodes of the link is rewired, 2) no self-loops are accepted after rewiring

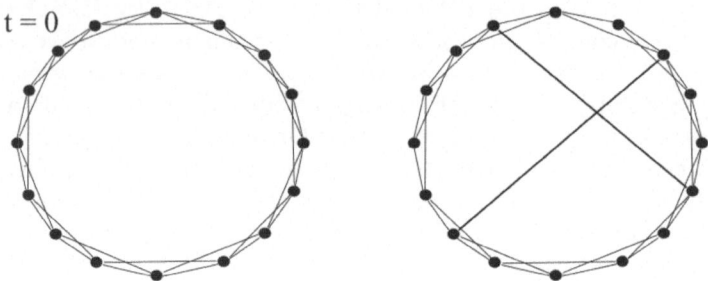

Fig. 3.11. Generation of W-S random networks. Starting from an initial regular lattice where each vertex has z neighbors, one end of the links is rewired with probability p_{rew} giving rise to shortcuts

and 3) only one link is allowed between each pair of vertices. Other methods have been later suggested relaxing some of these restrictions in order to simplify analytical work. In fact, analytical results of the scaling properties of the model are difficult since they depend not only on network size and density of connections but also on the given rewiring probability p_{rew}. However, Barrat and Weigt [17] did estimate an analytical expression for the scaling of the clustering coefficient (calculated using their own definition of clustering coefficient) to be:

$$C(p) = \frac{3(z-1)}{2(2z-1)}(1 - p_{\text{rew}})^3$$

In Sect. 3.5, we will display numerical comparison of the scaling properties for different p_{rew}.

Again, while the WS model captures some of the properties of real networks, it fails to resemble their degree distribution and correlations. Initially, as all vertices in the ring have the same degree, $p(k) = \delta(k - z)$. During the rewiring process, some vertices gain a few connections and others lose them making the distribution wider, creating in the end a binomial distribution exactly like the ER model (see Fig. 3.14).

3.4.3 Barabási-Albert (BA) Model of Scale-free Networks

During the 1990s, advances in computing power permitted for the first time the analysis of large real networks. Many of them exhibit scale-free (SF) degree distributions, $p(k) \sim k^{-\gamma}$, with exponents γ between 2 and 3. This discovery contrasted to the properties of the ER random model, extensively studied for decades as models for real networks. Barabási and Albert proved that two ingredients trigger the emergence of SF distributions in real networks: *growing of networks* adding new vertices in time and *preferential attachment* of new vertices with high degree vertices. A model including these ingredients was first introduced by Price in 1965 [18] trying to account for properties of citation networks, but it was a simpler version published in 1999 by Barábasi and Albert that became popular and provoked an avalanche of new papers and models.

The BA model, as depicted in Fig. 3.12, starts from an initially small and empty network of n_0 without connections. At every time step, a new vertex is included that makes $m \leq n_0$ connections to existing vertices, being the probability of connection proportional to their current degree:

$$\Pi(k_i) = \frac{k_i}{\sum_j k_j}$$

Older vertices tend to accumulate more and more connections and thus a higher probability to link to newly introduced vertices (rich-gets-richer phenomenon). Meanwhile, new vertices have few connections and thus a lower probability of gaining links in the future.

Networks generated by the BA model are found to have shorter average pathlength than ER and W-S networks of the same size and density with distance scaling as logarithm of size, $l \sim \ln(N)$. Analytical estimations predict even shorter correlation $l \sim \ln(N)/\ln(\ln(N))$ [19]. Clustering coefficient has been found to scale as $C \sim (\ln(N))^2/N$ [13], and it exhibits no degree-degree correlations. An interesting property of BA networks is that the procedure always generates connected networks in a single component, in large contrast to SF networks generated by the configuration model and similar methods where percolation processes are present. Another very important property of SF networks (not exclusive to the BA model) is that of robustness (or resilience) under attack or failures. As a few vertices accumulate most of the connections (hubs) and most vertices make few links, SF networks are very robust to the random removal of vertices, but selective attack of hubs produces large damage.

The BA is a very simple model intended to capture the main ingredients giving rise to SF degree distributions. This does not mean that it approximates

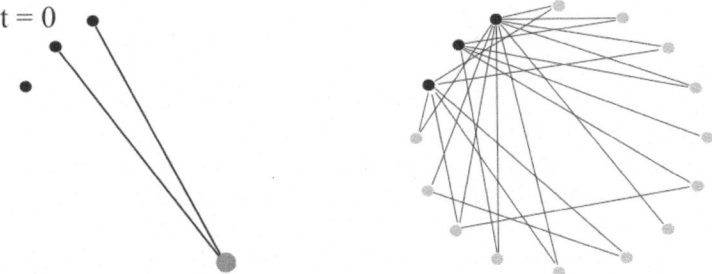

Fig. 3.12. Generation of BA scale-free random networks. Starting from an empty and small set of n_0 vertices, new vertices are included at each time step that make $m \leq n_0$ links to existing vertices with probability proportional to their current degree (*preferential attachment rule*)

other properties observed in real networks like degree-degree correlations and clustering coefficients. However, it is a flexible model and many modifications have been proposed to reproduce other properties. To only mention some of them, in [11] a mechanism to generate the degree-degree correlations of the WWW is presented (Fig. 3.4) and in [13, 20] mechanisms to introduce large C to evolving SF networks are discussed.

Another limitation of the BA model is that it generates networks with a unique exponent $\gamma = 3$ (when $N \to \infty$) whatever the initial conditions and for all allowed parameters $m \leq n_0$. Besides, in real networks the SF scaling does not generally happen over the whole range of degrees. Some networks display a SF distribution truncated by exponential decays due to "ageing" of vertices (receive no new connections after some time) or saturation (limited capacity for a vertex to host new connections) [21].

3.5 Comparison of Random Graph Models

The random models introduced in the previous section share some similarity properties. Besides, nomenclature makes differentiation confusing since the so-called "small-world networks" are not the only ones obeying the *small-world property*. Therefore, in this section we will study and compare the scaling characteristics of the three models in more detail.

3.5.1 Average Pathlength

All three random network models described in this chapter generate small-world networks. What is then the difference between them? Which model generates the smallest networks? In Sect. 3.4.2, W-S networks were defined as a transition between lattices and random graphs by a process of rewiring links. For small rewiring probabilities p_{rew}, the networks conserve high clustering while their average pathlength decays very fast approaching that of ER graphs. Thus, lattices have the longest l of all, increasing linearly with network size as $l \sim N/(4z) > \ln N$, faster than the "small-world upper limit". On the contrary, l of ER graphs scale in the large network limit as $l \sim \ln N/\ln z < \ln N$, where z is the average number of links per vertex. There is no analytical estimate for the scaling of W-S graphs, since this depends on N, z and p_{rew}.

In Figs. 3.13(a) and (b), this transition is depicted for generated networks of $N = 500$ vertices and different number of connections (represented here as connection densities instead of z). Figure 3.13(a) shows the fast decay of l: for a small rewiring probability of $p_{\text{rew}} = 0.01$, average pathlength of W-S lies very close to that of ER networks. Interestingly, it is observed that the differences between the models are only meaningful for rather sparse networks. At connection densities $\rho \approx 0.3$, lattices, random networks and W-S-networks have very similar average pathlengths; at $\rho \approx 0.5$, they are actually equal.

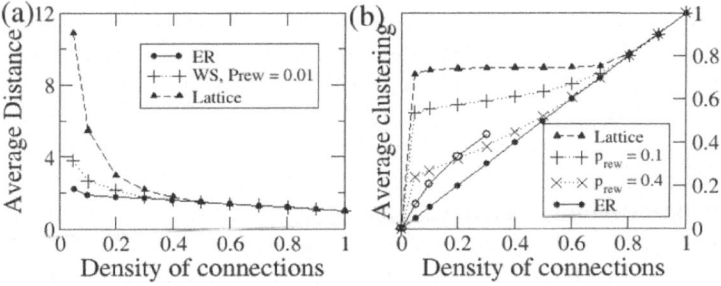

Fig. 3.13. W-S networks as a transition between regular lattices ($p_{\text{rew}} = 0$) and ER random graphs ($p_{\text{rew}} = 1$): (**a**) average pathlength l and; (**b**) average clustering coefficient C of generated networks with $n = 500$ vertices and increasing number of connections. BA model ($\circ$). All models differ at low densities. With increasing number of connections, average measures of all models become indistinguishable. Each data point is an average of 10 realizations

Scaling properties of l in SF networks are not trivial. Bollobás et al. [22] performed a strictly mathematical demonstration that the BA model scales as: $l \sim \ln N / \ln \ln N$. Independently, Cohen et al. [19] studied l of "random" SF networks, using the Molloy-Reed version of the configuration model. It consists of assigning each vertex a given degree probability $P(k_i)$ of getting a link and then introduce connections at random (The ER model is the special case where $P(k)$ is uniform: $P(k_i) = P(k_j)$ for all $i, j \in N$). They found that the scaling of the average pathlength depends on the exponent γ and for $\gamma < 3$, SF networks possess "ultra-small" diameter (again, in the large and sparse network limit).

$$2 < \gamma < 3 \quad \rightarrow \quad l \sim \ln \ln N$$
$$\gamma = 3 \quad \rightarrow \quad l \sim \frac{\ln N}{\ln \ln N}$$
$$\gamma > 3 \quad \rightarrow \quad l \sim \ln N$$

Note that the scaling found for $\gamma = 3$ in the Molloy-Reed SF networks coincides with that of the BA model. Although both methods generate graphs with scale-free degree distribution, they have different internal structure.

Summarizing, SF networks are the smallest of all while ER graphs are smaller than W-S graphs.

3.5.2 Clustering Coefficient

Figure 3.13(b) shows the transition of C in W-S networks, between lattices and ER graphs, to be slower with increasing p_{rew} than the transition of l (Fig. 3.13(a)). At small $p_{\text{rew}} = 0.01$, average pathlength of W-S networks is very close to that of ER graphs, while average clustering coefficient at $p_{\text{rew}} = 0.4$ starts to approach the curve of random graphs. The combination of these features, fast decrease of l while keeping high C, was the reason to coin the W-S graphs as "small-world networks" in analogy to social networks where the *"friends of my friends are also my friends"* effect causes high clustering and helps average pathlength between members of the network to be shorter.

Lattices have the largest C of all and W-S graphs have, for any $p_{\text{rew}} < 1.0$, larger C than ER graphs. Note that C of ER graphs equals its density of connections since this density is equal to the probability p of two vertices to be linked. Again, there is a critical density of connections ($\rho \approx 0.7 - 0.8$) where C of all models becomes indistinguishable. BA graphs have a relatively low C (solid line with open circles) but still larger than ER graphs. Unfortunately, it is not possible to generate BA graphs of larger densities because they are grown out of an initial empty graphs of finite size n_0 and new vertices make a maximum of $m \leq n_0$ connections. An analytical estimation of its scaling with network size was found by Klemm and Eguíluz in [13]. In Fig. 3.14, analytical estimations for the clustering coefficients of the different models are summarized.

3.5.3 Other Structural Differences

As mentioned in Sect. 3.3.1, the degree distribution $p(k)$, although important, does not explain very much about the internal structural organization of a network. Random graphs of the same size, connection density and degree distribution may possess very distinct internal structure. While W-S graphs should conserve much of the regularity of their parent lattices, degrees of ER and BA networks are uncorrelated, as depicted in Fig. 3.4(a) (even if BA networks are generated by a "preferential attachment" rule forcing new vertices to preferentially link to those with highest degree). This absence of degree-degree correlations is not a general property of scale-free networks, but intrinsic of the BA model.

Another important structural feature of real complex networks, not captured by any of the models presented here, is the presence of *modules* and *hierarchies*. In many real networks, and specially in biological ones, the combination of modular and hierarchical organization is believed to be a very important consequence of self-organisation: elements specialized in similar function are arranged into modules that hierarchically interact with each other. This requires a complex topology supporting the very rich range of functional capacities exhibited by living organisms. Imagine genetic networks where genes involved in similar regulatory processes form clusters, or in the cat cortex (Fig. 9.2 of Chap. 9) where cortical areas performing similar function form the visual, auditory, somatosensory-motor and frontolimbic clusters observed. After recognizing its importance, Ravasz et al. [23] studied such structures in metabolic networks and presented a simple model that generates networks with both scale-free degree distribution and hierarchical/modular architecture.

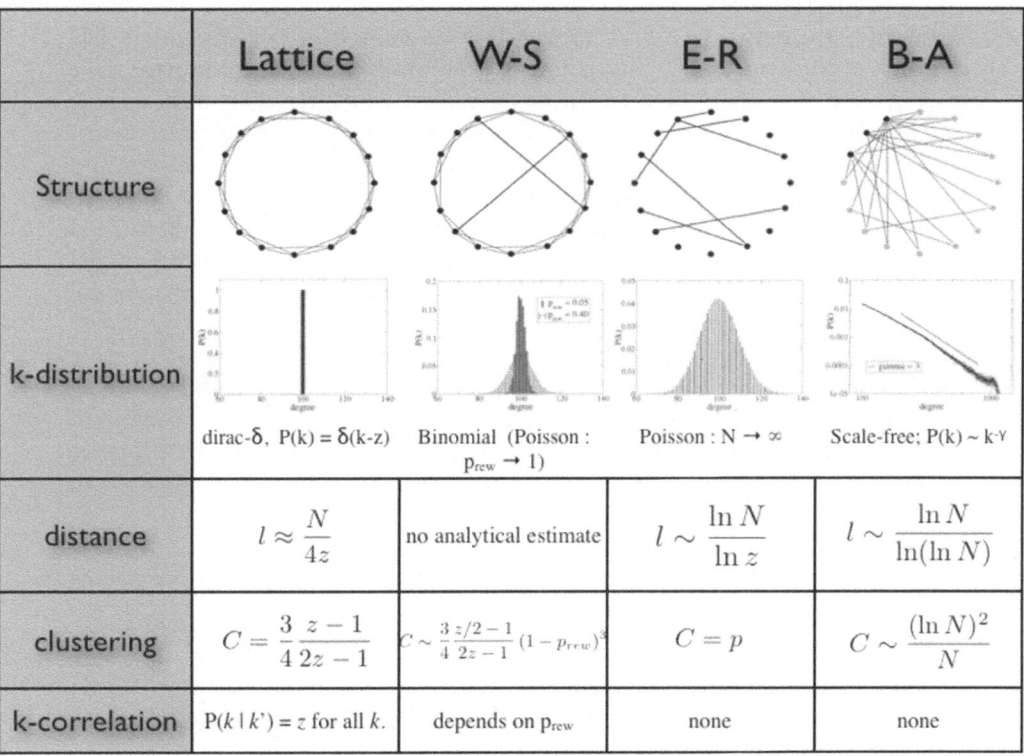

	Lattice	W-S	E-R	B-A
Structure				
k-distribution	dirac-δ, $P(k) = \delta(k-z)$	Binomial (Poisson : $p_{rew} \to 1$)	Poisson : $N \to \infty$	Scale-free; $P(k) \sim k^{-\gamma}$
distance	$l \approx \dfrac{N}{4z}$	no analytical estimate	$l \sim \dfrac{\ln N}{\ln z}$	$l \sim \dfrac{\ln N}{\ln(\ln N)}$
clustering	$C = \dfrac{3}{4}\dfrac{z-1}{2z-1}$	$C \sim \dfrac{3}{4}\dfrac{z/2-1}{2z-1}(1-p_{rew})^3$	$C = p$	$C \sim \dfrac{(\ln N)^2}{N}$
k-correlation	$P(k \mid k') = z$ for all k.	depends on p_{rew}	none	none

Fig. 3.14. Comparison of random graph models

3.6 Randomizing Networks and Comparison with Expected Properties

In Sect. 3.5, we compared random network models in terms of their scaling properties. In practice, however, scaling information is rarely available because usually we have one, and only one real network to be studied. In some cases (for example, the Internet or WWW), its structure might be known at different time stamps and thus its growing properties are available.

Then, how does one classify a real network? How expected, or surprising, are its properties and organization compared to those of random networks? Given any real network, a first approach is to generate a set of random ER networks of the same size N and number of connections M and compare their properties. But if the real network has a scale-free degree distribution, how representative is this comparison? A better approach is to generate a set of SF networks of the same size, number of connections and scaling factor γ and again, compare. In such a case, generation of static SF networks is suggested, using the configuration model and similar methods like the one presented in [24]. These methods are computationally more efficient than the BA model and, very importantly, the scaling factor γ is tunable while the BA model always returns $\gamma = 3$ exponent.

In reality, however, the degree distribution of real networks will rarely match that of any model, so, what should we expect the measures to be, given the degree distribution? By rewiring the connections, an ensemble of maximally random networks of the same size, density and degree sequence can be generated to be used as null hypotheses. The method is depicted in Fig. 3.15 and summarized as the following:

Select two connections at random, say, $(a_1 \rightarrow b_1)$ and $(a_2 \rightarrow b_2)$, and switch them if and only if:

1. *Neither the $(a_1 \rightarrow b_2)$ and $(a_2 \rightarrow b_1)$ links previously exist — otherwise double links would be introduced.*
2. *$b_2 \neq a_1$ and $b_1 \neq a_2$ exist — otherwise self-loops would be introduced.*

Fig. 3.15. Schematic representation of randomizing networks while conserving degree distribution. After randomly selecting two links, they are exchanged provided some restrictions are fulfilled. This procedure generates a maximally random network of same size, N, number of connections m and degree-distributions $P(k_{\text{in}})$ and $P(k_{\text{out}})$ as its parent network

When applied to matrices, the rewiring procedure will also conserve input intensity *in-S* (but not *out-S*). Given a digraph in adjacency list form, the pseudo-code for the algorithm might read like:

```
counter = 0
```

```
while counter < m_rew:
    1) Select one node at random, a₁, and one of its neighbors, b₁.
    2) Select another vertex a₂.
       Check that a₂ ≠ a₁, a₂ ≠ b₁ and NOT a₂ → b₁, otherwise
       start again.
    3) Select a neighbor of a₂, b₂.
       Check that b₂ ≠ a₁ and NOT a₁ → b₂, otherwise start again.
    4) include b₂ in the list net(a₁)      Swap the connections
       include b₁ in the list net(a₂)
       remove b₁ from net(a₁)              delete the old links
       remove b₂ from net(a₂)
       counter += 1
```

where m_{rew} is the maximum number of times the rewiring process will run (note that at each step two connections are rewired). The two links must be randomly chosen with uniform probability $1/M$ but the algorithm is taking first nodes a_1 and a_2 (with probability $1/N$) and then one of their neighbors. Individual links are thus selected with probability $p(b|a) = p(a)\,p(b) = \frac{1}{N}\frac{1}{out\text{-}k(a)}$. As a result, links of high degree nodes have a lower probability of being rewired. A very easy way to balance the situation is to initially generate a list of size M containing each node $out\text{-}k(a)$ times (as in the configuration model) and select every time the nodes a_1 and a_2 occur in this list. The probability of selecting any link is now:

$$p(b \,|\, a) = p(a)\,p(b) = \frac{out\text{-}k(a)}{M}\frac{1}{out\text{-}k(a)} = \frac{1}{M}$$

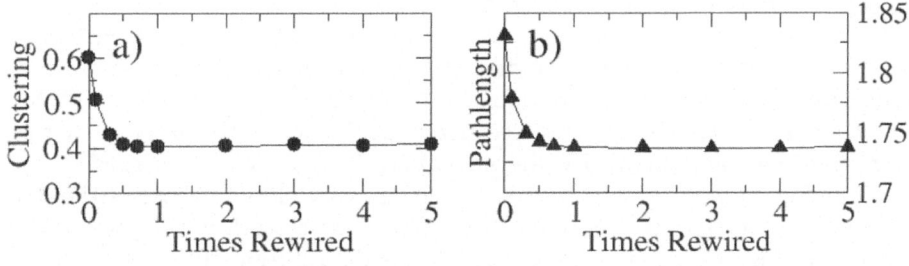

Fig. 3.16. Rewiring the cat cortical network towards generation of maximally random digraphs of same N, M and degree distributions: **a)** Average clustering coefficient and **b)** average pathlength of the rewired networks. Each data point is the average after 100 realizations

An important question one faces when rewiring networks is how long the process should run so that we are sure of having really randomized networks. In this case, as shown in Figs. 3.16(a) and (b), C and l of the cat cortex become stable after rewiring $2\,M$ connections.

The rewiring process will destroy the internal structure of the network: degree-degree correlations, clustering, communities, hierarchies etc. while conserving a basic property like the degree distribution. However, it is also possible to generate maximally random networks with given degree correlations or clustering coefficients, etc. Different levels of approximation are available and it is up to the readers to decide which level of detail is enough in their specific case.

3.7 Classification of the Cat Cortical Network

The cortical networks of cat and macaque are known to have clustered organization as well as small-world properties: high clustering coefficient and short average pathlength [4,5,9]. In an effort to understand how cortical organization has evolved towards such structures, Tononi and Sporns found that similar topology could emerge from the balance between *functional integration* and *segregation* [25].

In Table 3.2 the clustering coefficient C and average pathlength l of the cat cortex and of different generated random networks ($N = 53$, $M = 826$) are presented. C of cat cortex is much higher than that of ER random digraphs and thus, apparently very "surprising" based on what should be expected in a random network of its size and number of connections. l is always very small due to the high density of connections (~ 0.3). As pointed out in the previous section, we need to define a proper ensemble of random networks to be used as a "null hypothesis". Comparison to ER digraphs might be erroneous, as it is in this case. The degree distribution of the cat cortex in Fig. 3.17(b) shows no relation to any of the models introduced in this chapter. The distribution of ER networks is localized around the mean number of links per vertex z ($z_{cat} \approx 16$) but the distribution of cat is very wide, some areas have degree up to $k = 35$ while N is only 53. Using the rewiring procedure described in Sect. 3.6, a set of maximally random networks of the same size, number of connections *and* degree distribution was obtained. C and l of the cat cortex are still higher than those of rewired networks, but the differences are not so large as when compared to ER digraphs, and thus, not so "surprising".

C and l of the generated W-S networks are very close to those of the cat, suggesting that cat cortex is similar to a W-S network with p_{rew} between 0.05 and 0.1. In Fig. 3.17(a), the cumulative degree distribution of the cat cortex is shown together with that of W-S networks ($-+$ line) and "scale-free" networks generated by the configuration model with $\gamma = 3.5$ (solid line). As expected, the distribution of the W-S digraphs is very narrow (decays very quickly) and thus, very unlikely to be a representative model for cortical networks. Even

Table 3.2. Comparison of average clustering and average shortest path between the cat cortical network and random networks of same size and number of connections. "Rewired Cat" also possesses the same degree sequence. Each value is the average over 10 realizations

	Cat cortex	ER random	Rewired Cat	W-S ($p_{rew} = 0.05$)	W-S ($p_{rew} = 0.1$)
C	0.62	0.30	0.41	0.63	0.57
l	1.83	1.71	1.74	1.89	1.80

if the average properties of the W-S model closely reproduce those of the cat cortex, the internal structural organization is still very different. Besides, it is very difficult to imagine how cortical networks might have evolved as anything similar to a rewired regular lattice.

Wide range degree-distributions are typical of scale-free networks. Obviously, speaking of SF distribution in a network as small as 53 vertices is meaningless. However, the cumulative degree distribution of generated "SF networks" with $N = 53$ and $M = 826$ (solid line in Fig. 3.17(a)) closely follows the real distribution of the cat cortex. It is true that the existence of a SF distribution does not necessarily imply any mechanism of network construction, but SF properties usually emerge out of self-organized systems and the BA model presents an intuitive and likely evolutionary scenario. There is yet another observation suggesting that mammalian cortex has a SF nature: in [26], robustness of cortical networks is shown to behave like SF networks under random or selective attack.

Apart from its modular structure, mammalian cortical networks are also known to be hierarchically organized [6, 27]. Recently, various authors in this book found, by means of correlations in dynamical simulations, that functional connectivity of cat cortical network also follows a modular and

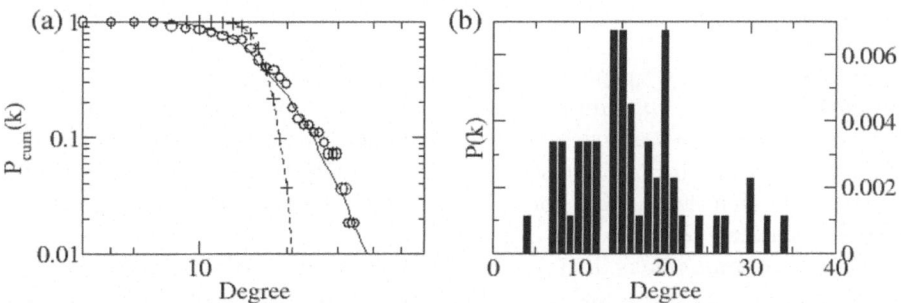

Fig. 3.17. Degree distribution of *in-k* of cat cortical network. (**a**) Cumulative degree distribution of cat cortex (○), generated W-S $p_{rew} = 0.3$ (−+ line) and scale-free $\gamma = 3.5$ (solid line) networks of same size and number of connections as the cat cortical network. Notice that W-S networks produce poor approximations; (**b**) Histogram of *in-k*

hierarchical pattern very close to that of its structural organization [8, 28]. Internal connections between areas within the same community (visual, auditory, somatosensory-motor and frontolimbic) are the first links being functionally expressed, while later the connections between areas in different communities are expressed. This separation unveils two hierarchical levels representing specialized information processing and integration of multisensorial information.

In the development of realistic models for the evolution of mammalian cortex, the idea of balancing functional integration and segregation proposed by Tononi and Sporns may be useful as a "macroscopic" driving force, yet more natural and detailed mechanisms need to be found to cause the system to self-organism into a modular and hierarchical structure. The broad degree distribution, suggesting a SF-type system, comes to the rescue since power laws are characteristic of self-organized systems. Although the model by Ravasz et al. [23], which generates hierarchical networks with a scale-free degree distribution by self-replication of structures, is an interesting starting point, it seems unlikely as a model for cortical evolution. A recipe for such an evolutionary model should at least reproduce the ingredients we have presented in this last section.

3.8 Further Reading

Readers looking for a first introduction to networks (or better said, graph theory) are encouraged to look for online resources. A web page called *An Interactive Introduction to Graph Theory* by Chris K. Caldwell provides an illustrative and educational introduction of basic concepts for the non-initiated (http://www.utm.edu/cgi-bin/caldwell/tutor/departments/math/graph /intro), while in his web page *Algorithmic Graph Theory*, Rashid Bin Muhammad provides a more concise introduction to mathematical graph theory: (http://www.personal.kent.edu/~rmuhamma/GraphTheory/graph Theory.htm)

Books on classical graph theory are usually difficult to read and far from our current interests in complex networks, but are a necessary reference if serious research on networks is to be done. Classical books treat the general case of directed graphs (digraphs) poorly, with the exception of a recent book by Jørgen Bang-Jensen and Gregory Gutin titled *Digraphs: Theory, Algorithms and Applications* [29].

There are some books dedicated to the '*new science*' of complex networks; however, we would recommend reading through some of the excellent reviews available. Newman [10] published the first major review dedicated to complex networks, their topology and random graph models. It lacks many of the current developments and challenges of complex networks research but it is still a "must" to anyone starting in the field. Two recent reviews [30, 31] come to the rescue and after going through structural properties of networks, they jump into dynamical processes occurring in networks. They also provide an overview of current fields where network theory has been applied, such as epidemiology, neuroscience, economics, etc. and summarize the challenges to be faced in each field.

Newman, Barabási and Watts recently published a compilation of selected papers which have been central in the development of complex networks theory [32].

References

1. O. Sporns, G. Tononi, K. Kötter: *The human connectome: A structural description of the human brain.* PLoS. Comput. Biol. 1(4), 0245–0251, 2005.
2. R. Kötter, F.T. Sommer: *Global relationship between anatomical connectivity and activity propagation in the cerebral cortex.* Phil. Trans. R. Soc. Lond. B, 355 (1393), 127–134, 2000.
3. J. W. Scannell, C. Blakemore, M. P. Young: *Analysis of connectivity in the cat cerebral cortex.* J. Neurosc. 15(2), 1463–1483, 1995.
4. J. W. Scannell, G. A. P. C. Burns, C. C. Hilgetag, M. A. O'Neill, M. P. Young: *The connectional organization of the cortico-thalamic system of the cat.* Cer. Cortex 9(3), (277–299), 1999.
5. C. C. Hilgetag, G. A. P. C. Burns, M. A. O'Neill, J. W. Scannell, M. P. Young: *Anatomical connectivity defines the organization of clusters of cortical areas in the macaque monkey and the cat.* Phil. Trans. R. Soc. Lond. B 355, 91–110, 2000.
6. C.C. Hilgetag, M.A. O'Neill, M.P. Young: *Hierarchical organization of macaque and cat cortical sensory systems explored with a novel network processor.* Phil. Trans. R. Soc. Lond. B 355, 71–89, 2000.
7. C.C. Hilgetag, M. Kaiser: *Clustered organization of cortical connectivity.* Neuroinf. 2, 353–360, 2004.
8. C.S. Zhou, L. Zemanová, G. Zamora, C. C. Hilgetag, J. Kurths: *Hierarchical organization unveiled by functional connectivity in complex brain networks.* Phys. Rev. Lett. 97, 238103, 2006.
9. O. Sporns, J.D. Zwi: *The small world of the cerebral cortex.* Neuroinf. 2, 145–162, 2004.
10. M. E. J. Newman: *The structure and function of complex networks.* SIAM Review 45(2), 167–256, 2003.
11. R. Pastor-Satorras, A. Vázquez, A. Vespignani: *Dynamical and correlation properties of the Internet.* Phys. Rev. Lett. 87, 258701, 2001.
12. A. Barrat, M. Barthélemy, R. Pastor-Satorras, A. Vespignani: *The architecture of complex weighted networks.* PNAS 101(11), 3747–3752, 2004.
13. K. Klemm, V. M. Eguíluz: *Highly clustered scale-free networks.* Phys. Rev. E 65, 036123, 2002.
14. E. A. Bender, E. R. Canfield: *The asymptotic number of labelled graphs with given degree sequences.* J. Combin. Theory A 24 (3),296–307, 1978.
15. M. Molloy, B. Reed: *A critical point for random graphs with given degree sequence.* Random. Structures Algorithms 6, 161–179, 1995.
16. M. E. J. Newman, S. H. Strogatz, D. J. Watts: *Random graphs with arbitrary degree distributions and their applications.* Phys. Rev. E 64, 026118 2001.

17. A. Barrat, M. Weigt: *On the properties of small-world network models*. Eur. Phys. J. B 13, 547–560, 2000.
18. D. J. de S. Price, *Networks of scientific papers*. Science, 149, 510–515, 1965.
19. R. Cohen, S. Havlin, D. ben-Avraham. *Structural properties of scale-free networks*. In Handbook of Graphs and Networks. Editors S. Bornholdt, H. G. Schuster, Wisley-VCH, Berlin, ISBN: 3-527-40336-1, 2002.
20. P. Holme, B.J. Kim: *Growing scale-free networks with tunable clustering*. Phys Rev. E 65, 026107, 2002.
21. L. A. N. Amaral, A. Scala, M. Barthélemy, E. Stanley: *Classes of small-world networks*. PNAS 97(21), 11149–11152, 2000.
22. B. Bollobás, O. Riordan, J. Spencer, G. Tusnády: *The degree sequence of a scale-free random graph process*. Random Structures and Algorithms 18, 279–290, 2001.
23. E. Ravasz, A. L. Somera, D. A. Mongru, Z. N. Ottvai, A.L. Barabási: *Hierarchical organization of modularity in metabolic networks*. Science 297, 1551–1555, 2002.
24. K.I. Goh, B. Kahng, D. Kim: *Universal behaviour of load in scale-free networks*. Phys. Rev. Lett. 87(27), 278701, 2001.
25. O. Sporns, G. Tononi: *Classes of network connectivity and dynamics*. Complexity 7(1), 28–38, 2001.
26. M. Kaiser, R. Martin, P. Andras, M. P. Young: *Simulation of robustness against lesions of cortical networks*. Eur. J. Neurosc. 25, 3185–3192, 2007.
27. L. da F. Costa, O. Sporns: *Hierarchical features of large-scale cortical connectivity*. Eur. Phys. J B 48, 567–573, 2005.
28. L. Zemanová, C.S. Zhou, J. Kurths: *Structural and functional clusters of complex brain networks*. Physica D 224, 202212, 2006.
29. J. Bang-Jensen, G. Gutin: *Digraphs: Theory, Algorithms and Applications*. Springer-Verlag, London. Springer Monographies in Mathematics, ISBN: 1-85233-268-9, 2000.
30. S. Boccaletti, V. Latora, Y. Moreno, M. Chavez, D.-U. Hwang: *Complex networks: structure and dynamics*. Physics Reports 424, 175–308, 2006.
31. L. da F. Costa, F. A. Rodrigues, G. Travieso, P. R. V. Boas: *Characterization of Complex Networks: A Survey of measurements*. arXiv:cond-mat 0505185, 2005.
32. M. E. J. Newman, A-L. Barabási, D. J. Watts: *The Structure and Dynamics of Networks*. Princeton University Press, 2006. ISBN: 0691113572.

Organization and Function of Complex Cortical Networks

Claus C. Hilgetag[1] and Marcus Kaiser[2]

[1] International University Bremen, School of Engineering and Science, Bremen, Germany
C.Hilgetag@iu-bremen.de,
http://www.iu-bremen.de/schools/ses/chilgetag/
[2] School of Computing Science, Newcastle University, Newcastle upon Tyne, United Kingdom
M.Kaiser@ncl.ac.uk,
http://www.biological-networks.org

Summary. This review gives a general overview of the organization of complex brain networks at the systems level, in particular in the cerebral cortex of the cat brain. We identify fundamental parameters of the structural organization of cortical networks, illustrate how these characteristics may arise during brain development and how they give rise to robustness of the cortical networks against damage. Moreover, we review potential implications of the structural organization of cortical networks for brain function.

4.1 Introduction

The network organization of the mammalian brain underlies its diverse stable and plastic functions. Experimental approaches from various directions have suggested that the specific organization of nerve fibre networks, particularly in the massively interconnected cerebral cortex of the brain, is closely linked to their function. However, the exact relationship between cortical network structure and brain function is still poorly understood, mainly due to the complexity of the available experimental data on brain networks, both with respect to their great volume and formidable intricacy. This complexity requires theoretical analyses as well as computational modeling to characterize the network organization and deduce functional implications of particular network parameters. In this review, we specifically focus on the structural and functional organization of brain networks, which are interconnecting neural elements at the large-scale systems level of the brain, and we use the cortico-cortical network of the cat as an example to illustrate different aspects of organization and function.

4.1.1 A Systems View of Brain Networks

The brain is a networked system of extraordinary complexity. Considering, for instance, the cellular organization of the human brain, one is faced with approximately 10^{10} single elements (e.g. [1], each of which on average is connected to more than 1,000 other elements [2]). To complicate matters, these elements are neither completely nor randomly connected, e.g. [3]. Fortunately, there is some regularity in the brain that might help to reduce the number of objects and interconnections that need to be considered. Neurons that possess similar connectional and functional features tend to group into large assemblies of several thousands to millions of cells that are regionally localized (forming cortical 'areas' or subcortical 'nuclei'). A first attempt to understand brain organization and function can, therefore, start with investigating the structure, connectivity and function of these assemblies of cells, rather than the complete cellular substance of the brain, e.g. [4]. It is this so-called systems level approach that we pursue here.

This review focuses on corticocortical connectivity at the systems level, specifically networks formed by long-range projections among cortical areas. There are two main reasons that motivate this approach. First, more extensive and reliable databases are currently available for systems level networks than for cellular neuronal circuits. There have been pioneering studies about the interconnections of different types of neurons at the level of cellular circuits, for instance, [3, 5–9]. However, detailed information about connectivity at the cellular level, based on systematic sampling of different cortical regions, is still largely missing. Second, systems connectivity data play an important role in many models of the brain (e.g. [4, 10–14]). These data have been used to derive conclusions about the global organization, development and evolution of the brain and to suggest modes of information processing. In particular, systems level connectivity may be responsible for important aspects of brain function, such as the neural activation patterns observed in functional imaging studies of perception and cognition (e.g. [15]), functional diversity and complexity [16], as well as other functional aspects reviewed here.

The systems level concept can be readily formalized and treated with the help of graph theoretical approaches, considering cortical areas as nodes and their interconnections by long-ranging nerve fibres as edges of directed or undirected graphs. In the theoretical systems level concept, areas and nuclei of the brain are well-defined, intrinsically uniform entities with sharp borders. It needs to be kept in mind, however, that experimental data present a more complicated picture. For instance, numerous, partly incongruent, mapping schemes exist for describing the parcellation of the cerebral cortex into different areas [17].

4.1.2 Types of Brain Connectivity

Different aspects of brain connectivity can be distinguished using the following, widely accepted classification [18]:

- Anatomical or *structural connectivity* denotes the set of physical connections linking neural units (cells and populations) at a given time. Structural connectivity data can range over multiple spatial scales, from area-intrinsic circuits to large-scale networks of inter-regional pathways. Anatomical connection patterns are relatively static over shorter time scales (seconds to minutes), but can be dynamic over longer time scales (hours to weeks); for example, during learning or development.
- *Functional connectivity* [19] captures patterns of statistical dependence between distributed neural units, which may be spatially remote, measuring their correlation/covariance, spectral coherence or phase-locking. Functional connectivity is time-dependent (typically using time series containing hundreds of milliseconds) and 'model-free', that is, it measures statistical interdependence (mutual information) without explicit reference to causal effects. Different methodologies for measuring brain activity may result in different statistical estimates of functional connectivity [20].
- *Effective connectivity* describes the set of causal effects one neural unit exerts over another [19]. Thus, unlike functional connectivity, effective connectivity is not 'model-free', but requires the specification of a causal model including structural connection parameters. Experimentally, effective connectivity can be inferred through network perturbations [21], or through the observation of the temporal ordering of neural events. Other measures estimating causal interactions can also be used (e.g. [22]). Effective connectivity is also time-dependent. Statistical interactions between brain regions change rapidly, reflecting the participation of varying subsets of brain regions and pathways in different cognitive tasks [23–26], behavioral or attentional states [24], and changes within the structural substrate related to learning [27].

Importantly, structural, functional and effective connectivity are mutually interrelated. Clearly, structural connectivity is an essential condition for the kinds of patterns of functional or effective connectivity that can be generated in a network. Structural inputs and outputs of a given cortical region, its connectional fingerprint [28], are major determinants of its functional properties. Conversely, functional interactions can contribute to the shaping of the underlying anatomical substrate [29], either directly through activity (covariance)-dependent synaptic modification, or, over longer time scales, through affecting an organism's perceptual, cognitive or behavioral capabilities, and thus its adaptation and survival.

4.1.3 A Case Study of Structural Connectivity: Inter-area Connections in the Cat Cerebral Cortex

The specific analyses described in this and other chapters in this volume are based on a global collation of cat cortical connectivity (892 interconnections of 55 areas) [30]. This collation of cat cortical data was developed from the data set described in [14] and forms part of a larger database of thalamocortical connectivity of the cat [31]. The database was created by the interpretation of a large number of reports of tract-tracing experiments from the anatomical literature. All tract-tracing experiments are based on the same general design. The investigated brain region of an anaesthetized animal is injected with a tracer substance (chemical or microrganismic), which is directly brought into the cells or taken up by neurons close to the injection site. The tracer, which is typically taken up at the axonal or dendritic branch endpoints, is then transported along the neuron's axon, *retrogradely* from the axonal terminals to the soma, or *anterogradely* in the opposite direction, or in both directions. After a method-dependent survival time, the animal is sacrificed and its brain is sectioned, histochemically processed, and analyzed under a microscope to show the distribution of transported tracer. Mathematically speaking, the experimental result of a tract-tracing experiment is a three-dimensional map of tracer concentration, contained in two-dimensional sections, at the moment of the animal's death. From the systematic analyses of the label distribution in the sections, in combination with a given parcellation of the cortical sheet into distinct areas, conclusions can be drawn about the specific interconnections of different areas by neural fiber projections. The invasive nature of these experiments explains why they cannot be applied to the human brain. As alternative non-invasive approaches with similar resolution and reliability are still missing, our knowledge about structural brain connectivity is largely restricted to non-human brains [32].

While the distribution of retrograde labelling may be quantified by systematic counts of labelled cells of projection origin (providing a numerical measure of the number of axons in a particular projection), the strength of anterograde label at a given location can often only be determined as an ordinal measure (e.g. 'sparse', 'moderate', 'dense'). The quantification of the number of projection origins in retrograde experiments is laborious as well; for this reason, the strength of fiber pathways is frequently only reported in ordinal terms (as for the specific cat data shown in Fig. 9.2 of Chap. 9: ones represent sparse, twos moderate and threes dense projections). It should be noted that the absent entries in a connectivity matrix can stand for connections that were investigated and were found to be absent, or potential projections that were not investigated in the first place. The potential impact of future additions to connectivity compilations needs to be carefully considered. However, previous simulations of connectivity matrices in which all entries with unknown information were assumed to exist did not result in principally different findings [33, 34].

In the remainder of this review, we discuss the organization of structural brain connectivity and its implication for the functions of the cerebral cortex, using as a particular example the complex network of interconnections between regions of the cerebral cortex of the cat. First, we explore the spatial layout of cortical networks in Sect. 4.2, and then their topological organization in Sect. 4.3, before presenting aspects of functional implications in Sect. 4.4. We conclude by summarizing the main conclusions and formulating open problems for future research.

4.2 Spatial Organization

The organization of neural systems is shaped by multiple constraints, ranging from limits placed by physical and chemical laws to diverse functional requirements. In particular, it is of interest to identify factors influencing the spatial layout of neural connectivity networks. Although spatial coordinates of cortical areas in the cat are not readily available at the moment, it may be expected that the spatial organization of the cat cortex is close to that in other mammals. In particular, we have analyzed detailed information about the spatial organization of the cerebral cortex in primates, such as the macaque monkey [35, 36].

One prominent guiding idea is that the establishment and maintenance of neural connections carries a significant metabolic cost that should be reduced wherever possible [37]. As a consequence, wiring length should be globally minimized in neural systems. A trend toward wiring minimization is apparent in the distributions of projection lengths for various neural systems, which show that most neuronal projections are short [2, 35, 36]. However, wiring length distributions also indicate a significant number of longer-distance projections, which are not formed between immediate neighbors in the network.

Alternatively, it has been suggested that wiring length reductions in neural systems are achieved not by minimal rewiring of projections within the networks, but by suitable spatial arrangement of the components. Under these circumstances, the connectivity patterns of neurons or regions remain unchanged, maintaining their structural and functional connectivity, but the layout of components is perfected such that it leads to the most economical wiring. In the sense of this 'component placement optimization' (CPO) [37], any rearrangement of the position of neural components, while keeping their connections unchanged, would lead to an increase of total wiring length in the network.

Using extensive connectivity datasets for systems and cellular neural networks combined with spatial coordinates for the network nodes, we found that optimized component rearrangements could substantially reduce total wiring length in all tested neural networks [36]. Specifically, total wiring length between 95 primate (macaque) cortical areas could be decreased by 32%, and wiring of neuronal networks in the nematode *Caenorhabditis elegans* could be shortened by 48% on the global level, and by 49% for neurons within frontal ganglia. The wiring length distribution before and after optimization as well as the reduction for the macaque cortical connectivity are shown in Fig. 4.1. Wiring length reductions were possible due to the existence of long-distance

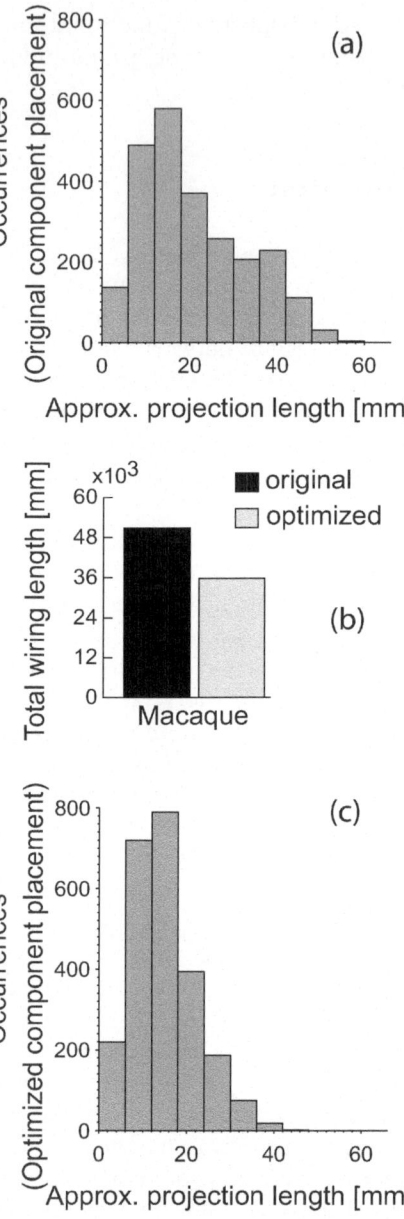

Fig. 4.1. Projection length distribution and total wiring length for original and optimally rearranged neural networks: (**a**) Approximated projection length distribution in the macaque monkey cortical connectivity network with 95 areas and 2,402 projections; (**b**) Reduction in total wiring length in rearranged layouts yielded by simulated annealing; (**c**) Approximated projection length distribution in neural networks with optimized component placement. The number of long distance connections is substantially reduced compared to the original length distribution in (**a**)

projections in the neural networks. Thus, biological neural networks feature shorter average pathlengths than networks lacking long-distance connections. Moreover, the average shortest pathlengths of neural networks, corresponding to the average number of processing steps along the shortest signalling paths, were close to pathlengths in networks optimized for minimal paths.

Minimizing average pathlength — that is, reducing the number of intermediate transmission steps in neural integration pathways — has several functional advantages. First, the number of intermediate nodes that may introduce interfering signals and noise is limited. Second, by reducing transmission delays from intermediate connections, the speed of signal processing and, ultimately, behavioral decisions is increased. Third, long-distance connections enable neighboring as well as distant regions to receive activation nearly simultaneously [35, 38] and thus facilitate synchronous information processing in the system (compare [39]). Fourth, the structural and functional robustness of neural systems increases when processing pathways (chains of nodes) are shorter. Each further node introduces an additional probability that the signal is not transmitted, which may be substantial (e.g. failure rates for transmitter release in individual synapses are between 50% and 90% [40]). Even when the signal survives, longer chains of transmission may lead to an increased loss of information. A similar conclusion, on computational grounds, was first drawn by John von Neumann [41] when he compared the organization of computers and brains. He argued that, due to the low precision of individual processing steps in the brain, the number of steps leading to the result of a calculation ('logical depth') should be reduced, and highly parallel computing would be necessary.

4.3 Topologic Organization

Various approaches can be used to investigate the topology of cortical networks, at the local node-based level (e.g. Koetter & Stephan, [42]), for small circuits of connected nodes (also called network motifs), or at the global level of the network. Several of these approaches are summarized in [43, 44] and are also reviewed elsewhere in this volume (Chap. 3). Moreover, generic approaches for the investigation of complex networks may also be applied to brain connectivity [45]. Methodologically, analyses have used either techniques from graph theory, or multivariate methods using clustering or scaling techniques to extract statistical structure.

All studies of cortical structural connectivity have confirmed that cerebral cortical areas in mammalian brains are neither completely connected with each other nor randomly linked; instead, their interconnections show a specific and characteristic organization. Various global connectivity features of cortical networks have been described and characterized with the help of multivariate analysis techniques, such as multidimensional scaling or hierarchical cluster analysis [43]. For example, clusters or *streams* of visual cortical areas have

(a)

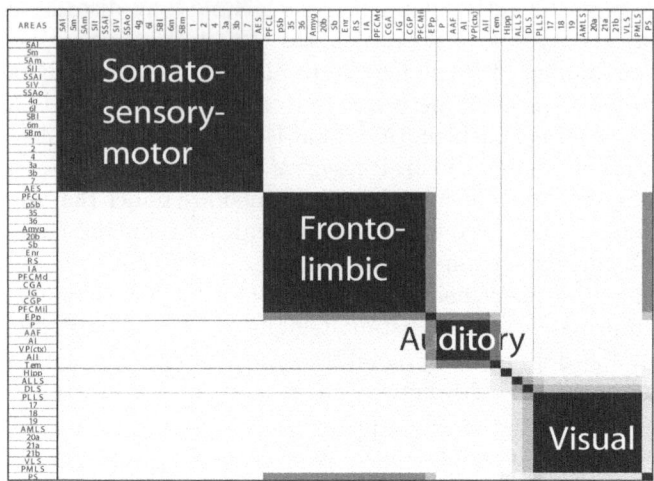

(b)

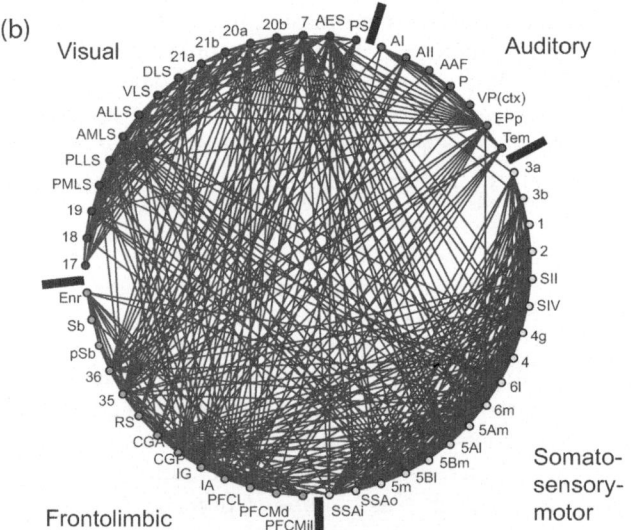

Fig. 4.2. Clustered organization of cat cortical connectivity: (**a**) Cluster count plot, indicating the relative frequency with which any two areas appeared in the same cluster, computed by stochastic optimization of a network clustering cost function [30]. Functional labels were assigned to the clusters based on the predominant functional specialization of areas within them, as indicated by the physiologic literature; (**b**) Cat cortical areas are arranged on a circle in such a way that areas with similar incoming and outgoing connections are spatially close. The ordering by structural similarity is related to the functional classification of the nodes, which was assigned as in (**a**)

been identified that are known to be segregated functionally [46] as well as in terms of their inputs, outputs and mutual interconnections [12]. Topological sequences of areas can be distinguished that might provide the layout for signaling pathways across cortical networks [47]. Alternatively, hierarchies of cortical areas can be constructed, based on the laminar origin and termination patterns of interconnections [48, 49].

Significantly, all large-scale cortical connection patterns examined to date, including global connectivity in cat and monkey brains as well as their subdivisions, exhibit small-world attributes with short pathlengths and high clustering coefficients [16, 30, 44]. These properties are also found in intermediate-scale connection patterns generated by probabilistic connection rules, taking into account metric distance between neuronal units [50]. The findings suggest that high clustering and short pathlengths can be found across multiple spatial scales of cortical organization.

To identify the clusters that are indicated by the high clustering coefficients of cortical networks, a computational approach based on evolutionary optimization can be used [30]. This stochastic optimization method delineated a small number of distinctive clusters in global cortical networks of cat and macaque [51] (Fig. 4.2). These clusters contained areas which were more frequently linked with each other than with areas in the remainder of the network, and the clusters followed functional subdivisions (e.g., containing predominantly visual or somatosensory-motor areas), as discussed in Sect. 4.6 below. The algorithm could also be tuned to identify clusters that no longer contained any known absent connections, and thus produced maximally dense clusters of areas, interpretable as network 'building blocks'.

A clustered organization of cortical networks was also indicated by application of the matching index, demonstrating distinct groups of cortical areas with similar input and output [43]. The matching index captures the pairwise similarity of areas in terms of their specific afferents and efferents from other parts of the network [43, 44] (as well as in Chap. 3), following one of the central assumptions of systems neuroscience that the functional roles of brain regions are specified by their inputs and outputs. In agreement with this concept, one finds that pairs of areas with high matching index also share functional properties [43].

The ubiquitous feature of a segregation of cortical networks into multiple, interconnected network clusters (or 'communities' in network analysis parlance) is explored from various perspectives in the following sections.

4.4 Network Development

The previously reviewed spatial and topologic analyses demonstrate that cortical networks are predominantly, but not exclusively, connected by short projections, and that cortical connections link areas into densely connected clusters. How does such an organization arise during the development of the brain?

It is known that neural systems on several scales show distance dependence for the establishment of projections. In systems ranging from connectivity between individual neurons in *Caenorhabditis elegans* [36] and rat visual cortex [52] to connectivity between mouse [2,53], macaque [35] (see Fig. 4.1(a)) or human [54] cortical areas, there is a higher tendency to establish short-distance than long-distance connections. This feature can be incorporated into models of cortical connection development.

4.4.1 Models for Network Development

Whereas various models exist for network development or evolution, only few of them consider spatial constraints. The standard model for generating scale-free networks, for example, uses growth and preferential attachment [55]. Starting with m_0 initial nodes, a new node establishes a connection with an existing node i with the probability

$$P_i = \frac{k_i}{\sum k},$$

that is, the number of edges of node i (k_i) divided by the total number of edges already established in the network. The resulting network consists of one cluster, and the degree distribution shows a power law. This model may result in degree distributions comparable with real biological networks but lacks multiple clusters or modules.

In order to generate scale-free networks with a modular organization, a hierarchical model for network development was designed [56, 57]. Starting with one root node, during each step two units are added that are identical to the network generated in the previous iteration. Then the bottom nodes of these two units are linked with the root of the network. While this algorithm resulted in a modular organization, it did not support a large variety of module sizes within the same network, as seen in real-world networks.

Waxman [58] proposed a connection establishment algorithm for the Internet in which the probability of a connection between two nodes decays exponentially with the spatial distance between them. In this way, the high costs for the wiring and maintenance of long-range connections can be represented. Initially, the nodes are distributed at random. Thereafter, edges are attached to the graph. The probability that an edge is established between two nodes decays exponentially with the distance between them. In contrast to the previous models, the location of the nodes is determined from the start, therefore, there was no growth in terms of the size of the network or the number of nodes.

A decay with distance, however, is also likely for growing and expanding biological systems, as the concentration of a chemical substance such as a growth factor decays with the distance from the place of production or emission [59]. We therefore explored different mechanisms for spatial and topological network development through computational modeling, considering (i) a simple dependence of projection formation on spatial distance and enclosing spatial borders, and (ii) dependence on distance as well as on developmental time windows. We also compared the algorithms with previously suggested topological mechanisms for network development.

4.4.2 Growth Depending on Spatial Distance

We simulated mechanisms of spatial growth, in such a way that connections between nearby nodes (i.e. areas) in the cortical network were more probable than projections to spatially distant nodes [58]. Such a distribution could, for instance, result from the concentration of unspecific factors for axon guidance decaying exponentially with the distance to the source [59].

At each step of the algorithm, a new area was added to the network until reaching the target number of nodes (55 areas for simulated cat cortical networks). New areas were generated at randomly chosen positions of the embedding space. The probability for establishing a connection between a new area u and existing areas v was set as

$$P(u,v) = \beta \, e^{-\alpha \, d(u,v)}, \qquad (4.1)$$

with $d(u,v)$ being the distance between the nodes and α and β being scaling coefficients. Areas that did not establish connections were disregarded. A more detailed presentation of the network growth model is given elsewhere [60,61].

We generated 50 networks of the size of the cat cortical network, through limited spatial growth in a fixed modeling space, and using parameters $\alpha = 5$ and $\beta = 2.5$. The spatial limits imposed during the simulations might represent internal restrictions of growth (e.g. by apoptosis [62]) as well as external factors (e.g. skull borders). The simulated networks yielded clustering coefficients and averaged shortest pathlength (ASP; see Chap. 3), shown in Table 4.1, similar to the cortical network. Moreover, the degree distribution of cortical and simulated limited growth networks showed a significant correlation (Spearman's rank correlation $\rho = 0.77$, $P < 3 \times 10^{-3}$).

Note that a small-world topology with similar ASP and clustering coefficient as in the biological networks could only be generated for limited spatial growth where the growing network quickly reached the borders of the embedding space. For unlimited growth, the ASP was much larger whereas the clustering coefficient was much lower than for the original cortical network.

Table 4.1. Comparison of cat cortical and simulated networks. Shown are the clustering coefficient C_{brain} and ASP_{brain} of the cat network as well as the average clustering coefficient and ASP of 50 generated limited and unlimited spatial growth networks with respective standard deviations

	C_{brain}	$C_{limited}$	$C_{unlimited}$	ASP_{brain}	$ASP_{limited}$	$ASP_{unlimited}$
cat	0.55	0.50 ± 0.02	0.29 ± 0.05	1.8	1.70 ± 0.04	3.86 ± 0.47

This is in accordance with experimental findings in which the lack of growth limits given by prohibiting apoptosis [62] resulted in a different layer architecture and network topology.

The presented spatial growth model proceeds independently of network activity. Such an approach is supported by experimental studies that show that activity is not necessary for the establishment of global connectivity. For example, after blocking neurotransmitter release and thus activity propagation during development, the global connectivity pattern of the brain remained unchanged [63].

We also investigated an alternative growth model, using a developmental mechanism of growth and preferential attachment, in which new nodes were more likely to establish links to existing nodes that already had many connections [55]. This model was also able to yield density and clustering coefficients similar to those in cortical networks. However, it failed to generate multiple clusters seen in the biological systems, as only one main cluster could be generated by this approach.

4.4.3 Growth Depending on Distance and Developmental Time Windows

Whereas the spatial growth model described above could replicate the small-world topology of the cat cortical network, there was no guarantee that multiple network clusters, as found in the cortical connectivity of the mammalian brain, would arise. Moreover, in cases where multiple clusters did occur, their size could not be controlled by the model parameters.

In order to explore the essential multiple-cluster feature of cortical connectivity, we modified the previous model and included one additional factor of cortical development, the formation of cortical areas and their interconnections during specific, overlapping *time windows*. Time windows arise during cortical development [64,65], as the formation of many cortical areas overlaps in time but ends at different time points, with highly differentiated sensory areas (for example, Brodmann area 17) finishing last. Based on this experimental finding, we explored a modified wiring rule in which network nodes were more likely to connect if they were (i) spatially close and (ii) developed during the same time window.

The following algorithm was used for network growth depending on distance combined with time windows (cf. Fig. 4.3(a)). First, three seed nodes were placed at spatially distant locations (cf. Fig. 4.3(b)). New nodes were placed randomly in space. The time window of a newly forming node was the same as that of the nearest seed node, as it was assumed to originate from, or co-develop with, that node. Second, the new node u established a connection with an existing node v with probability $P(u,v) = P_{temp}(u) \times P_{temp}(v) \times P_{dist}(u,v)$. The dependence P_{dist} decayed exponentially with the distance between the two nodes (cf. [60]). Third, if the newly formed node failed to establish connections, it was removed from the network.

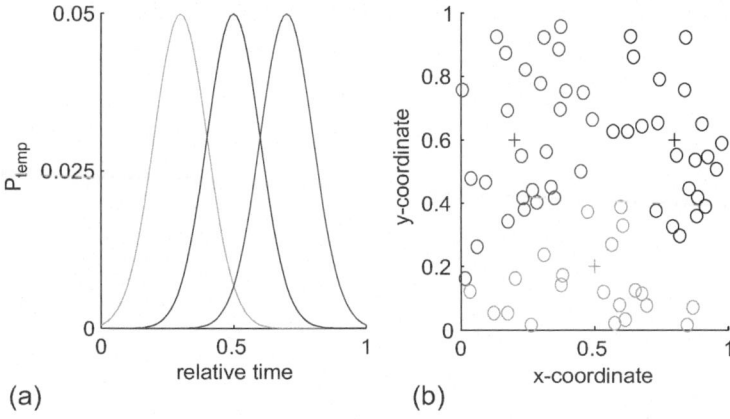

Fig. 4.3. Time windows and initial seed nodes: (**a**) Temporal dependence P_{temp} of projection establishment depending on node domain. Relative time was normalized such that '0' stands for the beginning of development and '1' for the end of network growth. The three seed nodes had different time windows, which were partially overlapping; (**b**) Two-dimensional projection of the 73 three-dimensional node positions. The gray level coding represents the time window corresponding to one of the three seed nodes (+)

Although the following results show networks with 73 nodes, comparable to the primate networks, networks similar to the cat cortical network also could be generated (not shown). The timed adjacency matrix shows the development of connections over time (Fig. 4.4(a)). Different gray levels represent the respective time windows of the nodes. The reordered matrix represents the original network

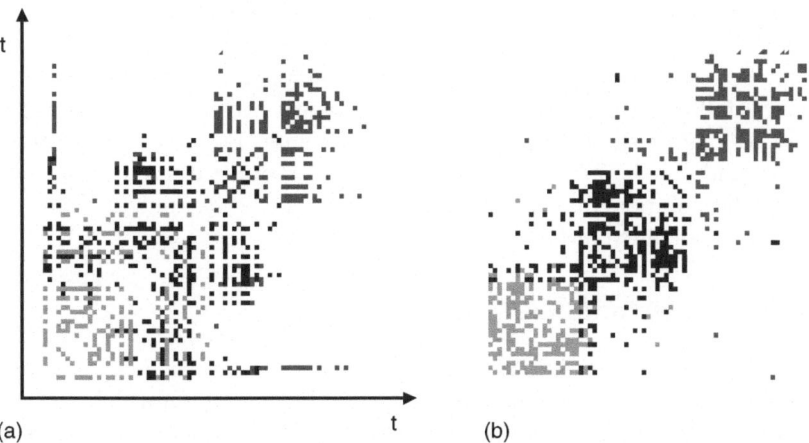

Fig. 4.4. (**a**) Timed adjacency matrix (the first nodes are in the left lower corner); (**b**) Clustered adjacency matrix. The matrix is the same as in (a), but nodes with similar connections are arranged adjacent in the node ordering

with different node order, in such a way that nodes with similar connectivity were placed nearby in the adjacency matrix(Fig. 4.4(b)).

Therefore, the inclusion of developmental time windows into the spatial growth algorithm generated multiple network clusters, with their number being identical to the number of different time windows that governed development. In addition to the number of clusters, the size of clusters could be varied by changing the width of the corresponding time window.

These results demonstrate that simple mechanisms of spatial growth, in combination with constraints by spatial borders or developmental time windows, can account for many of the structural features of corticocortical connectivity, in particular the formation of multiple network clusters.

4.5 Network Robustness

The brain can be remarkably robust against physical damage. Significant loss of neural tissue may be compensated in a relatively short time by large-scale adaptation of the remaining brain parts (e.g. [66–68]. For example, there has been a case in which almost an entire hemisphere was removed from an 11 year-old boy who had medically intractable seizures. After three years of rehabilitation training in the hospital, however, the patient was left with few remaining functional deficits [69]. Similarly, on a more local scale for damage within specific brain regions, Parkinson's disease only becomes apparent after half of the pigmented cells in the affected substantia nigra are lost [70]. In other cases, however, the removal of small amounts of tissue (e.g. in regions specialized for language functions) can lead to severe functional deficits. These findings provide a somewhat contradictory picture of the robustness of the brain and highlight several questions. Can we formally evaluate robustness given the variability in the effects of brain lesions? Are severity and nature of the effects of localized damage predictable? And finally, how can robustness against the loss of large amounts of tissue be explained?

In the following sections, we explore network robustness through the effects of lesions of structural network components. Lesions can affect nodes or edges, and can be applied either randomly or in a targeted way, in which case specific components are eliminated by target criteria which assess the perceived importance of the components.

4.5.1 Impact of Node Lesions

Following Barabási and Albert [71], we assessed the impact of lesions on network connectivity and integrity by measuring the average shortest path (ASP) or characteristic pathlength. As described in Chap. 3, the ASP between any two nodes in the network is the number of sequential connections that are required, on average, to link one node to another by the shortest possible route [72]. In case a network becomes disconnected in the process of removing edges or nodes, and no path exists between two particular nodes, this pair of nodes is ignored. If no connected nodes remain, the average shortest path is set to zero.

In two separate approaches for sequential node removal, we removed nodes either randomly or by targeted elimination. During random removal, nodes were selected randomly, with a uniform probability distribution, and deleted from the graph. In the case of targeted removal, the nodes with the highest node degree were subsequently eliminated. After each deletion, the ASP of the resulting graph was calculated, and the removal of nodes was continued until all nodes were removed from the network.

To provide benchmarks for comparisons, random, small-world and scale-free networks of the same size as the cat network were created and lesioned in an analogous way. We used rewiring to generate small-world networks [73] and a modified version of growth and preferential attachment for scale-free networks [55]. Fifty benchmark networks were created for each of the conditions. Moreover, the process of random removal of nodes in the cat cortical networks was repeated fifty times.

For the cat brain network (Fig. 4.5), the random and small-world benchmark networks show a different behavior for targeted node removal when compared to the cortical network. The cat network's response to targeted node removal is largely within the 95% confidence interval for the scale-free benchmark networks; however, the peak ASP value and the fraction of deleted nodes where the peak occurs are comparatively lower for the cat cortical network. Thus, in terms of random and targeted node lesion behavior, the cortical networks most closely resemble scale-free networks.

The decline in ASP at a later stage during the elimination process, as observed for the brain and scale-free networks, deserves special attention. It can be for two reasons. First, it could be that the network becomes fragmented into different disconnected components. Each of these is smaller, and likely to have a shorter ASP. Second, the overall decrease in network size with successive eliminations can lead to a decrease in shortest path. This is, however, likely to be a slow process, as it will usually be offset by an increase in ASP due to the targeted nature of the elimination.

In conclusion, this shows that structural properties of cortical networks are quite robust towards the random elimination of nodes from the network. In contrast, the targeted removal of nodes, by removing the most highly-connected nodes first, leads to a rapid fragmentation of the network. Therefore, cortical networks are similar to scale free networks in their response to the random or targeted removal of nodes. Indeed, as for scale-free networks, brain networks contain nodes which are almost connected to all other nodes of the system. Examples of such 'hubs' for the cat would be amygdala and hippocampus in the subcortical domain and anterior ectosylvian sulcus (AES), agranular insula (Ia), and area 7 for cortical regions.

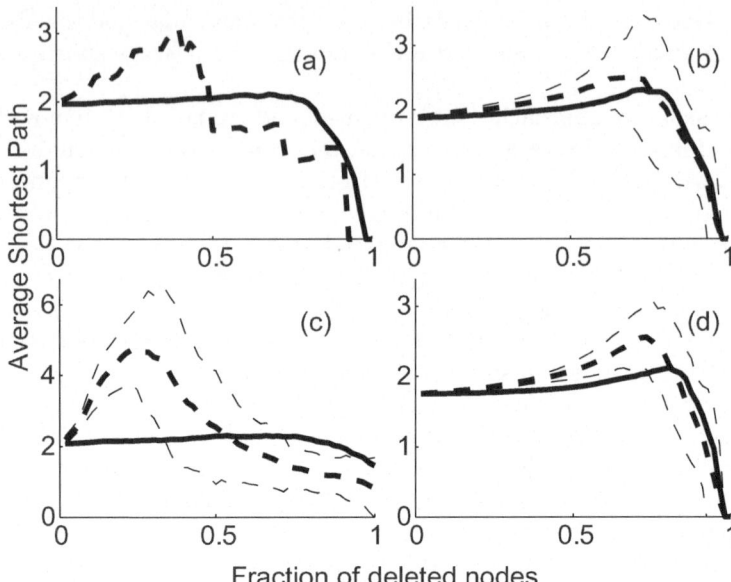

Fig. 4.5. Sequential node elimination in cat cortical and benchmark networks. The fraction of deleted nodes (zero for the intact network) is plotted against the average shortest path (ASP) after node removals. Nodes were removed randomly, or starting with the most highly connected nodes (targeted elimination): (**a**) Cat cortical network during targeted (*dashed*) and random (*solid line*) elimination. In the subsequent plots (**b**), (**c**) and (**d**), the dashed line shows the average effect of targeted elimination and the thin dashed lines the 95% confidence interval for the generated same-size benchmark networks. The solid line represents the average effect of random elimination; (**b**) Small-world benchmark network; (**c**) Scale-free benchmark network; (**d**) Random benchmark network

4.5.2 Impact of Edge Lesions

In many networks, the failure of single connections may be more likely than the extinction of entire nodes. We tested several measures for identifying vulnerable edges and compared their prediction performance for the cat cortical network. Among the tested measures, edge frequency in all shortest paths of a network yielded a particularly high correlation with vulnerability, and identified inter-cluster connections in biological networks [35].

Measures for Predicting Edge Vulnerability

We tested four candidate measures for predicting vulnerable edges in networks. First, the product of the degrees (PD) of adjacent nodes was calculated for each edge. A high PD indicates connections between two hubs which may represent potentially important network links. Second, the absolute difference in the adjacent node degrees (DD) of all edges was inspected. A large degree difference signifies connections between hubs and more sparsely connected network regions which may be important for linking central with peripheral regions of a network. Third, the matching index (MI) [43] was calculated as the number of matching incoming and outgoing connections of the two nodes adjacent to an edge, divided by the total number of the nodes' connections (excluding direct connections between the nodes [44]). A low MI identifies connections between very dissimilar network nodes which might represent important 'short cuts' between remote components of the network. Finally, edge frequency (EF), a measure similar to 'edge betweenness' [74, 75], indicates how many times a particular edge appears in all pairs shortest paths of the network. This measure focuses on connections that may have an impact on the characteristic path length by their presence in many individual shortest paths [35].

Prediction Performance

In the present calculation, both increase and decrease of ASP indicate an impairment of the network structure. Therefore, we took the deviation from the ASP of the intact network as a measure for structural impairments. We evaluated the correlation between the size of the prediction measures and the damage (shown in Table 4.2). While most of the local measures exhibited good correlation with ASP impact in real-world networks, the highest correlation was consistently reached by the EF measure. Also, the measures of matching index and difference of degrees show a high correlation.

After identifying a measure to predict *which* edges were most vulnerable, we looked at *where* in the network these edges resided. We generated 20 test networks; each consisting of three randomly wired clusters and six fixed inter-cluster connections (Fig. 4.6(a)). The inter-cluster connections (light gray) occurred in many shortest paths (Fig. 4.6(b)) leading to an assignment of the highest EF value, as no alternative paths of the same length were available. Furthermore, their elimination resulted in the greatest network damage as shown by increased ASP.

Table 4.2. Density, clustering coefficient CC, average shortest path ASP and correlation coefficients r for different vulnerability predictors of the analyzed networks (the index refers to the number of nodes). Tested prediction measures were the product of degrees (PD), absolute difference of degrees (DD), matching index (MI), and edge frequency (EF)

	Density	CC	ASP	r_{PD}	r_{DD}	r_{MI}	r_{EF}
Cat$_{55}$	0.30	0.55	1.8	0.08*	0.48**	−0.34**	0.77**

* Significant Pearson Correlation, 2-tailed 0.05 level.
** Significant Pearson Correlation, 2-tailed 0.01 level.

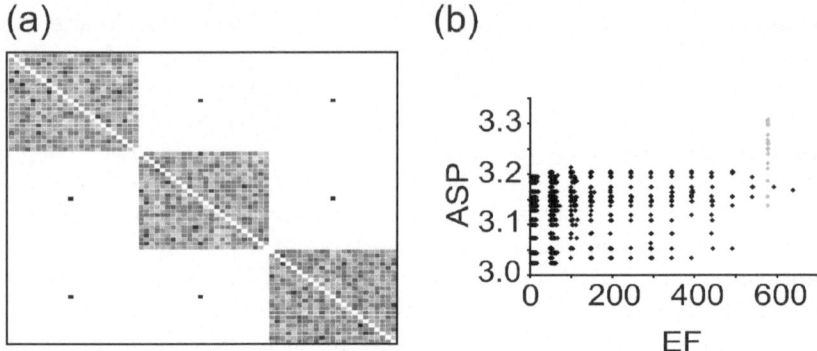

Fig. 4.6. Edges with high edge frequency form essential inter-cluster connections: (**a**) Connectivity of test networks with three clusters and six pre-defined inter-cluster connections. The network has comparable density as primate brain connectivity. The gray-level shading of a connection in the adjacency matrix indicates the relative frequency of an edge in 20 generated networks. White entries stand for edges absent in all networks; (**b**) Edge frequencies in the all-pairs shortest paths against ASP after elimination of edges. Light gray data points represent the values for the inter-cluster connections in all 20 test networks. Inter-cluster connections not only have the largest edge frequency, but also cause most damage after elimination

4.5.3 Conclusions of Network Lesion Studies

Cortical systems are known to consist of several distinct, linked clusters with a higher frequency of connection within than between the clusters (e.g. Fig. 4.2). Inter-cluster connections have also been considered important in the context of social contact networks, as 'weak ties' between individuals [76] and separators of communities [74]. We, therefore, speculated that connections between clusters might be generally important for predicting vulnerability. Whereas many alternative pathways exist for edges within clusters, alternative pathways for edges between clusters can be considerably longer. Interestingly, previously suggested growth mechanisms for scale-free networks, such as preferential attachment [55], or strategies for generating hierarchical networks [56] did not produce distributed, interlinked clusters. Consequently, the low predictive value of EF in the scale-free benchmark networks was attributable to the fact that scale-free networks grown by preferential attachment consisted of one central cluster, but did not possess a multi-cluster organization. This suggests that alternative developmental models may be required to reproduce the specific organization of biological networks (see Sect. 4.4 above).

An analysis of the similarity between the cat network and random, rewired, small-world, and scale-free benchmark networks shows that cat cortical connectivity is most similar to scale-free networks in several respects. Most importantly, cortical as well as scale-free networks show a huge disparity between random elimination of nodes or edges and targeted elimination in which the most vulnerable parts of the network are removed. Similarly, an analysis of edge vulnerability has shown that the targeted removal of connections between clusters has a large effect compared to the removal of connections within a cluster.

Is the brain optimized for robustness against random removal of nodes or edges? Although this question is highly debatable, there are several arguments why the robust architecture could be explained as a side effect of other functional constraints. For example, the formation of functional clusters is necessary to exclude signals with different modality and highly-connected areas could function as integrators of (multi-modal) information or spreaders of information to multiple clusters or many nodes within one cluster. Therefore, in the future it will be interesting to compare the function of areas with their number of projections and the function of their directly connected neighbors.

The effect of the structural network organization on the functional impact after lesions is also important for neurological lesion analysis [68]. The degree of connectedness of neural structures can affect the functional impact of local and remote network lesions, and this property might also be an important factor for inferring the function of individual regions from lesion-induced performance changes [21].

4.6 Functional Implications

Several experimental studies have demonstrated a close link between the organization of structural brain connectivity and functional connectivity. The analysis of neuronographic connection data for the monkey and cat cortex, for example, has revealed dense interconnections among visual, auditory and particularly somatosensory-motor areas, arranged in similar network clusters as those formed by structural connectivity [77]. Neuronographic data are produced by the disinhibition of local populations of cortical neurons through the application of strychnine, and by recording the resulting steady-state activation of remote areas. Thus, this kind of approach reveals eleptiform functional connectivity , which also can be described by a simple propagation model [78].

Similar findings were obtained for another type of functional connectivity independent from particular tasks or stimuli, the slow-frequency coupling among cortical areas in human fMRI resting state data [79]. These resting-state networks are very similar to the structural connectivity from the cat or monkey, in that most interactions proceed only across short distances and run in local clusters, which follow regional and functional subdivisions of the brain [80]. However, some functional interactions also exist across longer distances, particularly those between homotopic regions in the two hemispheres. However, the actual neural or metabolic mechanisms underlying resting-state coupling are still poorly understood. Thus, this type of connectivity currently needs to be interpreted with caution.

In the next subsections, we review particular functional implications of the clustered organization of cortical networks, based on various computational approaches.

4.6.1 Functional Motif Diversity

The approach of motif analysis [81] can be adapted to draw conclusions on functional diversity in cortical networks, based on an analysis of structural connectivity. This approach starts from a simple premise: areas need to be connected in order to interact, and the kind and number of different local circuits that can be formed within a given structural network may reflect the diversity of functions performed by this network. This idea was applied to a motif analysis of corticocortical connectivity, by distinguishing between structural and functional motifs. Structural motifs were defined in the conventional sense, as the set of different connection patterns involving $n = 2, 3, 4 \ldots$ nodes found in the given network, while functional motifs represented all possible subsets of identified structural motifs. By comparison with benchmark networks and through network evolution, Sporns and Koetter [82] found that, in cortical networks, the number of structural motifs is small, while the number of functional motifs is large, suggesting that cortical networks are organized as to achieve high functional diversity with a small number of different structural elements. However, this finding may be partly due to the global organization of the cortical networks. Since the studied networks are organized into densely connected clusters, the motif analysis resulted in a small number of structural motifs which are also completely, or almost completely, connected. Naturally, these structural motifs allow a large number of different functional submotives. Moreover, motifs are difficult to interpret in terms of building blocks of development or function. For example, current knowledge about the development or evolution of cortical networks makes it appear unlikely that brain networks develop by adding circuits of three or four areas. Moreover, the functional interactions within such ensembles may be too complex to represent a truly basic unit of cortical functioning.

4.6.2 Functional Complexity

Central aspects of cortical functioning are provided by the structural and functional specialization of cortical regions on the one hand, and their integration in distributed networks, on the other. This relationship has been formalized as an expression of functional complexity, with complexity being defined as

$$C(X) = H(X) - \sum_i H(\boldsymbol{x}_i | X - \boldsymbol{x}_i), \qquad (4.2)$$

where $H(X)$ is the entropy of the system and the second term on the right-hand side of the equation denotes the conditional entropy of each element, given the entropy of the rest of the system [16]. Thus, this measure describes the balance between the functional independence of a system's elements and their functional integration.

The complexity of cortical networks, as defined in (4.2), was explored through a simple functional model of the network, created by injecting the areas with white noise and investigating their functional coupling. Random rewiring of the actual cortical networks reduced their functional complexity, while graph evolution and selection for maximum complexity tended to produce small-world networks with a similar organization of multiple interconnected clusters as seen in the actual networks. Thus, the clustered organization of the networks appears to be well suited to support functional complexity, by facilitating high integration within clusters, yet high independence between clusters.

4.6.3 Critical Range of Functional Activations

In complex neural networks stable activation patterns within a critical functional range are required to allow function to be represented (cf. Chap. 7). In this critical range, the activity of neural populations in the network persists, falling between the extremes of quickly dying out or activating the whole network. We used a basic percolation model to investigate how functional activation spreads through a cortical network which has a clustered organization across several levels of organization. Specifically, cortical networks not only form clusters at the level of connected areas, but neural populations within areas are also more strongly connected with each other than they are with neurons in other cortical areas. Similar clustering can be observed at the even finer levels of hypercolumns and columns [83] (see Chap. 8). Our simulations demonstrated that hierarchical cluster networks were more easily activated than random networks, and that persistent and scalable activation patterns could be produced in hierarchically clustered networks, but not in random networks of the same size. This was due to the higher density of connections within the clusters facilitating local activation, in combination with the sparser connectivity between clusters which hindered the spreading of activity to the whole network. The critical range in the hierarchical networks was also larger than that in simple, same-sized small-world networks (Fig. 4.7). A detailed description of these results will be published elsewhere. The findings indicate that a hierarchical cluster architecture may provide the structural basis for the stable functional patterns observed in cortical networks.

While the reviewed results give ground for optimism that the structural organization of cortical networks provides the basis for their diverse, complex and stable functions, the actual mechanisms linking structural and functional connectivity are still speculative. Further computational modeling may be helpful in exploring some of the potential mechanisms, in particular by taking into account different types of organization at the level of area-intrinsic connectivity.

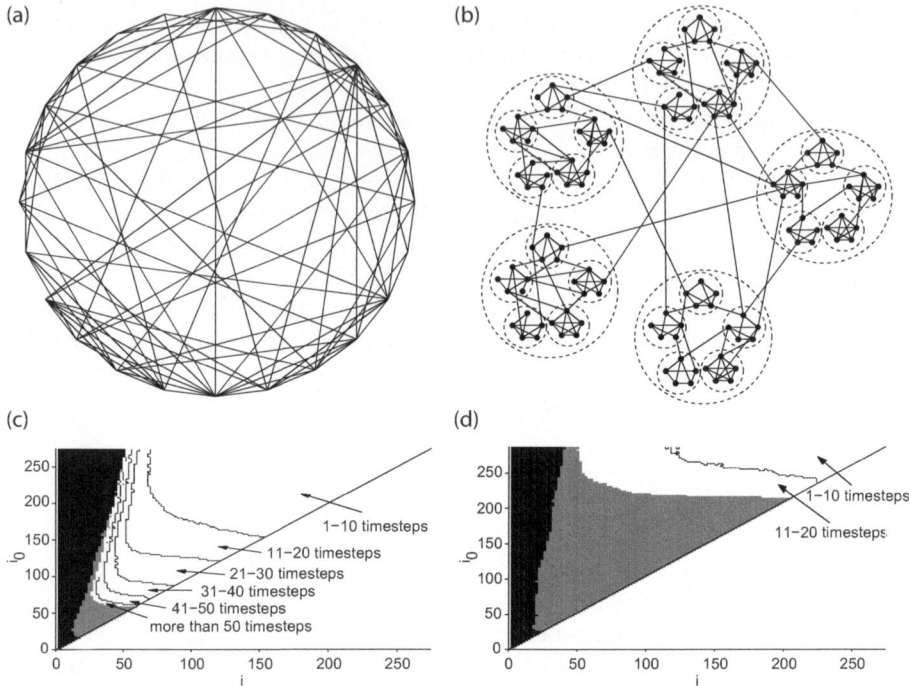

Fig. 4.7. Functional criticality in small-world and hierarchical cluster networks: Schematic views of (**a**) a small-world network and; (**b**) a hierarchical multiple-cluster network; (**c**) Critical range (gray) of persistent activity for small-world and; (**d**) hierarchical networks. In the critical parameter range, activity neither dies out nor spreads through the whole network

4.7 Conclusions and Open Questions

A number of preliminary conclusions can be drawn from the findings presented here. First, cerebral cortical fibre networks balance short overall wiring with short processing paths. This combination of desirable spatial and topologic network features results in high functional efficiency, providing a reduction of conduction delays along fibre tracts in combination with a reduction of transmission delays at node relays. Second, a central aspect of the topological organization of cortical networks is their segregation into distributed and interconnected multiple clusters of areas. Such a modular organization can also be observed at smaller levels of the cortical architecture, for instance, in densely intraconnected cortical columns. Experimental physiologic studies have demonstrated that the area clusters correspond to functional subdivisions of the cerebral cortex, suggesting a close relationship between global connectivity and function. Third, biologically plausible growth mechanisms for spatial network development can be implemented in simple computational models. Networks resulting from the simulations show the observed distance distribution of cortical connectivity as well its multi-cluster network organization. Fourth, the clustered organization of cortical connectivity may also be the reason for the robustness of cortical networks against damage, in particular for randomly inflicted impairments. In their response to damage and attack on nodes, cortical networks show behavior similar to scale-free networks. Due to this feature, they are robust to random lesions of nodes, but react critically to the removal of highly-connected nodes. Finally, the outlined structural organization of cortical networks has various functional implications: networks of hierarchically organized, inter-linked clusters provide the circuitry for diverse functional interactions and may lead to increased functional complexity as well as a wider critical range of activation behavior.

Nonetheless, many open questions remain. For example, what determines the specific layout of long-range cortical projections, given that their layout is not completely specified by spatial proximity? Is the organization of cortical networks into clusters already determined during ontogenetic development, or are they formed later on by activity-dependent rewiring of the networks? How are long-range connections integrated with the intrinsic micro-circuitry of cortical areas? What determines how widely brain functions are distributed in cortical networks? More broadly, what is the exact relationship between structural and functional connectivity at the systems level? Can general rules be formulated that describe the functional interactions of areas based on their structural connectivity? Progress in answering these questions will depend on a close and fruitful interaction between quantitative anatomical and physiological brain research and new approaches in data analysis and network modeling.

Acknowledgements

M.K. acknowledges financial support from the German National Academic Foundation and EPSRC (EP/E002331/1).

References

1. Braendgaard, H., Evans, S.M., Howard, C.V., Gundersen, H.J.: The total number of neurons in the human neocortex unbiasedly estimated using optical disectors. J. Microsc. **157**(Pt 3) (1990) 285–304.
2. Braitenberg, V., Schüz, A.: Cortex: Statistics and Geometry of Neuronal Connectivity. 2nd edn. Springer (1998).
3. Binzegger, T., Douglas, R.J., Martin, K.A.C.: A quantitative map of the circuit of cat primary visual cortex. J. Neurosci. **24** (2004) 8441–8453.
4. Young, M.P.: The organization of neural systems in the primate cerebral cortex. Phil. Trans. R. Soc. **252** (1993) 13–18.
5. Binzegger, T., Douglas, R.J., Martin, K.A.C.: Axons in cat visual cortex are topologically self-similar. Cereb Cortex **15**(2) (2005) 152–165.

6. Silberberg, G., Grillner, S., LeBeau, F.E.N., Maex, R., Markram, H.: Synaptic pathways in neural microcircuits. Trends. Neurosci. **28**(10) (2005) 541–551.
7. Markram, H., Toledo-Rodriguez, M., Wang, Y., Gupta, A., Silberberg, G., Wu, C.: Interneurons of the neocortical inhibitory system. Nat. Rev. Neurosci. **5**(10) (2004) 793–807.
8. Schubert, D., Kotter, R., Luhmann, H.J., Staiger, J.F.: Morphology, electrophysiology and functional input connectivity of pyramidal neurons characterizes a genuine layer va in the primary somatosensory cortex. Cereb. Cortex **16**(2) (2006) 223–236.
9. Thomson, A.M., Morris, O.T.: Selectivity in the inter-laminar connections made by neocortical neurones. J. Neurocytol. **31**(3–5) (2002) 239–246.
10. Zeki, S., Shipp, S.: The functional logic of cortical connections. Nature **335**(6188) (1988) 311–317.
11. Van Essen, D.C., Anderson, C.H., Felleman, D.J.: Information processing in the primate visual system: an integrated systems perspective. Science **255**(5043) (1992) 419–423.
12. Young, M.P.: Objective analysis of the topological organization of the primate cortical visual system. Nature **358**(6382) (1992) 152–155.
13. Crick, F., Koch, C.: Are we aware of neural activity in primary visual cortex? Nature **375**(6527) (1995) 121–123.
14. Scannell, J., Blakemore, C., Young, M.: Analysis of connectivity in the cat cerebral cortex. J. Neurosci. **15**(2) (1995) 1463–1483.
15. Friston, K.J.: Models of brain function in neuroimaging. Annu. Rev. Psychol. **56** (2005) 57–87.
16. Sporns, O., Tononi, G., Edelman, G.M.: Theoretical neuroanatomy: relating anatomical and functional connectivity in graphs and cortical connection matrices. Cereb. Cortex **10** (2000) 127–141.
17. Stephan, K.E., Zilles, K., Kotter, R.: Coordinate-independent mapping of structural and functional data by objective relational transformation (ORT). Philos. Trans. R. Soc. Lond. B Biol. Sci. **355**(1393) (2000) 37–54.
18. Sporns, O., Chialvo, D.R., Kaiser, M., Hilgetag, C.C.: Organization, development and function of complex brain networks. Trends Cogn. Sci. **8** (2004) 418–425.
19. Friston, K.J.: Functional and effective connectivity in neuroimaging: a synthesis. Hum. Brain Mapp. **2** (1994) 56–78.
20. Horwitz, B.: The elusive concept of brain connectivity. Neuroimage **19**(2 Pt 1) (2003) 466–470.
21. Keinan, A., Sandbank, B., Hilgetag, C.C., Meilijson, I., Ruppin, E.: Fair attribution of functional contribution in artificial and biological networks. Neural Comp. **16** (2004) 1887–1915.
22. Seth, A.K.: Causal connectivity of evolved neural networks during behavior. Network **16**(1) (2005) 35–54.
23. Varela, F., Lachaux, J.P., Rodriguez, E., Martinerie, J.: The brainweb: phase synchonization and large-scale integration. Nature Rev. Neurosci. **2** (2001) 229–239.
24. Büchel, C., Friston, K.J.: Modulation of connectivity in visual pathways by attention: cortical interactions evaluated with structural equation modelling and fMRI. Cereb. Cortex. **7** (1997) 768–778.
25. Bressler, S.L.: Large-scale cortical networks and cognition. Brain Res. Brain Res. Rev. **20**(3) (1995) 288–304.

26. Buchel, C., Friston, K.J.: Dynamic changes in effective connectivity character-
    ized by variable parameter regression and Kalman filtering. Hum. Brain. Mapp.
    6(5–6) (1998) 403–408 Clinical Trial.
27. Buchel, C., Coull, J.T., Friston, K.J.: The predictive value of changes in effective
    connectivity for human learning. Science **283**(5407) (1999) 1538–1541.
28. Passingham, R.E., Stephan, K.E., Kötter, R.: The anatomical basis of functional
    localization in the cortex. Nat. Rev. Neurosci. **3** (2002) 606–616.
29. Izhikevich, E.M., Gally, J.A., Edelman, G.M.: Spike-timing dynamics of neu-
    ronal groups. Cereb. Cortex **14** (2004) 933–944.
30. Hilgetag, C.C., Burns, G.A.P.C., O'Neill, M.A., Scannell, J.W., Young, M.P.:
    Anatomical connectivity defines the organization of clusters of cortical areas in
    the macaque monkey and the cat. Phil. Trans. R. Soc. Lond. B **355** (2000)
    91–110.
31. Scannell, J.W., Burns, G.A., Hilgetag, C.C., O'Neil, M.A., Young, M.P.: The
    connectional organization of the cortico-thalamic system of the cat. Cereb.
    Cortex **9**(3) (1999) 277–299.
32. Crick, F., Jones, E.: Backwardness of human neuroanatomy. Nature **361**(6408)
    (1993) 109–110.
33. Young, M.P., Scannell, J.W., O'Neill, M.A., Hilgetag, C.C., Burns, G.,
    Blakemore, C.: Non-metric multidimensional scaling in the analysis of neu-
    roanatomical connection data and the organization of the primate cortical visual
    system. Phil. Trans. R. Soc. **348** (1995) 281–308.
34. Hilgetag, C.C., O'Neill, M.A., Young, M.P.: Indeterminancy of the visual cortex.
    Science **271**(5250) (1996) 776–777.
35. Kaiser, M., Hilgetag, C.C.: Modelling the development of cortical networks.
    Neurocomputing **58–60** (2004) 297–302.
36. Kaiser, M., Hilgetag, C.C.: Nonoptimal component placement, but short pro-
    cessing paths, due to long-distance projections in neural systems. PLoS. Com-
    put. Biol. (2006) e95.
37. Cherniak, C.: Component placement optimization in the brain. J. Neurosci.
    14(4) (1994) 2418–2427.
38. Masuda, N., Aihara, K.: Global and local synchrony of coupled neurons in
    small-world networks. Biol. Cybern. **90** (2004) 302–309.
39. von der Malsburg, C.: The correlation theory of brain function. Technical report,
    Max-Planck-Institute for Biophysical Chemistry (1981).
40. Laughlin, S.B., Sejnowski, T.J.: Communication in neuronal networks. Science
    301 (2003) 1870–1874.
41. von Neumann, J.: The Computer and the Brain. Yale University Press (1958).
42. Kötter, R., Stephan, K.E.: Network participation indices: characterizing com-
    ponent roles for information processing in neural networks. Neural Networks **16**
    (2003) 1261–1275.
43. Hilgetag, C.C., Kötter, R., Stephan, K.E., Sporns, O.: Computational meth-
    ods for the analysis of brain connectivity. In: Computational Neuroanatomy.
    Humana Press, Totowa, NJ (2002) 295–335.
44. Sporns, O.: Graph theory methods for the analysis of neural connectivity pat-
    terns. In: Neuroscience Databases. Kluwer Academic, Dordrecht (2002) 169–183.
45. da Fontura Costa, L., Rodrigues, F.A., Travieso, G., Villas Boas, P.R.: Charac-
    terization of complex networks: A survey of measurements. cond-mat **0505185**
    (2005) v3.

46. Ungerleider, L., Mischkin, M.: Two cortical visual systems. In Ingle, M., Goodale, M., Mansfield, R., eds.: The New Cognitive Neurosciences. MIT Press, Cambridge, MA (1982).
47. Petroni, F., Panzeri, S., Hilgetag, C.C., Kötter, R., Young, M.P.: Simultaneity of responses in a hierarchical visual network. Neuroreport 12 (2001) 2753–2759.
48. Felleman, D.J., van Essen, D.C.: Distributed hierarchical processing in the primate cerebral cortex. Cereb. Cortex 1 (1991) 1–47.
49. Hilgetag, C.C., O'Neill, M.A., Young, M.P.: Hierarchical organization of macaque and cat cortical sensory systems explored with a novel network processor. Philos. Trans. R. Soc. Lond. Ser. B 355 (2000) 71–89.
50. Sporns, O., Zwi, J.D.: The small world of the cerebral cortex. Neuroinformatics 2 (2004) 145–162.
51. Hilgetag, C.C., Kaiser, M.: Clustered organisation of cortical connectivity. Neuroinformatics 2 (2004) 353–360.
52. Hellwig, B.: A quantitative analysis of the local connectivity between pyramidal neurons in layers 2/3 of the rat visual cortex. Biol. Cybern. 82 (2000) 111–121.
53. Schuz, A., Chaimow, D., Liewald, D., Dortenman, M.: Quantitative aspects of corticocortical connections: a tracer study in the mouse. Cereb. Cortex 16 (2005) 1474–1486.
54. Schuez, A., Braitenberg, V.: The human cortical white matter: quantitative aspects of corticocortical long-range connectivity. In Schuez, A., Miller, R., eds.: Cortical areas: unity and diversity. CRC Press, London (2002) 377–385.
55. Barabási, A.L., Albert, R.: Emergence of scaling in random networks. Science 286 (1999) 509–512.
56. Barabási, A.L., Ravasz, E., Vicsek, T.: Deterministic scale-free networks. Physica A 3–4 (2001) 559–564.
57. Ravasz, E., Somera, A.L., Mongru, D.A., Oltvai, Z.N., Barabási, A.L.: Hierarchical organization of modularity in metabolic networks. Science 297 (2002) 1551–1555.
58. Waxman, B.M.: Routing of multipoint connections. IEEE J. Sel. Areas Commun. 6(9) (1988) 1617–1622.
59. Murray, J.D.: Mathematical Biology. Springer, Heidelberg (1990).
60. Kaiser, M., Hilgetag, C.C.: Spatial growth of real-world networks. Phys. Rev. E 69 (2004) 036103.
61. Kaiser, M., Hilgetag, C.C.: Edge vulnerability in neural and metabolic networks. Biol. Cybern. 90 (2004) 311–317.
62. Kuida, K., Haydar, T.F., Kuan, C.Y., Gu, Y., Taya, C., Karasuyama, H., Su, M.S., Rakic, P., Flavell, R.A.: Reduced apoptosis and cytochrome c-mediated caspase activation in mice lacking caspase. Cell 94(3) (1998) 325–337.
63. Valverde, S., Cancho, R.F., Solé, R.V.: Scale-free networks from optimal design. Europhys. Lett. 60(4) (2002) 512–517.
64. Sur, M., Leamey, C.A.: Development and plasticity of cortical areas and networks. Nature Rev. Neurosci. 2 (2001) 251–262.
65. Rakic, P.: Neurogenesis in adult primate neocortex: an evaluation of the evidence. Nature Rev. Neurosci. 3 (2002) 65–71.
66. Spear, P., Tong, L., McCall, M.: Functional influence of areas 17, 18 and 19 on lateral suprasylvian cortex in kittens and adult cats: implications for compensation following early visual cortex damage. Brain Res. 447(1) (1988) 79–91.

67. Stromswold, K.: The cognitive neuroscience of language acquisition. In Gazzaniga, M., ed.: The New Cognitive Neurosciences. 2nd edn. MIT Press, Cambridge, MA (2000) 909–932.
68. Young, M.P., Hilgetag, C.C., Scannell, J.W.: On imputing function to structure from the behavioural effects of brain lesions. Phil. Trans. R. Soc. **355** (2000) 147–161.
69. Traufetter, G.: Leben ohne links. Spiegel Special (4) (2003) 30.
70. Pakkenberg, B., Moller, A., Gundersen, H.J., Mouritzen, D.A., Pakkenberg, H.: The absolute number of nerve cells in substantia nigra in normal subjects and in patients with parkinson's disease estimated with an unbiased stereological method. J. Neurol. Neurosurg. Psychiatry **54**(1) (1991) 30–33.
71. Albert, R., Jeong, H., Barabási, A.L.: Error and attack tolerance of complex networks. Nature **406** (2000) 378–382.
72. Diestel, R.: Graph Theory. Springer, New York (1997).
73. Watts, D.J., Strogatz, S.H.: Collective dynamics of 'small-world' networks. Nature **393** (1998) 440–442.
74. Girvan, M., Newman, M.E.J.: Community structure in social and biological networks. Proc. Natl. Acad. Sci. **99**(12) (2002) 7821–7826.
75. Holme, P., Kim, B.J., Yoon, C.N., Han, S.K.: Attack vulnerability of complex networks. Phys. Rev. E **65** (2002) 056109.
76. Granovetter, M.S.: The strength of weak ties. Am. J. Sociol. **78**(6) (1973) 1360–1380.
77. Stephan, K.E., Hilgetag, C.C., Burns, G.A.P.C., O'Neill, M.A., Young, M.P., Kötter, R.: Computational analysis of functional connectivity between areas of primate cerebral cortex. Phil. Trans. R. Soc. **355** (2000) 111–126.
78. Kötter, R., Sommer, F.T.: Global relationship between anatomical connectivity and activity propagation in the cerebral cortex. Phil. Trans. Roy. Soc. Lond. Ser. B **355** (2000) 127–134.
79. Achard, S., Salvador, R., Whitcher, B., Suckling, J., Bullmore, E.: A resilient, low-frequency, small-world human brain functional network with highly connected association cortical hubs. J. Neurosci. **26** (2006) 63–72.
80. Salvador, R., Suckling, J., Coleman, M.R., Pickard, J.D., Menon, D., Bullmore, E.: Neurophysiological architecture of functional magnetic resonance images of human brain. Cereb. Cortex **15**(9) (2005) 1332–1342.
81. Milo, R., Shen-Orr, S., Itzkovitz, S., Kashtan, N., Chklovskii, D., Alon, U.: Network motifs: simple building blocks of complex networks. Science **298** (2002) 824–827.
82. Sporns, O., Kötter, R.: Motifs in brain networks. PLoS Biol. **2** (2004) 1910–1918.
83. Buzsaki, G., Geisler, C., Henze, D.A., Wang, X.J.: Interneuron diversity series: circuit complexity and axon wiring economy of cortical interneurons. Trends Neurosci. **27**(4) (2004) 186–193.

5

Synchronization Dynamics in Complex Networks

Changsong Zhou, Lucia Zemanová and Jürgen Kurths

Institute of Physics, University of Potsdam PF 601553, 14415 Potsdam, Germany,
cszhou@agnld.uni-potsdam.de

Summary. Previous chapters have discussed tools from graph theory and their contribution to our understanding of the structural organization of mammalian brains and its functional implications. The brain functions are mediated by complicated dynamical processes which arise from the underlying complex neural networks, and synchronization has been proposed as an important mechanism for neural information processing. In this chapter, we discuss synchronization dynamics on complex networks. We first present a general theory and tools to characterize the relationship of some structural measures of networks to their synchronizability (the ability of the networks to achieve complete synchronization) and to the organization of effective synchronization patterns on the networks. Then, we study synchronization in a realistic network of cat cortical connectivity by modeling the nodes (which are cortical areas composed of large ensembles of neurons) by a neural mass model or a subnetwork of interacting neurons. We show that if the dynamics is characterized by well-defined oscillations (neural mass model and subnetworks with strong couplings), the synchronization patterns can be understood by the general principles discussed in the first part of the chapter. With weak couplings, the model with subnetworks displays biologically plausible dynamics and the synchronization pattern reveals a hierarchically clustered organization in the network structure. Thus, the study of synchronization of complex networks can provide insights into the relationship between network topology and functional organization of complex brain networks.

5.1 Introduction

Real-world complex networks are interacting dynamical entities with an interplay between dynamical states and interaction patterns, such as the neural networks in the brain. Recently, the complex network approach has been playing an increasing role in the study of complex systems [1]. The main research focus has been on the topological structures of complex systems based on simplified graphs, paying special attention to the global properties of complex networks, such as the small-world (small-world networks (SWNs) [2]) scale-free (scale-free networks (SFNs) [3]) features, or to the presence or absence of some very small subgraphs, such as network motifs [4]. Such topological studies have revealed important organizational principles in the structures of many realistic network systems [1]. The structural characterization of networks has been discussed in Chap. 3.

However, a more complete understanding of many realistic systems would require characterizations beyond the interaction topology. A problem of fundamental importance is the impact of network structures on the dynamics of the networks. The elements of many complex systems display oscillatory dynamics. Therefore, synchronization of oscillators is one of the widely studied dynamical behavior on complex networks [5]. It is important to emphasize that synchronization is especially relevant in brain dynamics [6]. Synchronization of neuronal dynamics on networks with complex topology thus has received significant recent attention [7–10].

Most previous studies have focused on the influence of complex network topology on the ability of the network to achieve synchronization. It has been shown that SWNs provide a better synchronization of coupled excitable neurons in the presence of external stimuli [7]. In pulse-coupled oscillators, synchronization becomes optimal in a small-world regime [8], and it is degraded when the degree becomes more heterogeneous with increased randomness [9]. Investigation of phase oscillators [11] or circle maps [12] on SWNs has shown that when more and more shortcuts are created at larger rewiring probability p, the transition to the synchronization regime becomes easier [11]. These observations have shown that the ability of a network to synchronize (synchronizability) is generally enhanced in SWNs as compared to regular chains. Physically, this enhanced synchronizability was attributed to the decreasing of the average network distance due to the shortcuts. On the other hand, it has been shown that the synchronizability also depends critically on the heterogeneity of the degree distribution [13]. In particular, random networks with strong heterogeneity in the degree distribution, such as SFNs, are more difficult to synchronize than random homogeneous networks [13], despite the fact that heterogeneity reduces the average distance between nodes [14]. The synchronizability in most previous studies is based on the linear stability of the complete synchronization state using spectral analysis of the network coupling matrix [5].

These studies focusing on the impacts of network topology assumed that the coupling strength is uniform. However, most complex networks in nature where synchronization is relevant are actually weighted, e.g. neural networks [15], networks of cities in the synchronization of epidemic outbreaks [16], and communication and other technological networks whose functioning relies on the synchronization of interacting units [17]. The connection weights of many real networks are often highly heterogeneous [18]. It has been shown that weighted coupling has significant effects on the synchronization of complex networks [19–22].

In this chapter, we discuss the synchronization of nonlinear oscillators coupled in complex networks. Our emphasis is to demonstrate how the network topology and the connection weights influence the synchronization behavior of the oscillators. The theory and method are mainly based on general models of complex networks, and we also study synchronization and the relationship between dynamical clusters and anatomical communities in a realistic complex network of brain cortex.

The chapter is organized as follows. In Sect. 5.2, we present the general dynamical equations and the linear stability analysis for the complete synchronization state when the oscillators are identical. Then, we demonstrate the leading parameters that universally control the synchronizability of a general class of random weighted networks in Sect. 5.3. In Sect. 5.4, we carry out simulations of hierarchical synchronization in SFNs outside the complete synchronization regimes. We demonstrate the influence of small-world connections in Sect. 5.5. Section 5.6 is devoted to synchronization analysis in cat cortical networks. We discuss the possible relevance of the analysis of dynamical complex neural networks and meaningful extensions in Sect. 5.7.

5.2 Dynamical Equations and Stability Analysis

The dynamics of a general network of N coupled oscillators is described by:

$$\dot{\boldsymbol{x}}_j = \tau_j \boldsymbol{F}(\boldsymbol{x}_j) + \sigma \sum_{i=1}^{N} A_{ji} W_{ji} [\boldsymbol{H}(\boldsymbol{x}_i) - \boldsymbol{H}(\boldsymbol{x}_j)] \qquad (5.1)$$

$$= \tau_j \boldsymbol{F}(\boldsymbol{x}_j) - \sigma \sum_{i=1}^{N} G_{ji} \boldsymbol{H}(\boldsymbol{x}_i), \quad j = 1, \ldots, N , \qquad (5.2)$$

where $\boldsymbol{x}_j$ is the state of oscillator j and $\boldsymbol{F} = \boldsymbol{F}(\boldsymbol{x})$ governs the dynamics of each individual oscillator. The parameter τ_j controls the time scales of the oscillators, which are not identical in general. $\boldsymbol{H} = \boldsymbol{H}(\boldsymbol{x})$ is the output function, and σ is the overall coupling strength. $A = (A_{ji})$ is the adjacency matrix of the underlying network of couplings, where $A_{ji} = 1$ if there is a link from node i to node j, and $A_{ji} = 0$ otherwise. Here, we assume that the coupling is bidirectional so that $A_{ij} = A_{ji}$, i.e., A is symmetric. The number of connections of a node, the degree k_j, is just the row sum of the adjacency matrix A, i.e., $k_j = \sum_i A_{ji}$. More details about network characterization are found in Chap. 3. W_{ji} is the weight of the incoming strength for the link from node i to node j. Note that the incoming and output weights can be in general asymmetric, $W_{ji} \neq W_{ij}$. Here $G = (G_{ji})$ is the coupling matrix combining both topology [adjacency matrix $A = (A_{ji})$] and weights [weight matrix $W =$

(W_{ji})]: $G_{ji} = -W_{ji}$ for $i \neq j$ and $G_{jj} = \sum_i W_{ji} A_{ji}$. By definition, the rows of matrix G have zero sum, $\sum_{i=1}^{N} G_{ji} = 0$.

As we mentioned in the introduction, much previous work characterizes the synchronizability of networks using graph spectral analysis. The framework of this analysis is based on the *master stability function*. The readers are referred to the references [5,23] for the details. Here, we outline the main idea, which is to consider the ideal case of identical oscillators, i.e., $\tau_1 = \tau_2 = \cdots = \tau_N = 1$. In this case, it is easy to see that the completely synchronized state, $x_1(t) = x_2(t) = \cdots = x_N(t) = s(t)$, is a solution of (5.2), i.e., all the oscillators follow the same trajectory in the phase space, and the trajectory belongs to the attractor of the isolated oscillator. So this solution is also called the invariant *synchronization manifold*. However, synchronization in the network can only be observed when the synchronization state is robust against desynchronizing perturbations. Now the crucial question is: Is this solution stable? And under what conditions is it stable?

To study the stability of the synchronization state, we consider small perturbations of the synchronization state s, $\delta \dot{x}_j = x_j - s$, which are governed by the linear variational equations

$$\delta \dot{x}_j = \mathbf{D}F(s)\delta x_j - \sigma \mathbf{D}H(s) \sum_{i=1}^{N} G_{ji}\delta x_i, \quad j = 1, \cdots, N, \tag{5.3}$$

where $\mathbf{D}F(s)$ and $\mathbf{D}H(s)$ are the Jacobians on s.

The main idea of the master stability function is to project δx into the eigenspace spanned by the eigenvectors v of the coupling matrix G. By doing so, (5.3) can be diagonalized into N decoupled blocks of the form

$$\dot{\xi}_l = [\mathbf{D}F(s) - \sigma \lambda_l \mathbf{D}H(s)]\xi_l, \quad l = 1, \cdots, N, \tag{5.4}$$

where ξ_l is the eigenmode associated to the eigenvalue λ_l of the coupling matrix G. Here, $\lambda_1 = 0$ corresponds to the eigenmode parallel to the synchronization manifold, and the other $N-1$ eigenvalues λ_l represent the eigenmodes transverse to the synchronization manifold.

Note that all the variational equations in (5.4) have the same form:

$$\dot{\xi} = [\mathbf{D}F(s) - \epsilon \mathbf{D}H(s)]\xi. \tag{5.5}$$

They differ only by the parameter $\epsilon = \sigma \lambda_l$. From this, we understand that the stability of each mode is determined by the property of the master stability of the normal form in (5.5) and the eigenvalue λ_l. In this chapter, we focus on the cases where G has real eigenvalues, ordered as $0 = \lambda_1 \leq \lambda_2 \cdots \leq \lambda_N$. The largest Lyapunov exponent $\Lambda(\epsilon)$ of (5.5) as a function of the parameter ϵ is called the *master stability function*. If $\Lambda(\epsilon) < 0$, it follows that $\xi(t) \sim \exp(\Lambda t) \to 0$ when $t \to \infty$ and the mode is stable, otherwise, small perturbations will grow with time t and the mode is unstable. For many oscillatory dynamical systems [23], (5.5) is stable (e.g., $\Lambda(\epsilon) < 0$) in a single, finite interval $\epsilon_1 < \epsilon < \epsilon_2$, where the thresholds ϵ_1 and ϵ_2 are determined only by $\boldsymbol{F}$, $\boldsymbol{H}$, and $\boldsymbol{s}$. A transverse mode is damped if the corresponding eigenvalue satisfies $\epsilon_1 < \sigma\lambda_i < \epsilon_2$, and the complete synchronization state is stable when all the transverse modes are damped, namely,

$$\epsilon_1 < \sigma\lambda_2 \le \sigma\lambda_3 \le \cdots \le \sigma\lambda_N < \epsilon_2 . \tag{5.6}$$

This condition can only be fulfilled for some values of σ when the eigenratio R meets

$$R \equiv \lambda_N/\lambda_2 < \epsilon_2/\epsilon_1 . \tag{5.7}$$

It is impossible to synchronize the network completely if $R > \epsilon_2/\epsilon_1$, since there is no σ value for whom the solution is linearly stable. The eigenratio R depends only on the network structure, as defined by the coupling matrix G. If R is small, in general the condition in (5.7) will be more easily satisfied. It follows that the smaller the eigenratio R the more synchronizable the network and vice versa [5], and we can characterize the synchronizability of the networks with R, without referring to specific oscillators. In some special cases, $\epsilon_2 = \infty$, and the synchronization state is stable when the overall coupling strength σ is larger than a threshold $\sigma_c = \epsilon_1/\lambda_2$. More detailed characterization of the synchronizability in this case can be found in [21].

In the following section, we characterize the synchronizability of weighted networks using only the eigenratio R.

5.3 Universality of Synchronizability

5.3.1 Universal Formula

How the synchronizability, the eigenratio R, depends on the structure of networks is one of the major questions in previous studies. Based on spectral graph theory, previous work has obtained bounds for the eigenvalues of *unweighted* networks ($W_{ji} = 1$) [13, 24]. For arbitrary networks [24], the eigenvalues are bounded as $\frac{4}{ND_{\max}} \le \lambda_2 \le \frac{N}{N-1}k_{\min}$ and $\frac{N}{N-1}k_{\max} \le \lambda_N \le \max(k_i + k_j) \le 2k_{\max}$, where $k_{\min}$ and $k_{\max}$ are the minimum and maximum degrees, respectively, k_i and k_j are the degrees of two connected nodes, and $D_{\max}$, the diameter of the graph, is the maximum of the distances between nodes. From this, one gets

$$k_{\max}/k_{\min} \le R \le ND_{\max}\max(k_i + k_j)/4 . \tag{5.8}$$

The reader can find more details concerning how these bounds are obtained in [24]. Such bounds can provide some insights into the synchronizability of the networks. For example, networks with heterogeneous degrees have low synchronizability, since R is bounded away from 1 by $k_{\max}/k_{\min}$. However, such bounds are not tight. The upper bound, as a function of the network size N, can be orders of magnitude larger than the lower one even for small networks (especially for random networks), thus providing limited information about the actual synchronizability of the networks.

In the following, we present tighter bounds for more general weighted networks, including unweighted networks as special cases. We restrict ourselves to sufficiently random networks. Our analysis is based on the combination of a mean field approximation and new graph spectral results [21].

First, in random networks with a large enough minimal degree $k_{\min} \gg 1$, (5.1) $(\tau_j = 1, \forall j)$ can be approximated as

$$\dot{x}_j = F(x_i) + \sigma \frac{S_j}{k_j} \sum_{i=1}^{N} A_{ji}[H(x_i) - H(x_j)] . \tag{5.9}$$

The reason is that each oscillator i receives signals from a large and random sample of other oscillators in the network and x_i is not affected directly by the individual output weights W_{ji}. Thus, we may assume that W_{ji} and $H(x_i)$ are statistically uncorrelated and the following approximation holds

$$\sum_{i=1}^{N} W_{ji} A_{ji} H(x_i) \approx \frac{1}{k_j} \sum_{i=1}^{N} A_{ji} H(x_i) \sum_{i=1}^{N} W_{ji} A_{ji} = \bar{H}_j S_j \tag{5.10}$$

if $k_i \gg 1$. Here, $\bar{H}_j = (1/k_j) \sum_{i=1}^{N} A_{ji} H(x_i)$ is the local mean field, and S_j is the intensity of node j. Defined as the total input weight of node,

$$S_j = \sum_{i=1}^{N} A_{ji} W_{ji} , \tag{5.11}$$

the intensity S_j is a significant measure integrating the information of connectivity and weights [18].

Now, if the network is sufficiently random, the local mean field $\bar{H}_j$ can be approximated by the global mean field of the network, $\bar{H}_j \approx \bar{H} = (1/N) \sum_{i=1}^{N} H(x_i)$, since the information that a node obtains from its $k_j \gg 1$ connected neighbors is well distributed in the network and the averaged signal is close enough to the average behavior of the whole network. Moreover, close to the synchronized state s, we may assume $\bar{H}_j \approx H(s)$, and the system is approximated as

$$\dot{x}_j = F(x_j) + \sigma S_j[H(s) - H(x_j)], \quad j = 1, \ldots, N . \tag{5.12}$$

We call this the *mean field approximation*, which indicates that the oscillators are decoupled and forced by a common oscillator s with a forcing strength proportional to the intensity S_j (cf. Chap. 1). The variational equations (5.12) have the same form as (5.4), except that λ_l is replaced by S_l. If there is some σ satisfying

$$\epsilon_1 < \sigma S_{\min} \leq \cdots \leq \sigma S_l \leq \cdots \leq \sigma S_{\max} < \epsilon_2 , \tag{5.13}$$

then all the oscillators are synchronizable by the common driving $\boldsymbol{H}(\boldsymbol{s})$, corresponding to a complete synchronization of the whole network. These observations suggest that the eigenratio R can be approximated as

$$R \approx S_{\max}/S_{\min} \, , \tag{5.14}$$

where $S_{\min}$, $S_{\max}$ are the minimum and maximum intensities S_j in (5.11), respectively.

Next, we present tight bounds for the above approximation. Equation (5.9) means that the coupling matrix G is replaced by the new matrix $G^a = (G_{ji}^a)$, with $G_{ji}^a = \frac{S_j}{k_j}(\delta_{ji}k_j - A_{ji})$. G^a can be written as $G^a = S\hat{G} = SD^{-1}(D - A)$, where $S = (\delta_{ji}S_j)$ and $D = (\delta_{ji}k_j)$ are the diagonal matrices of intensities and degrees, respectively, and $\hat{G}$ is the normalized Laplacian matrix [25]. Importantly, now the contributions from the topology and weight structure are separated and accounted for by $\hat{G}$ and S, respectively. We can show that in large enough complex networks, such as SFNs in many realistic complex systems, the largest and smallest nonzero eigenvalues of the matrix G^a are bounded by the eigenvalues μ_l of $\hat{G}$ as

$$S_{\min}\mu_2 c \leq \lambda_2 \leq S_{\min}c', \quad S_{\max} \leq \lambda_N \leq S_{\max}\mu_N \, , \tag{5.15}$$

where c and c' can be approximated by 1 for most large complex networks of interest, such as realistic SFNs. The upper bound of λ_N in the inequality (5.15) follows from

$$\lambda_N = \max_{||v||=1} ||S\hat{G}v|| \leq \max_{||v||=1} ||Sv|| \max_{||v||=1} ||\hat{G}v|| = S_{\max}\mu_N \tag{5.16}$$

where $|| \cdot ||$ is the Euclidean norm and v denotes the normalized eigenvector of G^a. The other bounds in (5.15) are obtained in a similar spirit. If the network is sufficiently random, the spectrum of $\hat{G}$ tends to the *semicircle law* for large networks with arbitrary expected degrees [25]. The semicircle law says that if $k_{\min} \gg \sqrt{K}$, where K is the mean degree of the network, the distribution function of the eigenvalues of $\hat{G}$ follows a semicircle with the center at 1.0 and the radius $r = 2/\sqrt{K}$ when $N \to \infty$. In particular, we have $\max\{1 - \mu_2, \mu_N - 1\} = [1 + o(1)]\frac{2}{\sqrt{K}}$ for $k_{\min} \gg \sqrt{K}\ln^3 N$. From these, it follows that

$$\mu_2 \approx 1 - 2/\sqrt{K}, \quad \mu_N \approx 1 + 2/\sqrt{K} \, , \tag{5.17}$$

which we find to provide also a good approximation under the weaker condition $k_{\min} \gg 1$, regardless of the degree distribution of the random network. From (5.15) and (5.17), we have the following approximation for the bounds of R:

$$\frac{S_{\max}}{S_{\min}} \leq R \leq \frac{S_{\max}}{S_{\min}} \frac{1 + 2/\sqrt{K}}{1 - 2/\sqrt{K}} \, . \tag{5.18}$$

We stress that this result applies to unweighted networks ($S_i = k_i$) and the upper bound in the inequality (5.18) is much tighter than that in (5.8).

The bounds in (5.18) show that the contribution of the network topology is mainly captured by the mean degree K. Therefore, for a given K, the synchronizability of random networks with a large $k_{\min}$ is expected to be well approximated by the following universal formula:

$$R = A_R \frac{S_{\max}}{S_{\min}} , \qquad (5.19)$$

where the pre-factors A_R are expected to be close to 1. In the case of uniform intensity ($S_i = 1 \; \forall i$), they are given by the upper bounds, $A_R = \frac{1+2/\sqrt{K}}{1-2/\sqrt{K}}$, and $A_R \to 1$ in the limit $K \to \infty$. Equation (5.19) is consistent with the approximation in (5.14) and indicates that the synchronizability of these networks is primarily determined by the heterogeneity of the intensities, regardless of the degree distribution.

In simpler words, the universal formula says that if you give me a random enough but arbitrary network, we then can measure the intensities and tell you whether or not the network is synchronizable for a given oscillator (which specifies ϵ_1 and ϵ_2). If $S_{\max}/S_{\min}$ is clearly smaller than ϵ_2/ϵ_1, then we can predict that the suitable coupling strength for complete synchronization is $\epsilon_1/S_{\min} < \sigma < \epsilon_2/S_{\max}$.

5.3.2 Numerical Confirmation of the Universality

In the following, we present some results of numerical simulations of the synchronizability of various weighted and unweighted networks. These results confirm the above universal formula obtained based on physical arguments and graph spectral analysis. For example, let us consider the following weighted coupling scheme:

$$W_{ji} = S_j/k_j , \qquad (5.20)$$

in which the intensities S_j follow an arbitrary distribution not necessarily correlated with the degrees k_j. This means that the intensities S_j of a node j are equally distributed into the k_j input connections of this node, which serves as an approximation of realistic networks. However, nonuniform weights of the input links does not change our conclusion according to the approximation in (5.9). In [21], we presented results for more realistic weighted networks and showed that the universal formula also applies.

With the weighted coupling in (5.20), (5.9) and (5.1) are identical and $G^a = G$. This weighted coupling scheme includes many previously studied systems as special cases. If $S_j = k_j \; \forall j$, it corresponds to the widely studied case of unweighed networks [5, 13, 26]. In the case of fully uniform intensity

$(S_j = 1 \; \forall j)$, it accommodates a number of previous studies about synchronization of coupled maps [27,28]. The weighted scheme studied in [19], $W_{ji} = k_j^\theta$, is another special case of (5.20) where $S_j = k_j^{1+\theta}$. We have applied the weighted scheme to various network models:

(i) *Growing SFNs with aging* [29]. This model of complex networks extends the Barabási-Albert (BA) model [3]. Starting with $2m + 1$ fully connected nodes, at each time step we connect a new node to m existing nodes according to the probability $\Pi_i \sim k_i \tau_i^{-\alpha}$, where τ_i is the age of the node. The minimum degree is then $k_{\min} = m$ and the mean degree is $K = 2m$. For the aging exponent $-\infty < \alpha \leq 0$, this growing rule generates SFNs with a power-law tail $P(k) \sim k^{-\gamma}$ and the scaling exponent in the interval $2 < \gamma \leq 3$ [29], as in most real SFNs. For $\alpha = 0$, we recover the usual Barabási-Albert (BA) model [3], which has $\gamma = 3$.

(ii) *Random SFNs* [30]. Each node is assigned to have a number $k_i \geq k_{\min}$ of "half-links" according to the distribution $P(k) \sim k^{-\gamma}$. The network is generated by randomly connecting these half-links to form links, prohibiting self- and repeated links.

(iii) *K-regular random networks.* Each node is randomly connected to K other nodes.

We now present results for two different distributions of intensity S_i that are uncorrelated with the distribution of the degree k_i: (1) a uniform

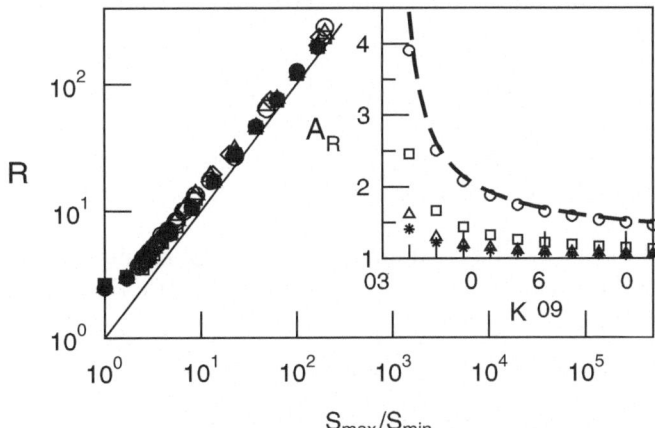

Fig. 5.1. R as a function of $S_{\max}/S_{\min}$. Filled symbols: uniform distribution of $S_i \in [S_{\min}, S_{\max}]$. Open symbols: power-law distribution of S_i, $P(S) \sim S^{-\Gamma}$ for $2.5 \leq \Gamma \leq 10$. Different symbols are for networks with different topologies: BA growing SFNs (*circles*), growing SFNs with aging exponent $\alpha = -3$ (*squares*), random SFNs with $\gamma = 3$ (*diamonds*), and K-regular random networks (*triangles*). The number of nodes is $N = 2^{10}$ and the mean degree is $K = 20$. Inset: A_R as a function of K for $S_{\max}/S_{\min} = 1$ (*circles*), 2 (*squares*), 10 (*triangles*), and 100 (*stars*), obtained with a uniform distribution of S_i in K-regular networks. The dashed lines are the bounds. Solid line: (5.19) with $A_R = 1$.

distribution in $[S_{min}, S_{max}]$; and (2) a power-law distribution, $P(S) \sim S^{-\Gamma}$, $S \geq S_{min}$, where S_{min} is a positive number. Consistently with the prediction of the universal formula, if $k_{min} \gg 1$, the eigenratio R collapses into a single curve for a given K when plotted as a function of S_{max}/S_{min}, irrespective of the distributions of k_j and S_j, as shown in Fig. 5.1. The behavior of the fitting parameter A_R is shown in the inset of Fig. 5.1. For a uniform intensity, it is very close to the upper bounds. It approaches 1 very quickly when the intensities become more heterogeneous ($S_{max}/S_{min} > 3$). Therefore, (5.19) with $A_R = 1$ (Fig. 5.1, solid line) provides a good approximation of the synchronizability for any large K if the intensities are not very homogeneous.

5.4 Effective Synchronization in Scale-Free Networks

So far, we have presented an analysis of the stability of the complete synchronization state and the synchronizability of the networks based on the spectrum of the weighted graphs. The main conclusion is that, for random networks, the ability of the network to achieve complete synchronization is determined by the maximal and minimal values of the intensities S_j. The intensity of a node in (5.11), defined as the sum of the strengths of all input connections of that node, incorporates both topological and weighted properties of the network.

Now we carry out simulations on concrete dynamical systems. We would like to demonstrate that the intensities S_j are still the important parameter for the organization of *effective synchronization* on the network outside the complete synchronization regime, e.g., when the network is perturbed by noise, or when the oscillators are non-identical, which are typical cases in more realistic systems.

In the following we give a summary of the main results (for more details, see [31]). For this purpose, we analyze the paradigmatic Rössler chaotic oscillator $\boldsymbol{x} = (x, y, z)$:

$$\dot{x} = -0.97x - z, \qquad (5.21)$$

$$\dot{y} = 0.97x + 0.15y, \qquad (5.22)$$

$$\dot{z} = x(z - 8.5) + 0.4. \qquad (5.23)$$

The oscillations are chaotic, and its time-dependent phase can be defined as $\phi = \arctan(y/x)$ [32, 33], as illustrated in Fig. 5.2.

Without loss of generality, we consider SFNs generated with the BA model [3] and use the coupling scheme in (5.20). For simplicity, we also sorted the label of the nodes according to the degrees, $k_1 \geq k_2 \geq \cdots \geq k_N$. We compare unweighted networks (UN) ($W_{ij} = 1$) by taking $S_j = k_j$ and weighted networks (WN) ($W_{ji} = 1/k_j$) by taking uniform intensities $S_j = 1$ for all the

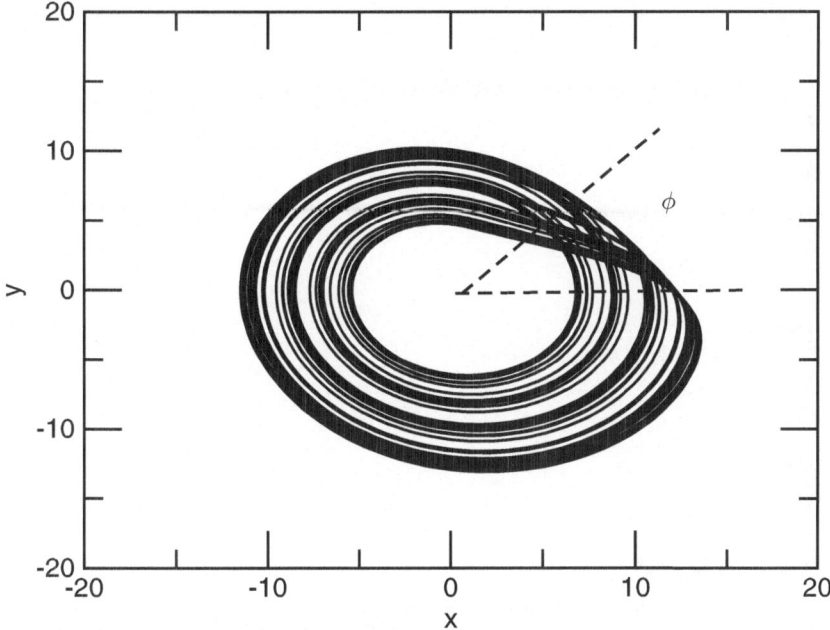

Fig. 5.2. Chaotic attractor of the Rössler oscillator. ϕ indicates the phase of the oscillation

nodes. Here, we use the output function $\boldsymbol{H}(\boldsymbol{x}) = \boldsymbol{x}$ in our simulations. In this case, $\epsilon_1 = \Lambda_F$, the largest Lyapunov exponent of the isolated Rössler chaotic oscillator, and $\epsilon_2 = \infty$.

Numerical simulations confirm that synchronization is achieved when $\sigma > \sigma_c = \Lambda_F/\lambda_2$. The transition to synchronization is shown in Fig. 5.3 as a function of the normalized coupling strength $g = \sigma\langle S\rangle$, where $\langle S\rangle$ is the average intensity of the networks. Here, we have plotted the average synchronization error $E = (1/N)\sum_{j=1}^{N}\Delta X_j$, where $\Delta X_j = \langle|x_j - X|\rangle_t$ is the time-averaged distance between the oscillator x_j and the mean activity of the whole network, $X = (1/N)\sum_{j=1}^{N}x_j$. When complete synchronization is achieved at $g > g_c$, one has $E = 0$ after a sufficiently long transient. Additionally, we show the oscillation amplitude A_X of the mean field X, calculated as the standard deviation of X over time. As expected, the WN with uniform intensity achieves complete synchronization at a critical coupling strength g_c smaller than that of the UN having the same mean degree K. It is important to emphasize that the network already maintains collective oscillations when the coupling strength g is much smaller than the threshold value g_c for complete synchronization. This is manifested by an amplitude A_X of the collective oscillations, which has the same level as that of the completely synchronized state at $g > g_c$.

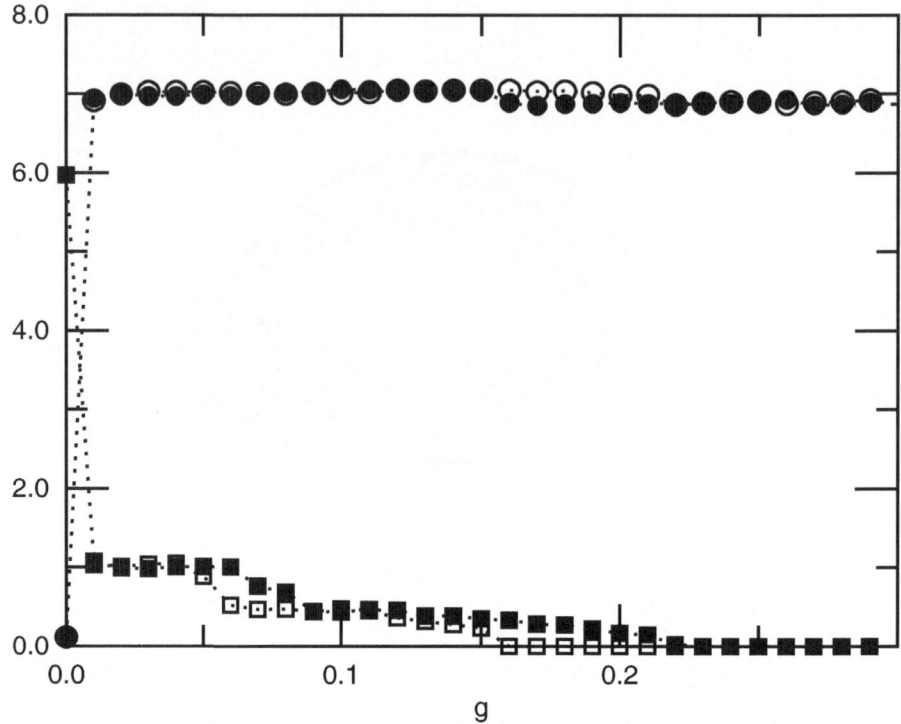

Fig. 5.3. Transition to synchronization in the UN and WN, indicated by the synchronization error E (*squares*) and the amplitude A_X of the mean field X (*circles*). The filled symbols are for the UN and the open symbols for the WN. In both networks, $N = 1000$ and $K = 10$ ($k_{\min} = 5$)

5.4.1 Hierarchical Synchronization

Now we look into the different behavior of the two networks outside the complete synchronization regime, when the coupling is too weak ($g < g_c$) or when the synchronization state at $g > g_c$ is perturbed by noise. Noise is simulated by adding independent Gaussian random perturbations $D\eta_j(t)$ with a standard deviation D to the variables of the oscillators, i.e., $\langle \eta_j(t)\eta_i(t - \tau) \rangle = \delta_{ji}\delta(\tau)$. We examine the synchronization difference ΔX_j of an individual oscillator with respect to the collective oscillations X of the whole network. The typical results are as shown in Figs. 5.4(a) and (b) for weak coupling and noise perturbation, respectively.

It is seen that ΔX_j is almost the same for the nodes in the WN (only slightly smaller for nodes with larger degrees), since in this case, the intensity is fully uniform ($S_j = 1$) and independent of the degrees. In sharp contrast, the synchronization difference is strongly heterogeneous in the UN and is negatively correlated with the intensity ($S_j = k_j$) of the nodes. To get a clear

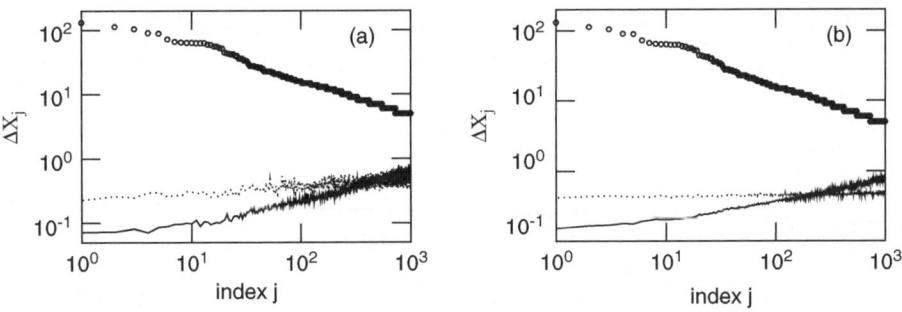

Fig. 5.4. Synchronization difference ΔX_j of the oscillators with respect to the global mean field X in the UN (*solid line*) and WN (*dotted line*). The symbol (○) denotes the degree k_j of the nodes. Note the log-log scales of the plots: **(a)** The coupling strength is weak ($g = 0.1$); **(b)** The synchronized state ($g = 0.3$) is perturbed by noise ($D = 0.5$)

dependence of ΔX on the degree k, we calculate the average value $\Delta X(k)$ among all nodes with degree k, i.e.,

$$\Delta X(k) = \frac{1}{N_k} \sum_{k_j = k} \Delta X_j \,, \tag{5.24}$$

where N_k is the number of nodes with degree $k_j = k$ in the SFN network. Now a pronounced dependence can be observed for the UN, as shown in Fig. 5.5. For both weak couplings and noise perturbations, the dependence is characterized by a power-law scaling

$$\Delta X(k) \sim k^{-\alpha} \,, \tag{5.25}$$

with the exponent $\alpha \approx 1$. These results demonstrate that in the UN, where the intensities ($S_j = k_j$) are heterogeneous due to the power-law distribution of the degrees, a small portion of nodes with large intensities synchronize more

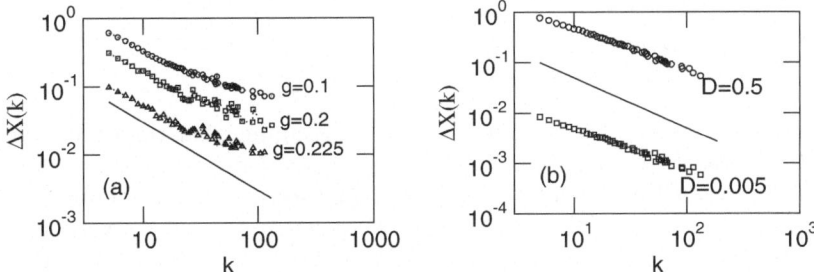

Fig. 5.5. The average values $\Delta X(k)$ as a function of k at various coupling strength g in the UN: **(a)** The coupling strength is weak; **(b)** The synchronized state ($g = 0.3$) is perturbed by noise. The solid lines with slope -1 are plotted for reference

closely to the mean field X, while most of the nodes with small intensities are still rather independent of X outside the synchronization regime. The effective synchronization patterns of the networks are controlled by the distribution of the intensities S_j, while complete synchronization is determined by the maximal and minimal values.

5.4.2 Effective Synchronization Clusters

We have shown that the synchronization behavior of the individual oscillators in the UN is highly nonuniform; in particular, the nodes with large degrees, i.e., the hubs, are close to the mean field. As a result, the synchronization difference between them should also be relatively small. We define an effective synchronization cluster for those oscillators that synchronize to each other within some threshold. For this purpose, we have calculated the pairwise synchronization difference $\Delta X_{ij} = \langle |x_i - x_j| \rangle_t$. A pair of oscillators ($i \neq j$) is considered to be synchronized effectively when their synchronization difference is smaller than a threshold: $\Delta X_{ij} \leq \Delta_{th}$. Since the synchronization difference is heterogeneous, there is no unique choice of the threshold value Δ_{th}. What we can expect is that with smaller values of Δ_{th}, the size of the effective cluster is smaller. The effective synchronization clusters for different values of the threshold Δ_{th} are shown in Figs. 5.6(a) and (b). The same clusters are also represented in the space of degrees (k_i, k_j) in Figs. 5.6(c) and (d), respectively. Note that almost all the oscillators forming the clusters have a degree $k_j > k_{th}$, where k_{th} is the threshold degree satisfying $\Delta X(k_{th}) = \Delta_{th}$; or correspondingly, the effective cluster is formed by nodes with $j < J_{th}$, where J_{th} is the mean index of nodes with degree $k_j = k_{th}$. The triangular shape of the effective clusters in Fig. 5.6 is well described by the relation $i + j \leq J_{th}$. Above the solid line $i + j = J_{th}$, those oscillators ($i \leq J_{th}$ and $j \leq J_{th}$) having large enough degrees, i.e., $k_i \geq k_{th}$ and $k_j \geq k_{th}$, are close to the mean field with $\Delta X_i \leq \Delta_{th}$ and $\Delta X_j \leq \Delta_{th}$, but the pairwise distance is large, $\Delta X_{ij} > \Delta_{th}$. These results demonstrate clearly that the nodes with the largest intensities are the dynamical core of the networks.

5.4.3 Non-identical Oscillators

Now we consider *non-identical* oscillators by assuming that the time scale parameters τ_j are heterogeneous in (5.1), so that the oscillators have different mean oscillation frequencies Ω_j. In our simulations, we use a uniform distribution of τ_j in an interval $[1 - \Delta\tau, 1 + \Delta\tau]$, with $\Delta\tau = 0.1$. We define the phases of the oscillations as indicated in Fig. 5.2. The average frequency can be computed as $\Omega_j = \langle \dot{\phi}_j \rangle$.

Let us first examine the collective oscillations in the network. Fig. 5.7 shows the amplitude A_X of the mean field X as a function of the coupling

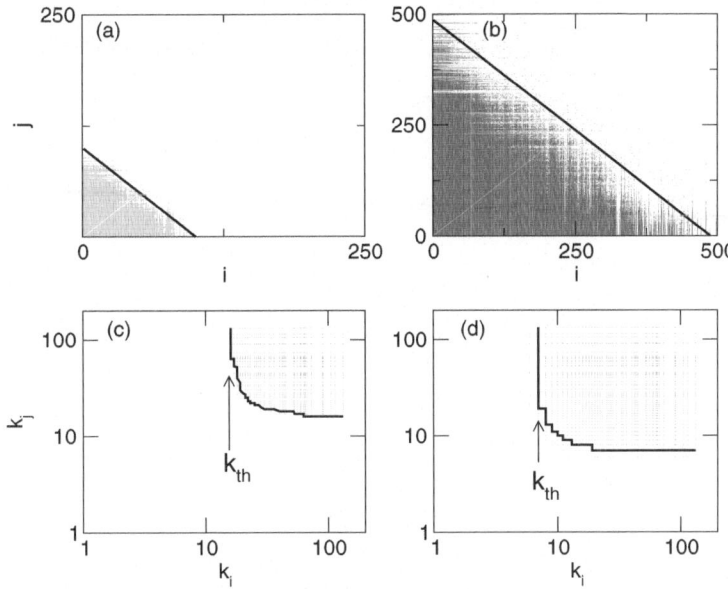

Fig. 5.6. The effective synchronization clusters in the synchronized UN in the presence of noise ($g = 0.3$, and $D = 0.5$), represented simultaneously in the index space (i, j) (**a**, **b**) and in the degree space (k_i, k_j) (**c**, **d**). A dot is plotted when $\Delta X_{ij} \leq \Delta_{th}$. (**a**) and (**c**) for the threshold value $\Delta_{th} = 0.25$, and (**b**) and (**d**) for $\Delta_{th} = 0.50$. The solid lines in (**a**) and (**b**) denote $i + j = J_{th}$ and are also plotted in (**c**) and (**d**) correspondingly. Note the different scales in (**a**) and (**b**) and the log-log scales in (**c**) and (**d**)

strength g for the UN and WN. It is seen that both networks generate a coherent collective oscillation when the coupling strength is larger than a critical value $g_{cr} \approx 0.08$. However, the UN generates a weaker degree of collective synchronization as indicated by a smaller amplitude A_X of the mean field.

Now we study in more detail synchronization behavior in the weak, intermediate and strong coupling regimes, indicated by the three vertical dashed lines in Fig. 5.7.

I. Weak Coupling: Non-synchronization Regime

We start with the weak coupling regime with $g = 0.05$. Here, neither the UN nor the WN display significant collective oscillations. The frequencies of the oscillators are still distributed and the phases of the oscillators are not locked. However, interesting dynamical changes can be already expected in the UN. Based on the mean field approximation described in Sect. 3.1, we have

$$\dot{x}_j = \tau_j F(x_j) + g \frac{S_j}{\langle S \rangle}(X - x_j), \quad k_j \gg 1 . \tag{5.26}$$

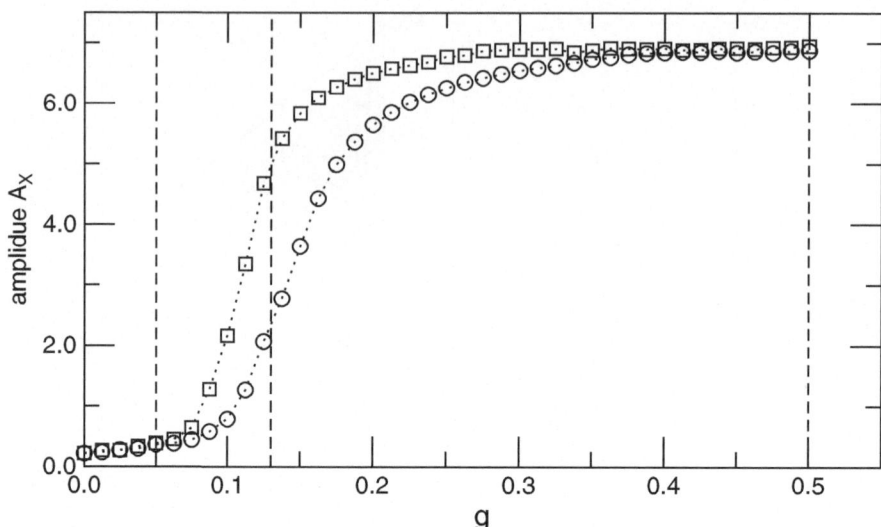

Fig. 5.7. The amplitude of the mean field as a function of the coupling strength g in the UN ($\circ$) and WN ($\square$). The networks have mean degree $K = 10$ and size $N = 1000$

For the UN, $S_j = k_j$, and this approximation means that the oscillators are forced by a common signal, the global mean field $\boldsymbol{X}$, with the forcing strength being proportional to their degree k_j. Now even though the overall coupling strength g is still small, the oscillators with large degrees are already strongly forced by the common signal $\boldsymbol{X}$. In this weak coupling regime, the global mean field displays some small fluctuations around the unstable fixed point of the isolated oscillator, $\boldsymbol{X} \approx \boldsymbol{x}_F$ ($\boldsymbol{F}(\boldsymbol{x}_F) = \boldsymbol{0}$), thus it has only a very small amplitude. These oscillators should somewhat synchronize to $\boldsymbol{X}$. As shown in Figs. 5.8(a) and (b), oscillators with $k > 10$ already display some degree of synchronization, indicated by a decreasing ΔX for larger k, while all the oscillators in the WN are distant from X, since $S_j = 1$ for all of them. A small distance of an oscillator j from X, which has an almost vanishing amplitude, shows that the oscillation amplitude A_j of the oscillator is small. We have calculated A_j as the standard deviation of the time series x_j. We can see from Fig. 5.8(c) that A_j indeed displays almost the same behavior as ΔX_j. This becomes even more evident when we compare the average value $A(k)$, similar to (5.24), (Fig. 5.8(d)), with $\Delta X(k)$, (Fig. 5.8(b)). The changes in the amplitudes can be understood as follows: taking $\boldsymbol{X} \approx \boldsymbol{x}_F$, from (5.29) one gets for the UN

$$\dot{\boldsymbol{x}}_j = \tau_j \boldsymbol{F}(\boldsymbol{x}_j) - g\frac{k_j}{K}(\boldsymbol{x}_j - \boldsymbol{x}_F), \quad k_j \gg 1 , \qquad (5.27)$$

which yields that hubs (nodes with the largest degrees) are experiencing a strong negative self-feedback, so that the trajectory is stabilized at the

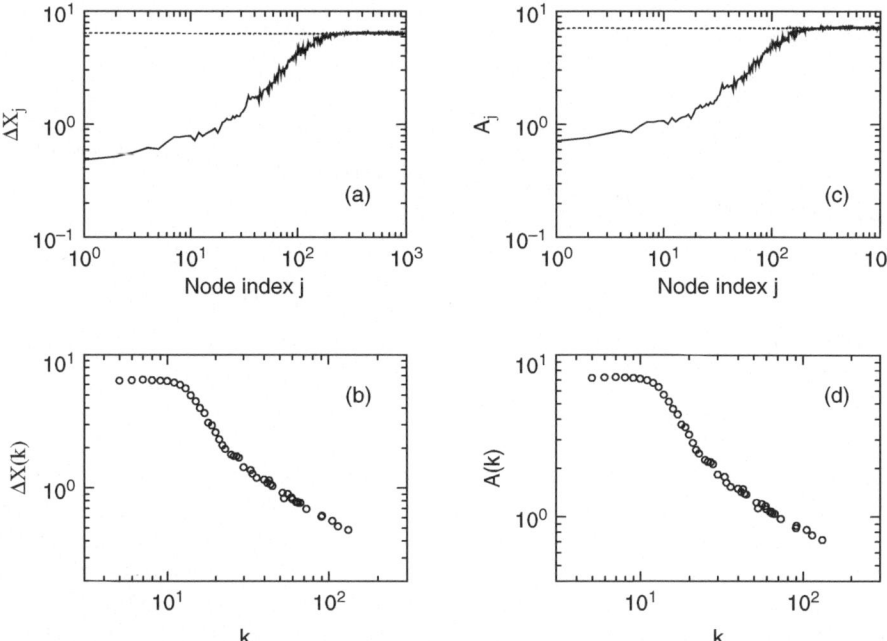

Fig. 5.8. (a) Synchronization difference ΔX_j of the oscillators with respect to the global mean field X in the UN (*solid line*) and WN (*dotted line*); (b) The average values $\Delta X(k)$ as a function of k in the UN; (c) and (d): as in (a) and (b), but for the oscillation amplitude A_j and the average value $A(k)$ of the oscillators. The results are averaged over 50 realizations of the random time scale parameters τ_j. The coupling strength is $g = 0.05$

originally unstable fixed point x_F, but with some fluctuations due to small non-vanishing perturbations from the mean activity of the neighbors.

To summarize: in the weak coupling regime, where even frequency and phase synchronization are not yet established, the heterogeneous UN already displays a form of hierarchical synchronization expressed by a change in the oscillation amplitudes.

II. Intermediate Coupling: Phase Synchronization

Next, we take an intermediate coupling strength $g = 0.13$, where both networks are in the regime of transition to strong collective oscillations (Fig. 5.7). In this regime, frequency and phase synchronization become evident, while the absolute distance ΔX is still large. In Fig. 5.9, we show the mean oscillation frequencies Ω_j of all oscillators. In the UN, we find that about 70% of the nodes are locked to a common frequency $\Omega = 0.99$, forming a frequency

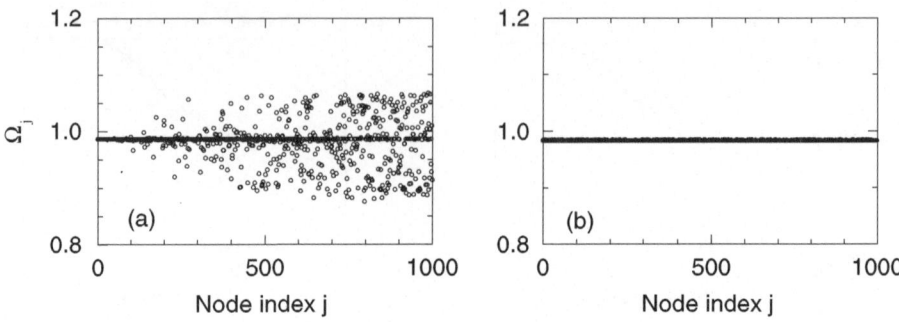

Fig. 5.9. The mean oscillation frequencies Ω_j of the oscillators in the UN: (**a**) and the WN; (**b**) at the coupling strength $g = 0.13$

synchronization cluster. Note that almost all nodes with the largest degrees k_j are synchronized in frequency, while many nodes with small degrees are not yet locked. In the WN, on the contrary, the frequencies of all nodes are locked so that the network is globally synchronized in frequency. The nodes that are not frequency locked in the UN are largely uncorrelated with each other and they do not generate significant contributions to the collective oscillations, while all the nodes in the WN have a significant contribution; as a result, the amplitude of the collective oscillation is much smaller in the UN.

Now we examine phase synchronization of the nodes with respect to $\boldsymbol{X}$. We measure phase synchronization by the time-averaged order parameter (Kuramoto parameter)

$$r_j = \langle \sin(\Delta\phi_j) \rangle^2 + \langle \cos(\Delta\phi_j) \rangle^2 , \qquad (5.28)$$

where $\Delta\phi_j = \phi_j - \phi_X$ is the difference of the phases of an individual oscillator j and the mean field $\boldsymbol{X}$. Here, the phases are defined as $\phi_j = \arctan(y_j/x_j)$ and $\phi_X = \arctan(Y/X)$ for an individual oscillator j and the mean field, respectively. Note that $r_j \approx 0$ when there is no phase locking and $r_j \approx 1$ when the phases are locked with an almost constant phase difference. Consistent with Fig. 5.9, we find that $r_j = 1$ for all oscillators in the WN; while $r_j < 1$ for many nodes with small degrees in the UN (Fig. 5.10(a)). To get a clear dependence of r on the degree k, we again calculate the average value $r(k)$ between all nodes with degree k. Now there is a more pronounced dependence between $r(k)$ and k (Fig. 5.10(b)).

We also calculate the absolute distances ΔX_j. They are not small on average in both networks in spite of phase synchronization (Fig. 5.10(c)), because phase locked oscillators may have significant (but bounded) phase differences. However, ΔX_j again displays the hierarchical structure in the UN (Fig 5.10(d)). The nodes with large degrees are not only locked in frequency, but also have small phase differences. So, in this regime, the hierarchical synchronization is manifested by different degrees of frequency and phase locking.

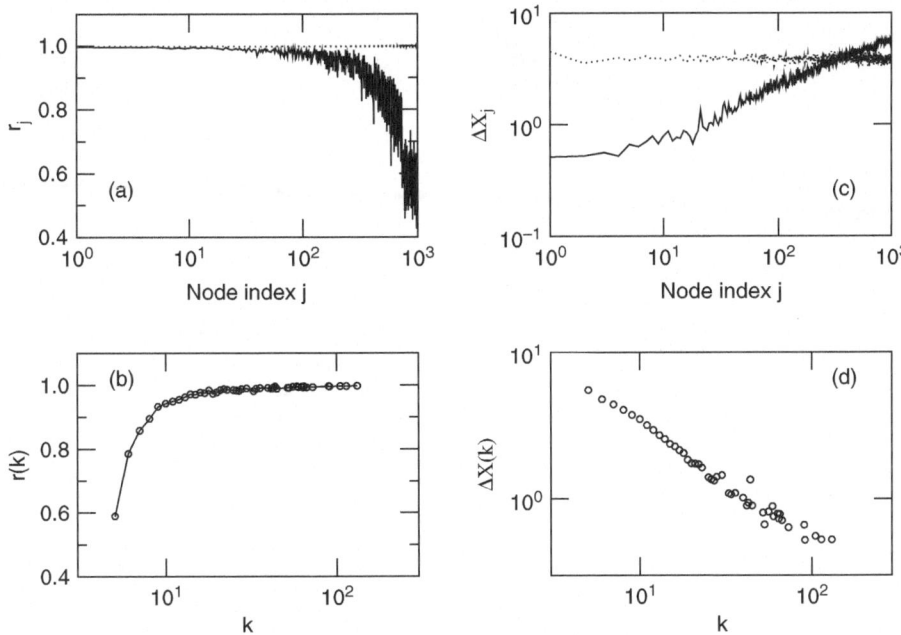

Fig. 5.10. (a) Phase synchronization order parameter r_j of node j with respect to the mean field $\boldsymbol{X}$ in the UN (*solid line*) and WN (*dotted line*). (b) Average value $R(k)$ of nodes with degree k as a function of k in the UN; (c) and (d) as in (a) and (b), but for the distance ΔX_j and its average value $\Delta X(k)$, respectively. The results are averaged over 50 realizations of random distribution of the time scale parameter τ_j. The coupling strength is $g = 0.13$

III. Strong Coupling: Almost Complete Synchronization

Now we consider the strong coupling regime where both networks have a large and saturated amplitude in their collective oscillations (Fig. 5.7).

We take $g = 0.5$, at which the amplitude of X is almost the same for both networks. The frequencies of all the oscillators are locked mutually as well as locked to the mean field; as a result, the phase synchronization order parameter is $r_j = 1$ for all oscillators in both UN and WN, i.e., the networks are globally phase synchronized. In the WN network, the phase difference $\Delta\phi_j$ between an oscillator and the mean field, averaged over time and over different realizations of random distribution of the time scale parameters τ_j, is small and on average rather homogeneous for all the oscillators (Fig. 5.11(a)). This implies that the oscillators are almost completely synchronized in the sense that $\Delta X \approx A_X \sin(\Delta\phi) \approx A_X \Delta\phi$ is also small and uniform on average (Fig. 5.11(c)). In the UN, however, many nodes with a degree smaller than the mean value K is not as strongly connected to the

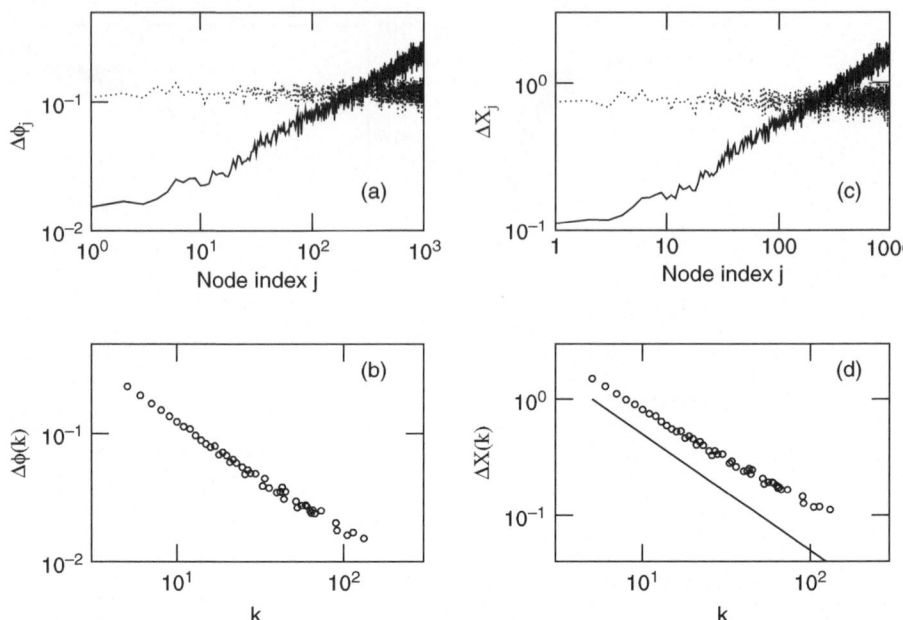

Fig. 5.11. (a) Averaged phase difference $\Delta\phi_j$ between a node j and the mean field X in the UN (*solid line*) and WN (*dotted line*); (b) Average value $\Delta\phi(k)$ of nodes with degree k as a function of k in the UN; (c) and (d) as in (a) and (b), but for the absolute difference ΔX_j and its average value $\Delta X(k)$, respectively. The solid line in (d) with slope $\alpha = 1$ is plotted for reference. The results are averaged over 50 realizations of the random time scale parameters τ_j. The coupling strength is $g = 0.5$

mean field, and on average they have phase differences larger than that of the WN (Fig. 5.11(a)), as is shown evidently by the average value $\Delta\phi(k)$ over nodes with degree k (Fig. 5.11(b)). Consequently, the synchronization difference ΔX_j is still heterogeneous (Fig. 5.11(c)) and $\Delta X(k) \sim k^{-\alpha}$ with $\alpha \approx 1$ (Fig. 5.11(d)).

5.4.4 Analysis of Hierarchical Synchronization

We have shown with unweighted SFNs that the effective synchronization displays a hierarchical organization according to the intensities, when the coupling is not strong enough, when there is noise or when the oscillators are non-identical. Here, we present an analysis of this hierarchical synchronization. The analysis is based on a mean field approximation of (5.1) with (5.26). For the case of UN, it reads

$$\dot{x}_j = \tau_j F(x_j) + \frac{gk_j}{K}(X - x_j), \quad k_j \gg 1 . \tag{5.29}$$

This approximation means that the oscillators are forced by a common signal $\boldsymbol{X}$, with the forcing strength being proportional to their degree k_j.

For identical oscillators ($\tau_j = 1, \forall j$), the linear variational equations of (5.29) are

$$\dot{\xi}_j = \left[\mathbf{D}\boldsymbol{F}(\boldsymbol{X}) - \frac{g}{K}k_j\mathbf{I} \right] \xi_j, \quad k_j \gg 1 , \tag{5.30}$$

which have the same form as (5.4), except that λ_j is replaced by k_j and $\mathbf{D}\boldsymbol{H}(\boldsymbol{s})$ is replaced by the identity matrix $\mathbf{I}$ when $\boldsymbol{H}(\boldsymbol{x}) = \boldsymbol{x}$. The largest Lyapunov exponent $\Lambda(k_j)$ (master stability function) of this linear equation is a function of k_j, i.e.,

$$\Lambda(k_j) = \Lambda_F - gk_j/K . \tag{5.31}$$

Remember that Λ_F is the largest Lyapunov exponent of the isolated oscillator $\boldsymbol{F}(\boldsymbol{x})$. $\Lambda(k_j)$ becomes negative for $\frac{g}{K}k_j > \Lambda_F$. For large values of k satisfying $\frac{g}{K}k \gg \Lambda_F$, we have $\Lambda(k) \approx -\frac{g}{K}k$.

Now suppose that the network is close to being completely synchronized, when the coupling strength g is below the threshold g_c, or when there is noise present in the system. For nodes with a large degree k so that $\Lambda(k) \approx -\frac{g}{K}k$ is sufficiently negative, the dynamics of the averaged synchronization difference $\Delta X(k)$ over large time scales can be expressed as

$$\frac{\mathrm{d}}{\mathrm{d}t}\Delta X(k) = \Lambda(k)\Delta X(k) + c , \tag{5.32}$$

where $c > 0$ is a constant denoting the level of perturbation with respect to the complete synchronization state, which depends on the noise level D or the coupling strength g. For the case of non-identical oscillators, the perturbation level (constant c) is due to the disorder in the time scale τ_j of the oscillators. From this, we get the asymptotic result $\Delta X(k) = c/|\Lambda(k)|$, giving

$$\Delta X(k) \sim k^{-1} , \tag{5.33}$$

which explains qualitatively the numerically observed scaling in Figs. 5.5, 5.10 and 5.11. The slight deviation of the scaling exponents from the linear result $\alpha = 1$ may result from the mean field approximation and significant nonlinearity, since the linear analysis in (5.32) is only a first order approximation.

For a general weighted random network, the degree k in (5.33) should be replaced by the intensity S, and we have

$$\Delta X(S) \sim S^{-1} . \tag{5.34}$$

5.5 Phase Synchronization in Small-World Networks of Oscillators

So far, we have analyzed networks that are random, where the nodes do not have spatial properties. However, in many realistic networks, the oscillators are arranged in space, and this spatial arrangement has a significant impact on the connection patterns. A good example of this type is the network of interacting neurons in local areas of the brain cortex. Here, the neurons have sparse connections to neighboring neurons, neither in a fully regular, nor in a completely random manner, but somewhere in between. A simple model describing this type of network is the SWN model proposed in [2, 34]. It is based on a regular array of oscillators, each coupled to its k nearest neighbors. With a probability p, a link is added (rewired) to a randomly selected pair of the oscillators, i.e., some long-range connections are introduced. For small p values, the resulting networks display both the properties of regular networks (high clustering) and of random networks (short pathlength), i.e., they are SWNs.

In this section, we demonstrate the important impact of such random long-range interactions on the synchronization of non-identical oscillators. We start with a regular ring of N nodes, each connected to its two nearest neighbors, i.e., $k = 2$. Shortcuts are then added between randomly selected pairs of nodes, with probability p per link of the basic regular ring, so that typically there are pN shortcuts in the resulting networks. In this way, the total number of connections also increases with p. Again, we use the Rössler chaotic oscillators $F(x)$ and output function $H(x)$ as in Sect. 4. The dynamical equation is

$$\dot{x}_j = \tau_j F(x_j) + \frac{g}{k_j} \sum_{i=1}^{N} A_{ji}(x_i - x_j) \,, \quad j = 1, \ldots, N \,, \qquad (5.35)$$

Note that the coupling strength g is normalized by the degree k_j of each node, so that we can scale out effects of the increasing average degree $K = \langle k_j \rangle$ when more and more shortcuts are added in the network at larger probability p. As in Sect. 4.3, we consider a uniform distribution of τ_j in the interval $[1 - \Delta\tau, 1 + \Delta\tau]$, and we fix $\Delta\tau = 0.4$ in the simulations.

We now discuss the synchronization behavior of (5.35) for networks with different shortcut probabilities p. The degree of synchronization is quantified by the amplitude A_X of the mean field oscillation $X = (1/N)\sum_{j=1}^{N} x_j$ as a function of the coupling strength g (Fig. 5.12(a)). We also examine the variation of the oscillation amplitudes in individual oscillators with respect to g by measuring the average value of the standard deviation of $x_j(t)$, $\langle A_j \rangle$ (Fig. 5.12(b)). We observe the following types of synchronous behavior:

i) When the shortcut probability p is very small ($p = 0.01$), the network is still dominated by local coupling and it does not display obvious collective synchronization effects over a broad range of g, as indicated by an almost vanishing mean field X. However, the oscillation amplitude of individual oscillators changes with g.

ii) With a larger number of shortcuts at $p = 0.1$, the network starts to synchronize and generates a coherent collective oscillation at a strong enough coupling strength.

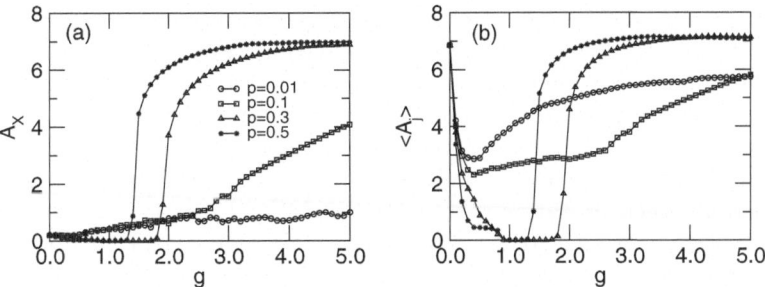

Fig. 5.12. Transition to oscillation death and synchronization in various SWNs of chaotic Rössler oscillators: (**a**) The amplitude A_X of the mean field X as a function of the coupling strength g; (**b**) The average value of the amplitudes A_j of all the individual oscillators. The network size is $N = 1024$

iii) At even larger values of p, e.g., $p = 0.3$ and $p = 0.5$, the networks display three dynamical regimes: (1) When the coupling strength is increased from very small values, the trajectory of each oscillator draws closer and closer to the unstable steady state $\boldsymbol{x}_F$ ($\boldsymbol{F}(\boldsymbol{x}_F) = \boldsymbol{0}$), as seen by a rapid decrease of the amplitude $\langle A_j \rangle$ of individual oscillators (Fig 5.12(b)). The oscillation frequencies Ω_j are still distributed in this regime (Fig 5.13(a,b)). (2) When a critical value g_1 is reached, all oscillators become stable at the same steady

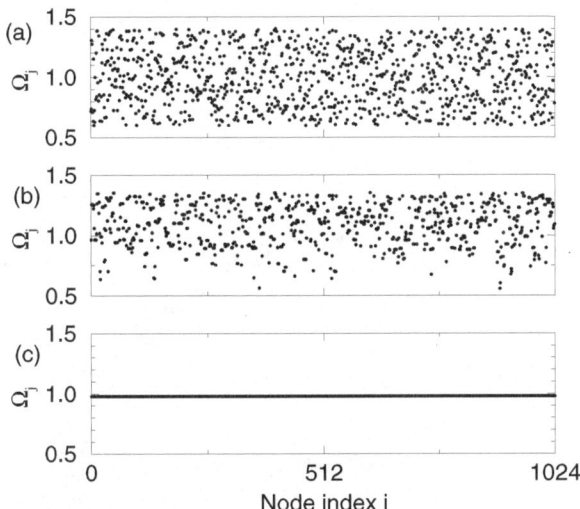

Fig. 5.13. Oscillation frequency Ω_j of the oscillators in SWNs with the shortcut probability $p = 0.5$ for different coupling strength: (**a**) $g = 0$; (**b**) $g = 0.1$; and (**c**) $g = 1.5$

state $\boldsymbol{x}_F$, so that $A_X = 0$ and $\langle A_j \rangle = 0$, and we observe oscillation death in SWNs, i.e., all the oscillators stop oscillating (Fig 5.12(b)). (3) When g is further increased to exceed another critical value g_2, the steady state $\boldsymbol{x}_F$ becomes unstable again, and the oscillation is restored. Importantly, the whole network is now in a global synchronization regime: the frequencies and phases of all oscillators are locked (Fig 5.13(c)). Comparing the critical value g_2 of the coherent synchronization regime for $p = 0.3$ and $p = 0.5$, one can see that networks with more shortcuts achieve this coherent synchronization with a smaller coupling strength.

For a fixed value of the coupling strength g, the two regimes of oscillation death and global synchronization can also be obtained by adding a sufficient number of shortcuts (Fig. 5.14). The system behavior is not sensitive to increasing p when $p < 0.02$. With a further increase of p, the oscillation amplitudes of the oscillators are reduced and finally the regime of oscillation death is reached, which is stable for networks in a certain range of p, and afterwards a coherent collective oscillation is observed due to global synchronization, which becomes more pronounced as more shortcuts are added to the network.

We have shown that the coupling topology in the SWNs has significant effects on the synchronization of strongly non-identical nonlinear oscillators. Compared to regular networks with local coupling ($p \approx 0$), SWNs with many shortcuts display enhanced synchronization as expressed by the regimes

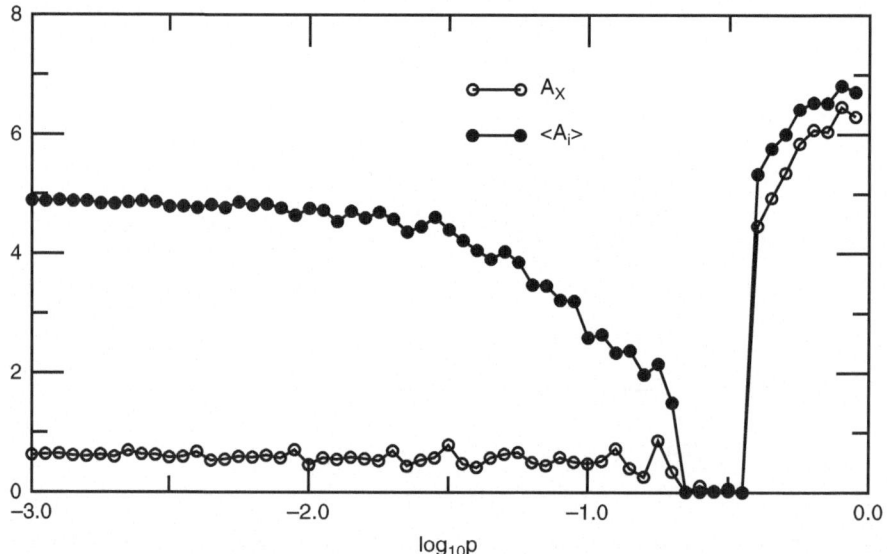

Fig. 5.14. Synchronization behavior vs. the shortcut probability p for a fixed coupling strength ($g = 1.5$)

of oscillation death and global synchronization similar to globally coupled networks [35].

5.6 Hierarchical Synchronization and Clustering in Complex Brain Networks

Synchronization of distributed brain activity has been proposed as an important mechanism for neural information processing [6]. Experimentally observed brain activity, characterized by synchronization phenomena over a wide range of spatial and temporal scales [36], reflects a hierarchical organization of the dynamics. Such an organization arises through a hierarchy of complex cortical networks: the microscopic level of interacting neurons, the mesoscopic level of mini-columns and local neural circuits, and the macroscopic level of nerve fiber projections between brain areas [15]. While details at the first two levels are still largely missing, extensive information has been collected about the latter level in the brain of animals, such as the cat and the macaque monkey [37]. The complex topology of cortical networks has been the subject of many recent analyses [37]. See Chaps. 3, 4, or 9 for a complete review. Analyses of the anatomical connectivity of the mammalian cortex [37] and the functional connectivity of the human brain [38] have shown that the two share typical features of many complex networks. However, the relationship between anatomical and functional connectivities remains one of the major challenges in neuroscience [6].

Conceptually modeling the dynamics of the neural system based on a realistic network of corticocortical connections and investigating the synchronization behavior should provide meaningful insights into this problem. Here we consider the cortical network of the cat. The cortex of the cat can be parcellated into 53 areas, linked by about 830 fibers of different densities [15] into a weighted complex network as shown in Fig. 5.15(a). This network displays typical small-world properties, i.e., short average pathlength and high clustering coefficient, indicating an optimal organization for an effective inter-area communication and for achieving high functional complexity [39, 40]. The degrees of the nodes are heterogeneous, for example, some nodes have only two or three links, while some others have up to 35 connections. Due to the small number of areas, it is difficult to claim a scale-free distribution [40], nevertheless, analyses comparing this network to scale-free network models with the same size and connectivity density does suggest a scale-free distribution (see Sect. 3.7 of Chap. 3).

Different from random networks models, the cortical network of the cat exhibits hierarchically clustered organization [40, 41]. There are a small number of clusters that broadly agree with the four functional cortical sub-divisions, i.e., visual cortex (V, 16 areas), auditory (A, 7 areas), somatosensory-motor (SM, 16 areas) and frontolimbic (FL, 14 areas). To distinguish these from the dynamical clusters in the following discussion, we refer to the topological

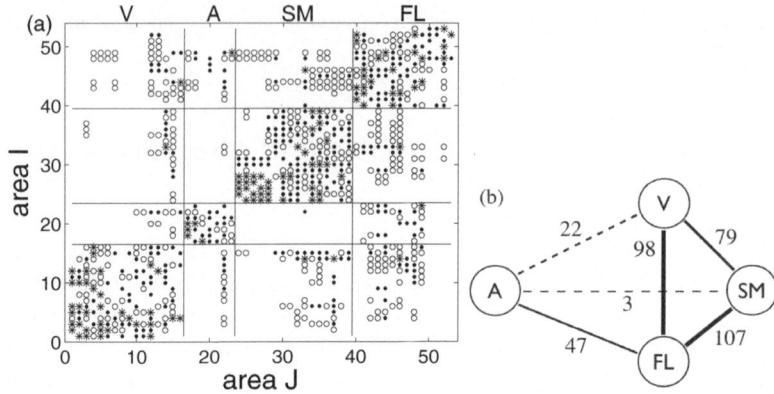

Fig. 5.15. (a) Connection matrix M^A of the cortical network of the cat brain. The different symbols represent different connection weights: 1 (• *sparse*), 2 (○ *intermediate*) and 3 (* *dense*). The organization of the system into four topological communities (functional sub-systems, V, A, SM, FL) is indicated by the dashed lines; (b) The number of connections between the four communities

clusters as *communities* [42]. The inter-community connections in Fig. 5.15(b) show that A is much less connected while V, SM and FL are densely connected with each other.

Next, we analyze synchronization dynamics of this network by simulating each cortical area with (i) periodic neural mass oscillators for modeling neural rhythms [43, 44] and (ii) a subnetwork of interacting excitable neurons [45, 46]. While the model with neural mass oscillators typically displays the synchronization behavior explained in Sect. 5.4, the model with subnetworks shows that the dynamics is also hierarchically organized and reveals different scales in the hierarchy of the network topology. In particular, in the biologically plausible regime, the most prominent dynamical clusters coincide closely with anatomical communities that agree broadly with the functional sub-divisions V, A, SM and FL.

5.6.1 Neural Mass Model

The mean activities of a population of neurons in the brain often exhibit rhythmic oscillations with well defined frequency bands, as seen in EEG measurements (cf. Chaps. 7 and 8). Such oscillations can be captured by realistic macroscopic models of EEG generation proposed in the early 1970s [43] (see Chap. 1 for a discussion). In this section, we use the neural mass model and parameters presented in [44]. A population of neurons contains two subpopulations: subset 1 consists of pyramidal cells receiving excitatory or inhibitory feedback from subset 2. Subset 2 is composed of local interneurons receiving excitatory input. The neural mass model describes the evolution of the macroscopic variables, i.e., mass potentials v^e, v^i and v^d for the excitatory, inhibitory and interneurons, respectively. A static nonlinear sigmoid function $f(v) = 2e_0/(1 + e^{r(v_0 - v)})$ converts the average membrane potential into an average pulse density of action potentials. Here, e_0 is the firing rate at the mass potential v_0 and r is the steepness of the activation. The input from another group of neurons and from external signals is fed into the population of interneurons. The dynamical equations for $I = 1, \ldots, N$ multiple coupled populations read

$$\ddot{v}_I^e = Aaf(v_I^d - v_I^i) - 2a\dot{v}_I^e - a^2 v_I^e , \tag{5.36}$$

$$\ddot{v}_I^i = BbC_4 f(C_3 v_I^e) - 2b\dot{v}_I^i - b^2 v_I^i , \tag{5.37}$$

$$\ddot{v}_I^d = Aa\left[C_2 f(C_1 v_I^e) + p_I(t) + \frac{g}{\langle S \rangle} \sum_J^N W_{IJ} f(v_J^d - v_J^i) \right] \tag{5.38}$$

$$-2a\dot{v}_I^d - a^2 v_I^d ,$$

where v_I^e, v_I^i and v_I^d are the mass potentials of the area I. Here, A and B are the average synaptic gain, a and b are the characteristic time constants of the EPSP and IPSP, respectively; C_1 and C_2, C_3 and C_4 are the average number of synaptic contacts, for the excitatory and inhibitory synapses, respectively. More detailed interpretation and standard values of these model parameters can be found in [44]. The coupling strength g is normalized by the mean intensity $\langle S \rangle$ as in Sect. 5.4.

Here we model the cat cortical network by simulating each cortical area (a large ensemble of neurons) by such a macroscopic neural mass oscillator, i.e., by taking the cortical network in Fig. 5.15(a) as the coupling matrix W_{IJ} in (5.39). As in [44], in our simulations we take $p_I(t) = p_0 + \xi_I(t)$ where $\xi_I(t)$ is a Gaussian white noise with standard deviation $D = 2$. We fix $p_0 = 180$ so that the system is in the periodic regime corresponding to alpha waves. A typical time series of the output, the average potential $V_I = v_I^d - v_I^i$, is shown in Fig. 5.16(a). Synchronization between the areas is measured by the linear correlation coefficient $R(I, J)$ between the outputs V_I and V_J. The average correlation $\langle R \rangle$ among all pairs of areas is shown in Fig. 5.16(b) as a function of the coupling strength g.

According to our analysis in Sects. 5.3 and 5.4, in a sufficiently random network, each oscillator is influenced by the mean activity of the whole network with a coupling strength proportional to the intensity $S_I = \sum_{J=1}^{N} W_{IJ}$. Not much direct relationship between the pair-wise coupling strength W_{IJ} and the strength of synchronization $R(I, J)$ is expected. We find that this still roughly holds for the cat cortical network although it is not very random due to the clustered organization. To demonstrate this, we distinguish three cases for any pair of nodes in the network: reciprocal projections (P2), uni-directional couplings (P1) and non-connection (P0), and compute the distribution of the correlation $R(I, J)$ for these cases separately. As seen in Fig. 5.17, when the coupling is weak (e.g., $g = 2$), the distributions for P0, P1 and P2 pairs

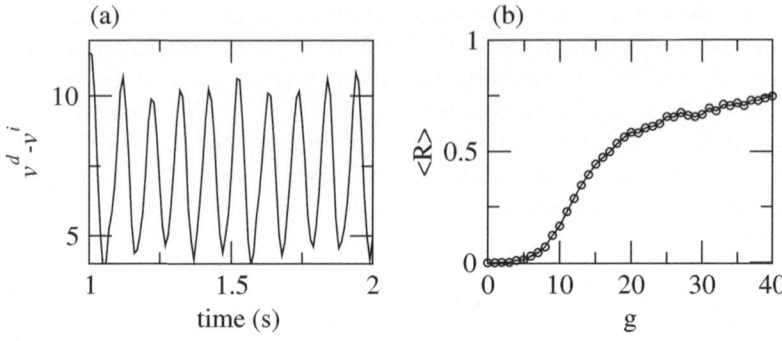

Fig. 5.16. (a) Typical activity $V = v^d - v^i$ of the uncoupled neural mass model; (b) The average correlation coefficient $\langle R \rangle = \frac{1}{N(N-1)} \sum_{I \neq J} R(I, J)$ ($N = 53$) vs. the coupling strength g in (5.39)

coincide and display a Gaussian shape around zero and no significant correlation is established. At a stronger coupling (e.g., $g = 5$), the P2 pairs have slightly stronger correlation than P1, however, the distributions still overlap significantly, as for strong coupling.

The dynamical pattern is not structured with very weak coupling, but with stronger coupling ($g \geq 5$), the system forms a major cluster including most of the areas from V, SM and FL, while the auditory system A remains relatively independent (Fig. 5.18). This is consistent with the inter-community connectivity shown in Fig. 5.15(b). Also, some other areas with the smallest degrees and intensities are also relatively independent. The correlation coefficient R_X between the activity V_I of an area and the global mean field $\bar{X} = (1/N) \sum^N V_I$ is shown in Fig. 5.19. It is roughly an increasing function of S, which basically reproduces the behavior presented in Sect. 5.4 (e.g., Figs. 5.5, 5.10 and 5.11).

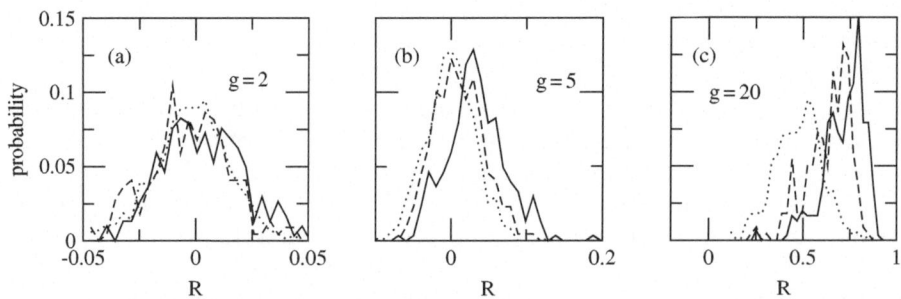

Fig. 5.17. (a) Distribution of the correlation R for P2 (*solid line*), P1 (*dashed line*) and P0 (*dotted line*) pairs at various values of the coupling strength g, (a) $g = 2$, (b) $g = 5$ and (c) $g = 20$

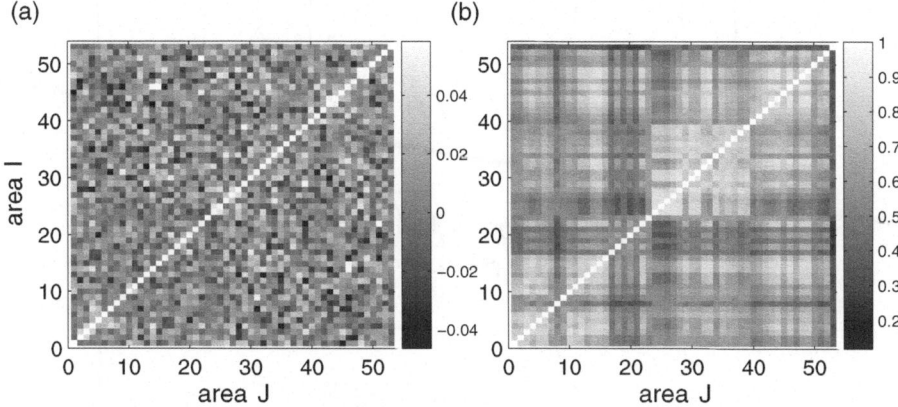

Fig. 5.18. Correlation matrices $R(I, J)$ for (**a**) weak coupling $g = 2$; (**b**) For strong coupling $g = 20$. Note the different gray-scales in the colorbars

5.6.2 Subnetworks of Interacting Neurons

Each brain area is composed of a large ensemble of neurons coupled in a complex network topology having several levels of organization; the detailed connectivity, however, is still largely unclear. In the following section, we model each cortical area with a sub-network of N_a interacting neurons. We use the SWN model [2] to couple the neurons. Specifically, a regular array of N_a neurons with a mean degree k_a is rewired with a probability p. Such a topology incorporates the basic biological feature that neurons are mainly connected to their spatial neighbors, but also have a few long-range synapses [47]. The small-world topology has been shown to improve the synchronization of interacting neurons [7–9]. Our model also includes other realistic, experimentally observed features, i.e., 25% of the N_a neurons are inhibitory and only a small number of neurons (about 5%) of one area receive excitatory synapses from another connected area [48]. To our knowledge, no information is available about the output synapses of each area, and for simplicity, we assume that the output signal from one area to another one is the mean activity of the output area. Individual neurons are described by the FitzHugh-Nagumo (FHN) excitable model [49] with non-identical excitability (cf. Chap. 1). A weak Gaussian white noise (with strength $D = 0.03$) is added to each neuron to generate sparse, Poisson-like irregular spiking patterns in isolated FHN neurons, as in realistic neurons.

Thus, our model of the neural network of a cat cortex is composed of a large ensemble of noisy neurons connected in a *network of networks*, and the dynamics of the neuron i in the area I is specified as:

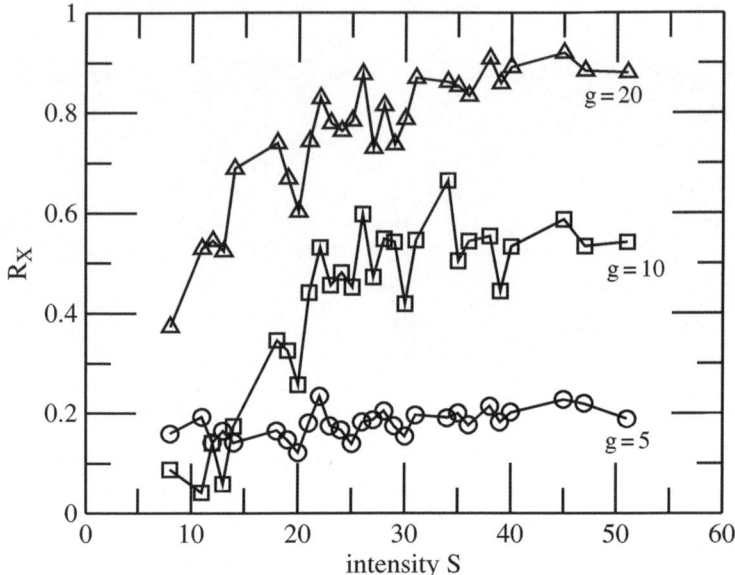

Fig. 5.19. Correlation between an area and the global mean field, as a function of the intensity S (averaged over nodes with the same intensity S) at various coupling strengths g

$$\epsilon \dot{x}_{I,i} = f(x_{I,i}) + \frac{g}{k_a} \sum_{j}^{N_a} M_I^L(i,j)(x_{I,j} - x_{I,i})$$

$$+ \frac{g}{\langle w \rangle} \sum_{J}^{N} M^A(I,J) L_{I,J}(i)(V_J - x_{I,i}) , \qquad (5.39)$$

$$\dot{y}_{I,i} = x_{I,i} + a_{I,i} + D\xi_{I,i}(t) , \qquad (5.40)$$

where

$$f(x_{I,i}) = x_{I,i} - \frac{x_{I,i}^3}{3} - y_{I,i} . \qquad (5.41)$$

Here, the matrix M^A represents the corticocortical connections in the cat network as in Fig. 5.15(a). M_I^L denotes the local SWN of the I-th area $(M_I^L(i,j): (i,j = 1, \ldots, N_a))$. A neuron j is inhibitory if $M_I^L(i,j) = -1$ for all of its connected neighbors. The label $L_{I,J}(i) = 1$ if the neuron i is among the 5% within the area I receiving the mean field signal $V_J = (1/N) \sum_l^{N_a} x_{J,l}$ from the area J, otherwise, $L_{I,J}(i) = 0$. The diffusive coupling, describing electrical synapses (gap junctions; see Chap. 2) and not being the most typical case in mammalian cortex, is mainly used for the simplicity of simulation at this stage. Normalized by the mean degree k_a of the SWNs within the areas, and normalized by the average weight $\langle w \rangle$ of inter-area connections, g represents the average coupling strength between any pair of neurons and is the control parameter in our simulations. (Note that we assume g to be equal for couplings within and between subnetworks).

The system is simulated with $N_a = 200$, $k_a = 12$ and $p = 0.3$ for the subnetworks. Our focus is to study the synchronization behavior at the systems level, i.e., the synchronization behavior between the mean activity V_I of the subnetworks and its relationship with the underlying cortical network in Fig. 5.15(a). The behavior demonstrated below does not depend critically on the parameters of the subnetworks, while the detailed synchronization behavior *within* the subnetwork does depend on them [7–9].

The coupling strength g controls the mutual excitation between neurons. At small g (e.g. $g = 0.06$), a neuron is not often excited by the noise-induced spiking of its connected neighbors, so the synchronization within and between the subnetworks is weak. This is shown by small fluctuations of the mean activity V_I of each area (Fig. 5.20(a)) and a small average correlation coefficient $\langle R \rangle$ among V_I (Fig. 5.20(d)). Weak synchronization in the subnetwork of an area is manifested by some clear peaks in V_I (Fig. 5.20(a)). When we increase g, the synchronization becomes stronger with more frequent and larger peaks in V_I (Fig. 5.20(b)) and at large enough g, the neurons are mutually excited achieving both strongly synchronized and regular spiking behavior (Fig. 5.20(c)). $\langle R \rangle$ approaches 1 (Fig. 5.20(d)), indicating an almost global synchronization of the network.

The patterns of the correlation matrix $R(I, J)$ are shown in Fig. 5.21. The behavior at strong couplings is very similar to that of the neural mass model in Fig. 5.18, since both models display well defined oscillations. However, Fig. 5.21(a) suggests that the dynamics of the present model with weak coupling has a nontrivial organization and an intriguing relationship with the underlying network topology. The distribution of R over all pairs of areas

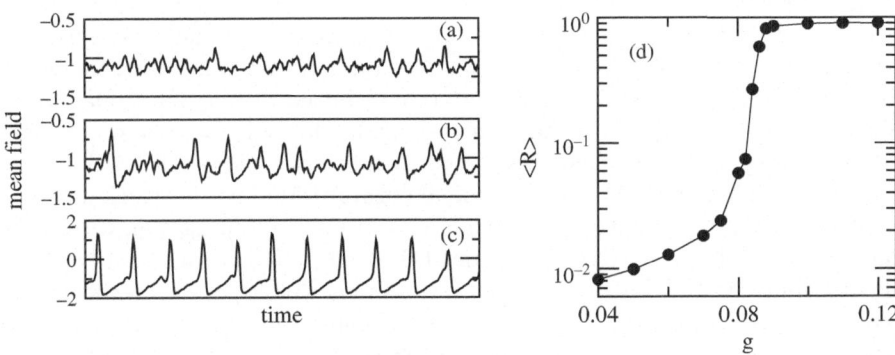

Fig. 5.20. Typical mean activity $V_I X$ of one area at various coupling strengths: (a) $g = 0.06$; (b) $g = 0.082$; (c) $g = 0.09$; (d) The average correlation coefficient $\langle R \rangle = \frac{1}{N(N-1)} \sum_{I \neq J} R(I, J)$ ($N = 53$) plotted vs. g

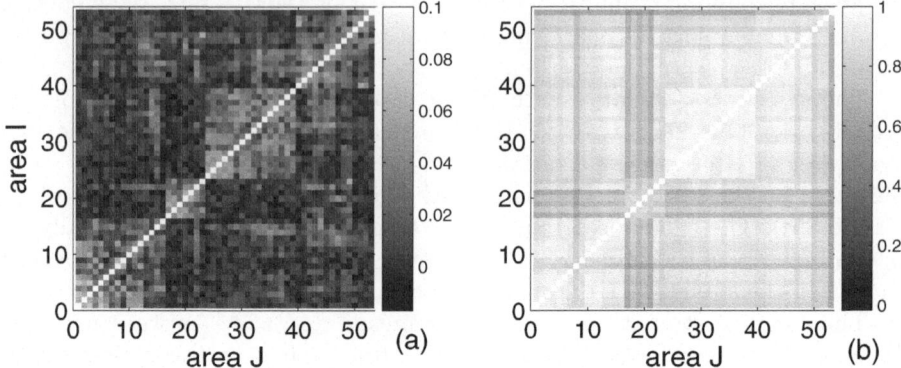

Fig. 5.21. Correlation matrices $R(I, J)$ at weak coupling $g = 0.06$: (**a**) and strong coupling $g = 0.12$; (**b**) Note the different gray-scales in the colorbars

displays a Gaussian peak around zero, but with a long tail for large values (Fig. 5.22(a), solid line). Although the correlations are relatively small, we find that the large values are significant when compared to the distribution of R of surrogate data by random shuffling of the time series V_I (Fig. 5.22(a), dash-dotted line). The weak coupling regime is biologically more realistic since here, the neurons only have a low frequency of irregular spiking and irregular mean activities (Fig. 5.20(a)), similar to those observed experimentally (e.g., EEG data [50]). The propagation of a signal between connected areas is mediated by synchronized activities (peaks in V) and a temporal correlation is most likely established when receivers produce similar synchronized activities from one input, or when two areas are excited by strongly correlated signals from common neighbors. Due to the weak coupling and the existence of subnetworks, such a synchronized response does not always occur and a local signal (excitation) does not propagate through the whole network. As a result, the correlation patterns are closely related to the network topology, although the values are relatively small due to infrequent signal propagation. With strong coupling, the signal can propagate through the whole network, corresponding to pathological situations, such as epileptic seizure [51].

Let us now characterize the dynamical organization and its relationship to the network topology. Based on an argument of signal propagation, we expect that the correlations for the P2, P1 and P0 areas should be different. Indeed, the distributions of R for these three cases display well-separated peaks in the weak coupling regime (Fig. 5.22(a)). Note especially that, all the P2 pairs have significant correlations compared to the surrogate data. With strong coupling (e.g. $g > 0.09$), where the excitation propagates through the whole network, the distribution is very similar to the neural mass model in Fig. 5.17(c), and the separation is no longer pronounced (not shown).

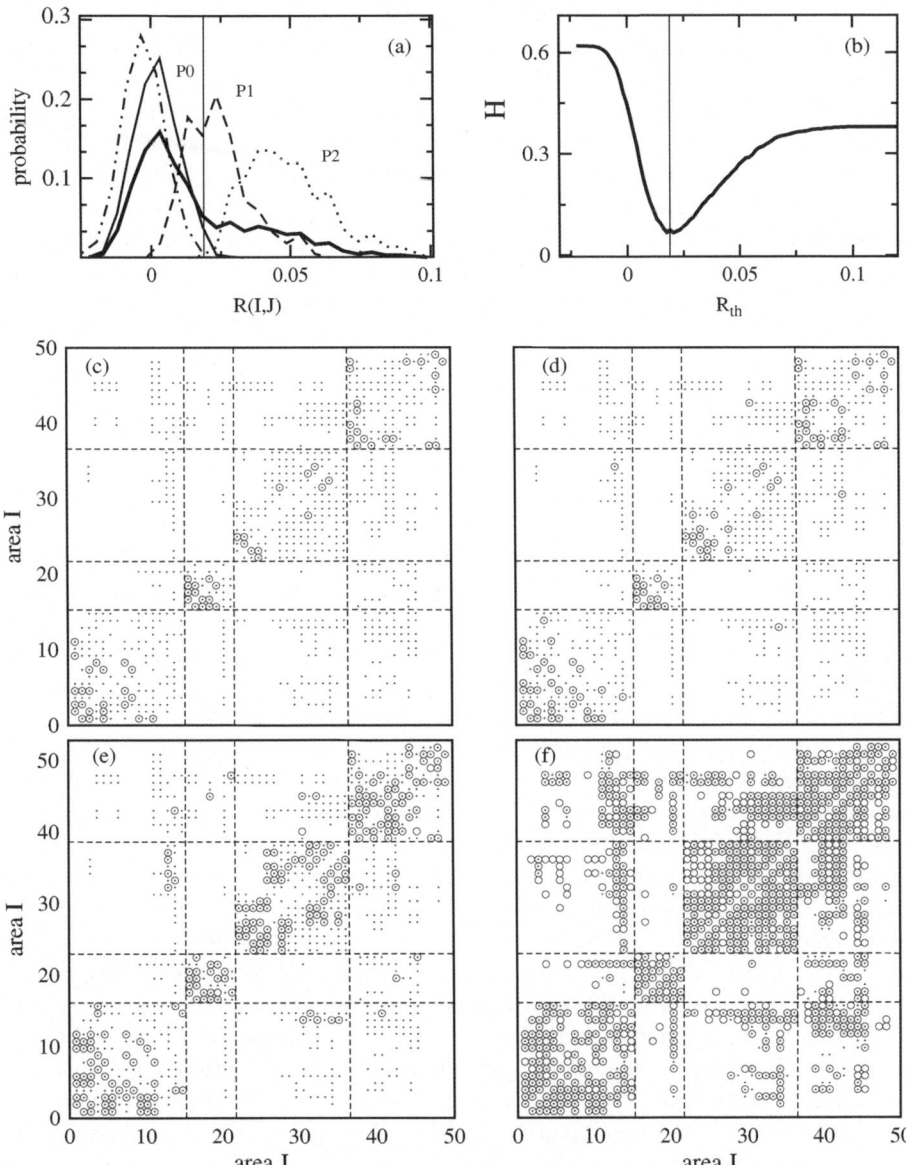

Fig. 5.22. (a) Distribution of the correlation R ($g = 0.07$) for all nodes (*solid line*), P2 (*light solid line*), P1 (*dashed line*) and P0 (*dotted line*). The dash-dotted line denotes the results for the surrogate data; (b) Hamming distance H vs. R_{th}. The vertical solid lines in (a) and (b) indicate the natural threshold $R_{th} = 0.019$. The functional networks ($\circ$) with thresholds $R_{th} = 0.070, 0.065, 0.055$, and 0.019 are shown in (c), (d), (e), and (f), respectively. The small dots indicate the anatomical connections

We extract a *functional network* M^F [38] by applying a threshold R_{th} to the correlation R, i.e., a pair of areas is considered to be functionally connected if $R(I, J) \geq R_{th}$ ($M^F(I, J) = 1$). We can then compare the topological structures of the anatomical network M^A and the functional networks M^F with varying R_{th} and examine how the various levels of synchronization reveal different scales in the network topology. focusing on the biologically meaningful weak coupling regime.

We focus on the biologically meaningful weak coupling regime and take $g = 0.07$ as the typical case. When R_{th} is very close to the maximal value of R, only a few P2 areas in the auditory system A are functionally connected, because of their strong anatomical links and sharing of many common neighbors. With lower values, e.g. $R_{th} = 0.07$, about 2/3 of the areas but only 10% of the P2 links are active in the functional network (Fig. 5.22(c)). Interestingly, within each anatomical community V, A, SM, and FL, a core subnetwork is functionally manifested in the form of connected components without inter-community connections. At lower values, e.g., $R_{th} = 0.065$, more areas from the respective communities are included into these components and a few inter-community connections appear to join the components from V, SM, and FL (Fig. 5.22(d)). This observation suggests that a core subnetwork coupled more strongly and communicating more frequently among the areas within the respective community is most likely to perform specialized functions of this community. Going to an even lower threshold, e.g., $R_{th} = 0.055$, all areas become involved and form a single connected functional network, but this network contains only about 1/3 of the anatomical P2 links and very few P1 links. However, the communication of the whole network is still mediated only by a small number of inter-community connections while most of the connections are within V, A, SM and FL, i.e., the functional network is highly clustered and agrees well with the anatomical communities. With further reduction of R_{th}, still more anatomical links are expressed as functional connections. For example, at $R_{th} = 0.019$, all P2 links are just fully expressed and about 70% of P1 links too. Meanwhile, about 4% of non-connected pairs (P0) establish significant functional connections (the significance level ≈ 0.004 at $R = 0.019$), since they have many common neighbors. Thus, the functional network reveals the anatomical network rather faithfully (Fig. 5.22(f)). To compare the matrices M^F (symmetrical) and M^A (asymmetrical) in a more quantitative way, we take the binary matrix of M^A, symmetrize all P1 links and compute the Hamming distance H, i.e., the percentage of elements between M^F and the binary and symmetrized M^A that are different. The closeness between them is confirmed by a very small Hamming distance $H = 0.074$, which is almost minimal for varying R_{th} (Fig. 5.22(b)). It is interesting to note that this value of the threshold R_{th} is exactly where the full distribution of R starts to deviate from the Gaussian and the distribution of P2 areas separates from that of the surrogate data (Fig. 5.22(a), solid line). We find that such a natural choice of R_{th} always reproduces the network topology well with $H \approx 0.06$ for different coupling strength $0.04 \leq g \leq 0.08$.

We have analyzed the dynamical clusters using the algorithm for hierarchical clustering in Matlab with the dissimilarity matrix $d = [d(I, J) = 1 - R(I, J)]$. Typical hierarchical trees for the weak and strong synchronization regimes are shown in Fig. 5.23. Figure 5.24 displays the most prominent clusters for the weak coupling regime. The functional clusters closely resemble the four communities obtained by using graphical tools based on anatomical structures [40, 41]. The four dynamical clusters sufficiently correspond to the functional sub-division of the cortex– C_1 (V), C_2 (A), C_3 (SM), C_4 (FL). However, it is also important to notice that there are a few nodes that belong to one anatomical community but join another dynamical cluster. For example, the area $I = 49$ (anatomically named as 36 in the cat cortex) of the fronto-limbic system is in the dynamical cluster C_2 mainly composed of areas from the auditory system (Fig. 5.24 (C_2)). A close inspection shows that these nodes bridging different anatomical communities and dynamical clusters are exactly the areas sitting in one anatomical community but in close connectional association with the areas in other communities [15]. In the strong synchronization regime, V, SM and FL join to form a major cluster (Fig. 5.25 (C_3)), while the auditory system A remains as a distinct cluster (Fig. 5.25 (C_2)). The formation of a cluster from community A both in the weak and the strong synchronization regimes is due to almost global connections within A. The cluster formation behavior in the strong coupling regime is also in good accordance with the inter-community connectivity shown in Fig. 5.15(b). There are also two single areas showing themselves as independent clusters. It turns out that these are the nodes with the minimal intensities in the network. In [46], we have shown that the clustering patterns remain almost the same in randomized networks that preserve the sequence of the

(a)   (b)

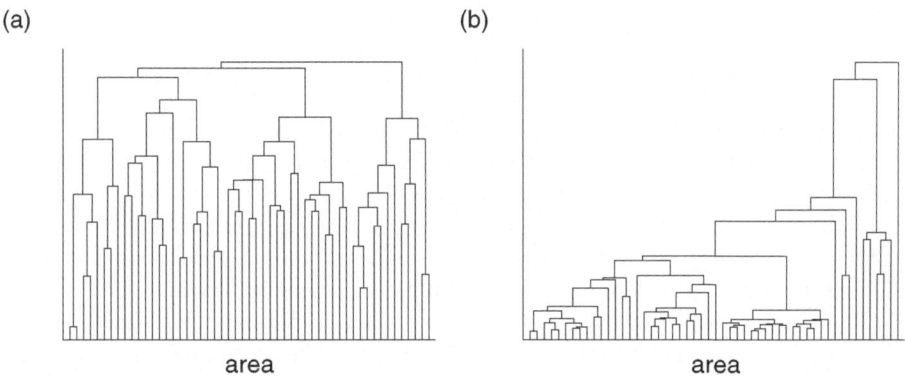

area   area

Fig. 5.23. Typical hierarchical tree of the dynamical clusters in the weak coupling regime; (**a**) $g = 0.07$ and strong coupling regime: (**b**) $g = 0.12$

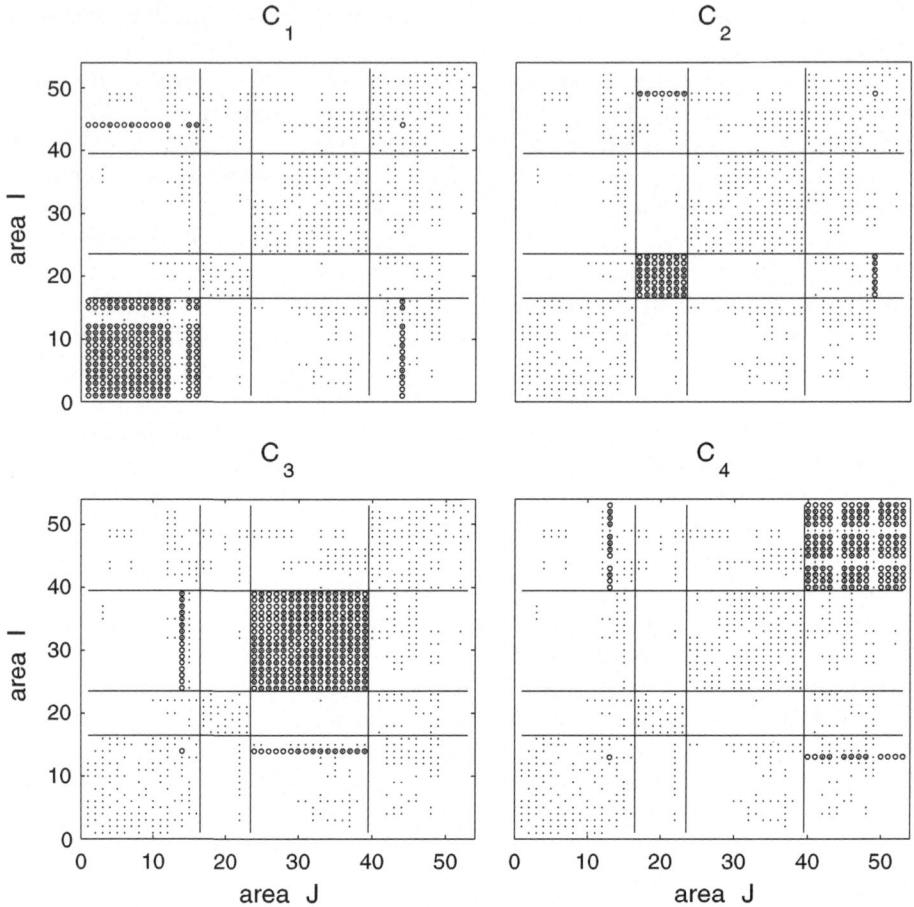

Fig. 5.24. Dynamical clusters (○) with weak coupling strength $g = 0.07$, overlaid on the underlying anatomical connections (·)

intensities S_I as in the cat cortical network; the auditory system A no longer forms a distinct cluster when the pronounced intra-community connections are destroyed in the randomized networks. This demonstrates that our understanding of synchronization based on weighted random networks in Sects. 5.3 and 5.4 can be applied when the node dynamics (mean activity of the subnetwork in this case) display a well-defined oscillatory behavior.

The comparison between models with subnetworks and those with neural mass oscillators indicates that self-sustained oscillator models may not be as appropriate for the understanding of the interplay between dynamics and structure in the brain as a hierarchical network of *excitable* elements.

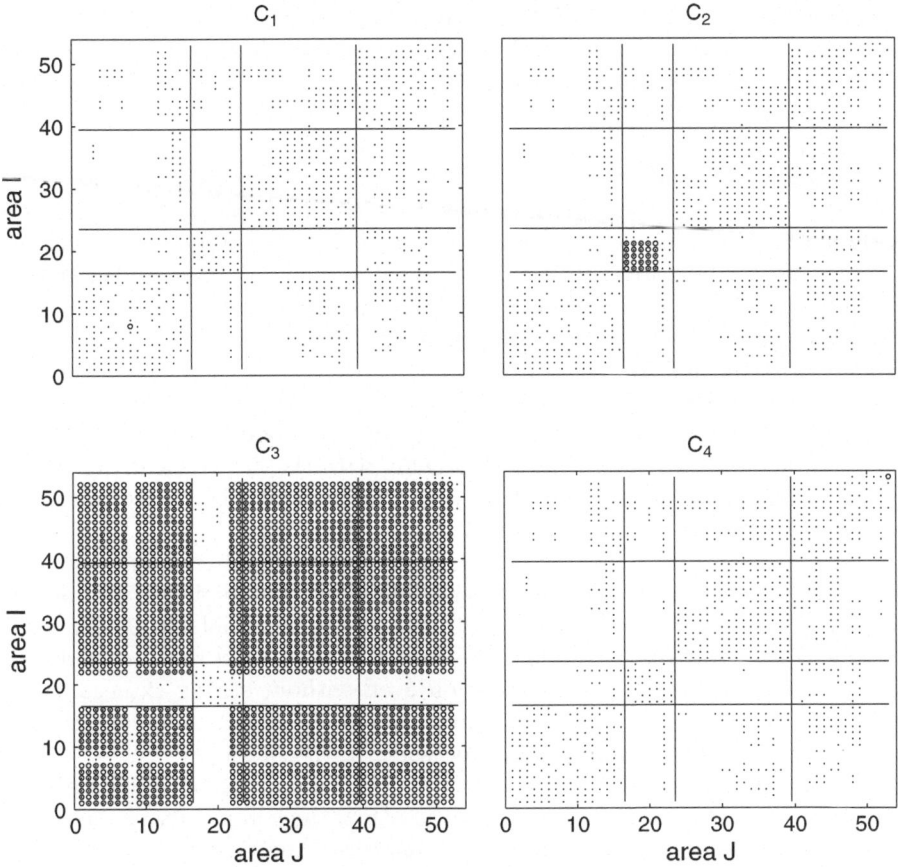

Fig. 5.25. Dynamical clusters ($\circ$) with strong coupling strength $g = 0.12$, overlaid on the underlying anatomical connections ($\cdot$).

5.7 Conclusion and Outlook

In this chapter, we discussed synchronization dynamics on complex networks. Firstly, we analyzed the relationship of some structural measures (such as degree and intensity of the nodes, fraction of random shortcuts etc.) in general network models to the synchronization behavior of the networks. Sections 5.2 and 5.3 considered the ideal case of complete synchronization allowing us to characterize the synchronizability of the network based solely on the spectral properties of the network. The main result is that the synchronizability, as measured by the ratio between the maximal and the minimal eigenvalues, is mainly determined by the maximal and minimal intensities in sufficiently random networks. In more general situations, the dynamics is perturbed away from the complete synchronization state, and we showed that the effective synchronization is hierarchically organized according to the distribution of the intensities. We also demonstrated that random long-range interactions in spatially extended networks (small-world networks) can induce different synchronization regimes, such as oscillation death and global synchronization with highly non-identical oscillators.

Secondly, we studied synchronization in a realistic network of cat cortical connectivity. We demonstrated that if well-defined oscillatory dynamics is assumed for the nodes (which represent cortical areas composed of large ensembles of neurons), for example, by employing a neural mass model, or by a subnetwork of rather strongly coupled neurons, the synchronization patterns can be understood by using general principles discussed in the first part of this chapter, i.e., the synchronization is mainly controlled by the global structural statistics (intensities) of the network. However, with weak coupling, the model with subnetworks displays biologically plausible dynamics and the synchronization pattern exhibits a close relationship with the hierarchically clustered organization in the network structure, i.e., the dynamics is mainly controlled by the local structures in the network.

Here, we have only focused on the highest structural level and modeled each large cortical area with one level of subnetwork and simple neuron dynamics. The maximal correlation (0.1–0.2) is low in a biologically plausible regime. The model displays a large region of frequent and regular spiking in the neurons and strong synchronization even without strong external influence. The model can be extended and improved in several ways, in order to address more realistic information processing in the brain:

(i) Biologically, a system of 10^5 neurons corresponding to a cubic millimeter of cortex is the minimal system size at which the complexity of the cortex can be represented, e.g., the number of synapses a neuron receives is 10^4 [52]. Thus, large sub-networks with other biologically realistic features, e.g., additional hierarchically clustered organization and more detailed spatial structure of neural circuits, should be considered. Such an extension is important for the modeling and simulation of experimentally observed hierarchical activity characterized by synchronization phenomena over a wide range of spatial and temporal scales.

(ii) Cortical neurons display rich dynamics which would require more subtle neuron models.

(iii) Biologically more realistic coupling by chemical synapses should be used and synaptic plasticity considered.

An extension of our model by additionally including the hierarchy of clustered structures reflecting the connectivity at the level of local neuronal circuits would allow localized and strong synchronization in some low-level clusters and naturally organize dynamics at higher scales. This will significantly broaden the biologically plausible regimes with stronger correlations, as observed experimentally [38].

Synchronization of distributed brain activity has been believed to be an important mechanism for neural information processing [6]. A carefully extended model could be used to investigate the relative contributions of network topology and task-related network activations to functional brain connectivity and information processing. The dynamics of the model could be then compared to the observed spread of activity in the cortex [53] and to the functional connectivity [38] at suitable spatio-temporal scales.

Simulations of such large, complex neural network models of the cortex and investigations of the relationship between network structure, dynamics organization and function of the system crossing various levels in the hierarchy would require significant developments both in neurophysics, in theory of dynamical complex networks and in algorithms of parallel computing [52].

Acknowledgements

The authors thank A.E. Motter and C.C. Hilgetag for helpful discussions.

References

1. See, e.g., reviews: S. H. Strogatz, Nature (London) **410**, 268 (2001); R. Albert and A.-L. Barabási, Rev. Mod. Phys. **74**, 47 (2002); S. Boccaletti et al., Phys. Rep. **424**, 175 (2006).
2. D. J. Watts and S. H. Strogatz, Nature (London) **393**, 440 (1998).
3. A.-L. Barabási and R. Albert, Science **286**, 509 (1999).
4. R. Milo, S. Shen-Orr, S. Itzkovitz, N. Kashtan, D. Chlkovskii and U. Alon, Science **298**, 824 (2002).
5. A partial list, e.g., P. M. Gade and C. K. Hu, Phys. Rev. E **62**, 6409 (2000); J. Jost and M. P. Joy, Phys. Rev. E **65**, 016201 (2001); M. Barahona and L. M. Pecora, Phys. Rev. Lett. **89**, 054101 (2002); A. E. Motter, C. S. Zhou and J. Kurths, Europhys. Lett. **69**, 334 (2005); Phys. Rev. E **71**, 016116 (2005); L. Donetti, P. I. Hurtado and M. A. Munoz, Phys. Rev. Lett. **95**, 188701 (2005); A. Arenas, A. Diáz-Guilera and C. J. Peréz-Vicentz, Phys. Rev. Lett. **96**, 114102 (2006).
6. E. Salinas and T. J. Sejnowski, Nature Neurosci. **2**, 539 (2001); P. Friés, Trends Cogn. Sci. **9**, 474 (2005); A. Schnitzler and J. Gross, Nature Neurosci. **6**, 285 (2005).
7. L. F. Lago-Fernández, R. Huerta, F. Corbacho and J. A. Sigüenza, Phys. Rev. Lett. **84**, 2758 (2000).
8. N. Masuda and K. Aihara, Biol. Cybern. **90**, 302 (2004).
9. X. Guardiola, A. Diaz-Guilera, M. Llas and C. J. Peréz, Phys. Rev. E **62**, 5565 (2000).
10. M. Timme, F. Wolf and T. Geisel, Phys. Rev. Lett. **92**, 074101 (2004); M. Denker, M. Timme, M. Diesmann, F. Wolf and T. Geisel, Phys. Rev. Lett. **92**, 074103 (2004); V. N Belykh, E. de Lange and M. Hasler, Phys. Rev. Lett. **94**, 188101 (2005).
11. H. Hong, M. Y. Choi and B. J. Kim, Phys. Rev. E **65**, 026139 (2002).
12. A. M. Batista, S. E. D. Pinto, R. L. Viana and S. R. Lopes, Physica A **322**, 118 (2003).

13. T. Nishikawa, A. E. Motter, Y.-C. Lai and F. C. Hoppensteadt, Phys. Rev. Lett. **91**, 014101 (2003).
14. F. Chung and L. Lu, Proc. Natl. Acad. Sci. U.S.A. **99**, 15879 (2002); R. Cohen and S. Havlin, Phys. Rev. Lett. **90**, 058701 (2003).
15. J. W. Scannell, G. A. P. C. Burns, C. C. Hilgetag, M. A. O'eil and M. P. Yong, Cereb. Cortex **9**, 277 (1999).
16. B. T. Grenfell, O. N. Bjornstad and J. Kappey, Nature (London) **414**, 716 (2001).
17. G. Korniss, M. A. Novotny, H. Guclu, Z. Toroczkai and P. A. Rikvold, Science **299**, 677 (2003).
18. A. Barrat, M. Barthélemy, R. Pastor-Satorras and A. Vespignani, Proc. Natl. Acad. Sci. U.S.A. **101**, 3747 (2004).
19. A. E. Motter, C. S. Zhou and J. Kurths, Europhys. Lett. **69**, 334 (2005); Phys. Rev. E **71**, 016116 (2005).
20. M. Chavez, D.-U. Hwang, A. Amann, H. G. E. Hentschel and S. Boccaletti, Phys. Rev. Lett. **94**, 218701 (2005).
21. C. S. Zhou, A.E. Motter and J. Kurths, Phys. Rev. Lett. **96**, 034101 (2006).
22. C. S. Zhou and J. Kurths, Phys. Rev. Lett. **96**, 164102 (2006).
23. L. M. Pecora and T. L. Carroll, Phys. Rev. Lett. **80**, 2109 (1998).
24. L. M. Pecora and M. Barahona, Chaos and Complexity Lett. **1**, 61 (2005).
25. F. Chung, L. Lu and V. Vu, Proc. Natl. Acad. Sci. U.S.A. **100**, 6313 (2003).
26. X. F. Wang, Int. J. Bifurcation Chaos Appl. Sci. Eng. **12**, 885 (2002).
27. J. Jost and M. P. Joy, Phys. Rev. E **65**, 016201 (2001).
28. S. Jalan and R. E. Amritkar, Phys. Rev. Lett. **90**, 014101 (2003).
29. S. N. Dorogovtsev and J. F. F. Mendes, Phys. Rev. E **62**, 1842 (2000).
30. M. E. J. Newman, S. H. Strogatz and D. J. Watts, Phys. Rev. E **64**, 026118 (2001).
31. C. S. Zhou and J. Kurths, Chaos **16**, 015104 (2006).
32. M. Rosenblum, A. Pikovsky and J. Kurths, Phys. Rev. Lett. **76**, 1804 (1996).
33. A. S. Pikovsky, M. Rosenblum and J. Kurths, *Synchronization – A universal concept in nonlinear sciences*, Cambridge University Press, 2001; S. Boccaletti, J. Kurths, G. Osipov, D. L. Valladares and C.S. Zhou, The Synchronization of Chaotic Systems, Phys. Rep. **366**, 1–101 (2002).
34. M. E. J. Newman, C. Moore and D. J. Watts, Phys. Rev. Lett. **84**, 3201 (2000).
35. G. V. Osipov, J. Kurths and C. S. Zhou, *Synchronization in Oscillatory Networks,* Spring, Berlin, 2007.
36. C. J. Stam and E. A. de Bruin, Hum. Brain Mapp. **22**, 97 (2004).
37. See a recent review: O. Sporns, D. R. Chialvo, M. Kaiser and C. C. Hilgetag, Trends Cogn. Sci. **8**, 418 (2004).
38. C. J. Stam, Neurosci. Lett. **355**, 25 (2004); V. M. Eguíluz, D. R. Chialvo, G. Cecchi, M. Baliki, and A. V. Apkarian, Phys. Rev. Lett. **94**, 018102 (2005); R. Salvador et al., Cereb. Cortex **15**, 1332 (2005).
39. O. Sporns and J. D. Zwi, Neuroinformatics **2**, 145 (2004).
40. C. C. Hilgetag and M. Kaiser, Neuroinformatics **2**, 353 (2004).
41. C. C. Hilgetag, G. A. Burns, M. A. O'Neill, J. W. Scannell and M. P. Young, Phil. Trans. R. Soc. Lond. B. **355**, 91 (2000).
42. M. E. J. Newman and M. Girvan, Phys. Rev. E. **69**, 026113 (2004).
43. F. H. Lopes da Silva, A. Hoeks, H. Smits and L. H. Zetterberg, Kybernetik **15**, 27 (1974).

44. F. Wendling, J. J. Bellanger, F. Bartolomei and P. Chauvel, Biol. Cybern. **83**, 367 (2000).
45. C. S. Zhou, L. Zemanová, G. Zamora, C. C. Hilgetag and J. Kurths, Phys. Rev. Lett. **97**, 238103 (2006).
46. L. Zemanová, C. S. Zhou, J. Kurths, Physica D **224**, 202 (2006).
47. G. Buzsaki, C. Geisler, D. A. Henze and X. J. Wang, Trends Neurosci. **27**, 186 (2004).
48. M. P. Young, Spat. Vis. **13**, 137 (2000).
49. R. FitzHugh, Biophys. J. **1**, 445 (1961).
50. E. Niedermeyer and F. Lopes da Silva, *Electroencephalography: Basic principles, clinical applications, and related fields*, Williams & Wilkins, 1993; R. Kandel, J. H., Schwartz, and T. M. Jessell, *Principles of Neural Science*, McGraw-Hill, 2000.
51. P. Kudela, P. J. Franaszczuk and G. K. Bergey, Biol. Cybern. **88**, 276 (2003).
52. A. Morrison, C. Mehring, T. Geisel, A. Aertsen and M. Diesmann, Neural Comput. **17**, 1776 (2005).
53. R. Kötter and F. T. Sommer, Phil. Trans. R. Soc. Lond. B **355**, 127 (2000).

6

Synchronization Analysis
of Neuronal Networks by Means
# of Recurrence Plots

André Bergner and Maria Carmen Romano, Jürgen Kurths and Marco Thiel

Nonlinear Dynamics Group, University of Potsdam
bergner@agnld.uni-potsdam.de

Summary. We present a method for synchronization analysis, that is able to handle large networks of interacting dynamical units. We focus on large networks with different topologies (random, small-world and scale-free) and neuronal dynamics at each node. We consider neurons that exhibit dynamics on two time scales, namely spiking and bursting behavior. The proposed method is able to distinguish between synchronization of spikes and synchronization of bursts, so that we analyze the synchronization of each time scale separately. We find for all network topologies that the synchronization of the bursts sets in for smaller coupling strengths than the synchronization of the spikes. Furthermore, we obtain an interesting behavior for the synchronization of the spikes dependent on the coupling strength: for small values of the coupling, the synchronization of the spikes increases, but for intermediate values of the coupling, the synchronization index of the spikes decreases. For larger values of the coupling strength, the synchronization index increases again until all the spikes synchronize.

6.1 Introduction

Networks are ubiquitous in nature, biology, technology and in the social sciences (see [1] and references therein). Much effort has been made to describe and characterize them in different fields of research. One key finding of these studies is that there are unifying principles underlying their behavior. In the past, two major approaches have been pursued to deal with networks. The first approach considers networks of regular topology, such as arrays or rings of coupled systems with nonlinear and complex dynamics on each node. The second approach concentrates on the topology of the network and sets aside the dynamics or at most considers a rather simple one at each node. Some of the prototypical types of network architectures that have been considered are random, small-world, scale-free and generalized random networks [2].

Recently, the study of complex dynamics on the nodes has been extended from regular to more complex architectures [3]. However, in most previous work, each node is still considered to be a phase oscillator (system with one predominant time scale), often pulse-coupled to each other. Much is left, however, to understand about network behavior with more realistic complex dynamics on the nodes of networks of complex architecture, such as chaotic and stochastic dynamics, which is found in many real application systems, such as in neural networks. The influence of the topology of the network on the dynamical properties of the complex systems is currently being investigated in the context of synchronisation [4–6].

Synchronization of complex systems has been intensively studied during the last years [7] and it has been found to be present in numerous natural and engineering systems [8]. Chaotic systems defy synchronization due to their sensitivity to slight differences in initial conditions. However, it has been demonstrated that these kind of systems are able to synchronize. In the case of two interacting non-identical chaotic systems (which is more likely to occur in nature than if they were identical), several types of synchronization might occur, dependent on the coupling strength between the systems. For rather weak coupling strength, phase synchronisation (PS) might set in. In this case, the phases and frequencies of the complex systems are locked, i.e. $|\phi_1(t) - \phi_2(t)| <$ const. and $\omega_1 \approx \omega_2$, whereas their amplitudes remain uncorrelated. If the coupling strength is further increased, a stronger relationship between the interacting systems might occur, namely generalized synchronization (GS). In this case, there is a functional relationship between both systems. Finally, for very strong coupling, both systems can become almost completely synchronized. Then, their trajectories evolve almost identically in time [7].

In the case of phase synchronization, the first step in the analysis is to determine the phases $\phi_1(t)$ and $\phi_2(t)$ of the two interacting systems with respect to the time t. If the chaotic systems have mainly one characteristic time scale, i.e. a predominant peak in the power spectrum, the phase can be estimated as the angle of rotation around one center of the projection of the trajectory on an appropriate plane. Alternatively, the analytical signal approach can be used [9]. However, for most of the complex systems found in nature, there is more than one characteristic time scale [10]. Hence, the approaches mentioned above to estimate the phase are not appropriate. Recently, a new method, based on the recurrence properties of the interacting systems [11], has been introduced to overcome this problem. By means of this technique, it is possible to analyze systems with a rather broad spectrum, as well as systems strongly contaminated by noise or subjected to non-stationarity [12].

In this chapter, we extend the recurrence based technique for phase synchronization analysis to systems with two predominant time scales, so that it is possible to obtain one synchronization index for each time scale. Moreover, we apply this method to large networks of different architectures with neuronal dynamics on their nodes.

The outline of this chapter is as follows: in Sect. 6.2, we introduce the concept of recurrence, as well as the synchronization index based on the recurrence properties of the system. In Sect. 6.2.2, we present the method to analyze the synchronization for two different time scales separately. In Sect. 6.3, we apply the method to complex networks of neurons and present the obtained results.

6.2 Phase Synchronization by Means of Recurrences

First, we show the problem of defining the phase in systems with rather broad power spectrum by using the paradigmatic system of two coupled non-identical Rössler oscillators:

$$
\begin{aligned}
\dot{x}_{1,2} &= -\omega_{1,2} y_{1,2} - z_{1,2} \\
\dot{y}_{1,2} &= \omega_{1,2} x_{1,2} + a y_{1,2} + \mu(y_{2,1} - y_{1,2}) \\
\dot{z}_{1,2} &= 0.1 + z_{1,2}(x_{1,2} - 8.5) \,,
\end{aligned}
\tag{6.1}
$$

where μ is the coupling strength and $\omega_{1,2}$ determine the mean intrinsic frequency of the (uncoupled) oscillators in the case of phase coherent attractors. In our simulations, we take $\omega_1 = 0.98$ and $\omega_2 = 1.02$. The parameter $a \in [0.15, 0.3]$ governs the topology of the chaotic attractor. When a is below a critical value a_c ($a_c \approx 0.186$ for $\omega_1 = 0.98$ and $a_c \approx 0.195$ for $\omega_2 = 1.02$), the chaotic trajectories always cycle around the unstable fixed point $(x_0, y_0) \approx (0, 0)$ in the (x, y) subspace (Fig. 6.1(a)). In this case, the rotation angle

$$
\phi = \arctan \frac{y}{x}
\tag{6.2}
$$

can be defined as the phase which increases almost uniformly. The oscillator has coherent phase dynamics, i.e. the diffusion of the phase dynamics is very low $(10^{-5}$–$10^{-4})$. In this case, other phase definitions, e.g. based on the Hilbert transform or on the Poincaré section, yield equivalent results [9]. However, beyond the critical value a_c, the trajectories no longer completely cycle around (x_0, y_0) – the attractor becomes the so-called funnel attractor. Such earlier returns in the funnel attractor happen more frequently with increasing a (Fig. 6.1(b)). It is clear that for the funnel attractors, usual (and rather simple) definitions of phase, such as (6.2), are no longer applicable [9].

Another problematic case arises if the systems under consideration have two predominant time scales, which is common in many real systems, e.g. neurons with spiking and bursting dynamics. In such cases, the definition of the phase given by (6.2) is also not appropriate.

Figure 6.2 shows these problems with the time series of a Hindmarsh-Rose neuron.[1]

Rosenblum et al. [13] have proposed the use of an ensemble of phase coherent oscillators that is driven by a non-phase-coherent oscillator in order to estimate the frequency of the latter and hence detect PS in such kind

---

[1] For the definition of the Hindmarsh-Rose neuron see Sect. 6.3 and also Chap. 1.

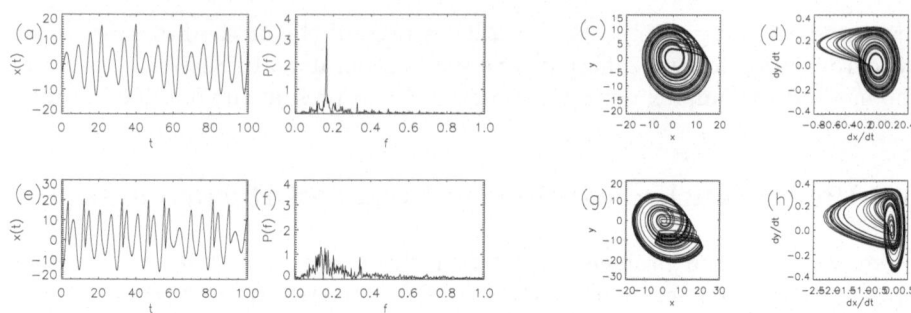

Fig. 6.1. (**a,e**): Segment of the x_1-component of the trajectory of the Rössler systems (6.1); (**b,f**): periodogram of the x-component of the trajectory; (**c,g**): projection of the attractor onto the (x, y) plane; (**d,h**): projection onto the $(\dot{x}, \dot{y})$ plane. Upper panel (**a,b,c,d**) computed for $a = 0.16$ and lower panel (**e,f,g,h**) computed for $a = 0.2925$

of systems. However, depending on the component one uses to couple the non-phase-coherent oscillator to the coherent ones, the result of the obtained frequency can be different.

Furthermore, Osipov et al. [10] have proposed another approach which is based on the general idea of the curvature of an arbitrary curve. For any two-dimensional curve $r = (u, v)$ they propose that the phase ϕ be defined as $\phi = \arctan \frac{\dot{v}}{\dot{u}}$. By means of this definition, the projection $\dot{r} = (\dot{u}, \dot{v})$ is a curve cycling monotonically around a certain point.

This definition of ϕ holds in general for any dynamical system if the projection of the phase trajectory onto some plane is a curve with a positive curvature. This approach is applicable to a large variety of chaotic oscillators, such as the Lorenz system [14], the Chua circuit [15] or the model of an ideal four-level laser with periodic pump modulation [16].

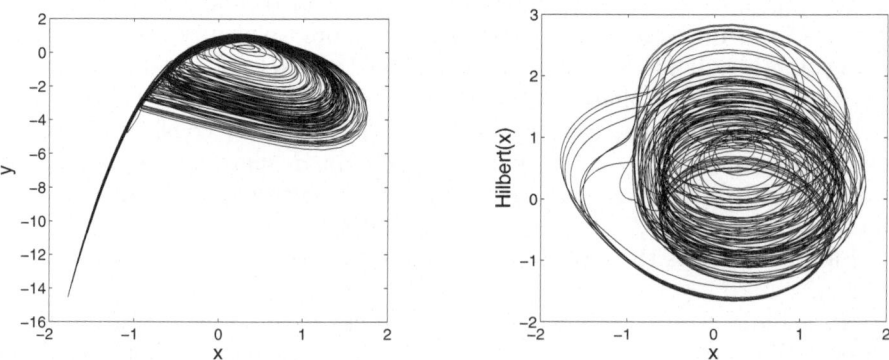

Fig. 6.2. Projection of the x-y-plane; (**a**) and the plot of the Hilbert Transform of x versus x; (**b**) for the Hindmarsh-Rose-neuron

This is clear for phase-coherent as well as funnel attractors in the Rössler oscillator. Here, projections of chaotic trajectories on the plane $(\dot{x}, \dot{y})$ always rotate around the origin (Figs. 6.1(c) and (d)) and the phase can be defined as

$$\phi = \arctan \frac{\dot{y}}{\dot{x}} . \qquad (6.3)$$

Although this approach works well in non-phase-coherent model systems, we have to consider that one is often confronted with the computation of the phase in experimental time series, which are usually corrupted by noise. In this case, some difficulties may appear when computing the phase given in (6.3), because derivatives are involved in its definition.

6.2.1 Cross-correlation of the Probability of Recurrence

We use a different approach, based on recurrences in phase space, to detect PS indirectly. We define a recurrence of the trajectory of a dynamical system $\{x_i\}_{i=1}^N$ in the following way: we say that the trajectory has returned at time $t=j$ to the former point in phase space visited at $t=i$ if

$$R_{i,j}^{(\varepsilon)} = \Theta(\varepsilon - \|x_i - x_j\|) = 1 , \qquad (6.4)$$

where ε is a pre-defined threshold and $\Theta(\cdot)$ is the Heaviside function. A "1" in the matrix at i,j means that x_i and x_j are neighboring, a "0" that they are not. The black and white representation of this binary matrix is called a recurrence plot (RP). This method has been intensively studied in the last years [11]: different measures of complexity have been proposed based on the structures obtained in the RP and have found numerous applications for example, in physiology and earth science [17]. Furthermore, it has been even shown that some dynamical invariants can be estimated by means of the recurrence structures [18].

Based on this definition of recurrence, one is able to tackle the problem of performing a synchronization analysis in the case of non-phase-coherent systems. We avoid the direct definition of the phase and use instead the recurrence properties of the systems in the following way: the probability $P^{(\varepsilon)}(\tau)$ that the system returns to the neighborhood of a former point x_i of the trajectory[2] after τ time steps can be estimated as follows:

$$P^{(\varepsilon)}(\tau) = \frac{1}{N-\tau} \sum_{i=1}^{N-\tau} \Theta(\varepsilon - \|x_i - x_{i+\tau}\|) = \frac{1}{N-\tau} \sum_{i=1}^{N-\tau} R_{i,i+\tau}^{(\varepsilon)} . \qquad (6.5)$$

This function can be regarded as a generalized autocorrelation function, as it also describes higher order correlations between the points of the trajectory

---

[2] The neighborhood is defined as a box of size ε centered at x_i, as we use the maximum norm.

dependent on the time delay τ. A further advantage with respect to the linear autocorrelation function is that $P^{(\varepsilon)}(\tau)$ is defined for a trajectory in phase space and not only for a single observable of the system's trajectory.

For a periodic system with period T, it can be easily shown that $P^{(\varepsilon)}(\tau)=1$ if $\tau = T$ and $P^{(\varepsilon)}(\tau) = 0$ otherwise. For coherent chaotic oscillators, such as (6.1) for $a=0.16$, $P^{(\varepsilon)}(\tau)$ has well-expressed local maxima at multiples of the mean period, but the probability of recurrence after one or more rotations around the fixed point is less than one (Fig. 6.3(b,d)).

Analyzing the probability of recurrence, it is possible to detect PS for non-phase-coherent oscillators as well. This approach is based on the following idea: Originally, a phase ϕ is assigned to a periodic trajectory $\boldsymbol{x}$ in phase space, by projecting the trajectory onto a plane and choosing an origin, around which the trajectory oscillates all the time. Then, an increment of 2π is assigned to ϕ when the point of the trajectory has returned to its starting position, i.e. when $\|\boldsymbol{x}(t + T) - \boldsymbol{x}(t)\| = 0$. Analogously to the case of a periodic system,

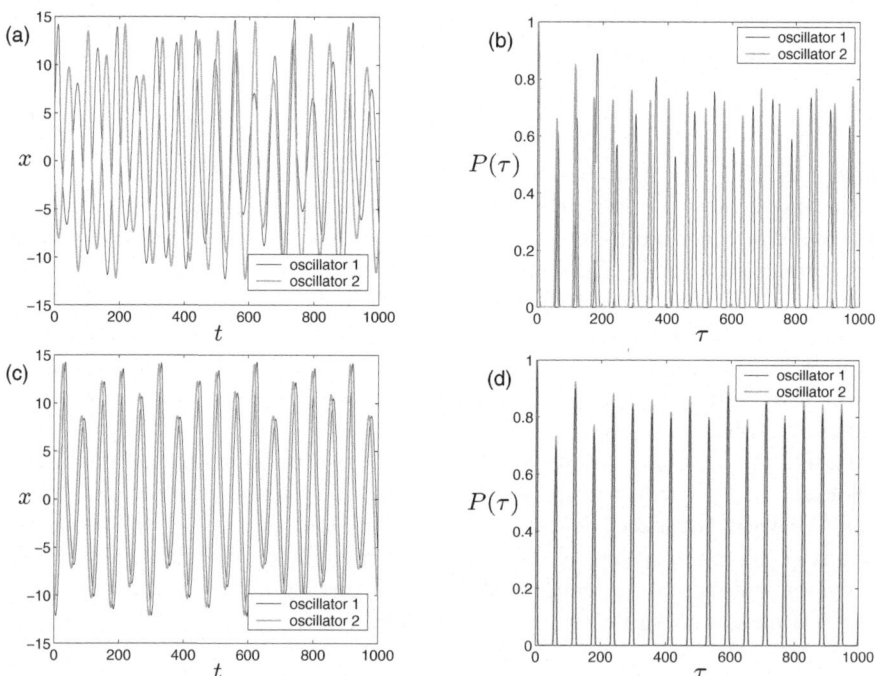

Fig. 6.3. Time series (**a** and **c**) and the probability of recurrence (**b** and **d**) of the Rössler system with parameters $a = 0.15$, $b = 0.2$, $c = 8.5$, $\omega_1 = 1$ and $\omega_2 = 1.05$. The coupling strength for the non-PS case (**a** and **b**) is $\mu_{\mathrm{nonPS}}=0.01$ and $\mu_{\mathrm{PS}}=0.07$ for the PS case (**c** and **d**), respectively. The values for CPR that have been calculated are $\mathrm{CPR}_{\mathrm{nonPS}} = 0.0102$ and $\mathrm{CPR}_{\mathrm{PS}} = 0.9995$. The figures show clearly how the peaks drift apart from each other in the absence of PS and coincide in the case of PS

we can assign an increment of 2π to ϕ for a complex non-periodic trajectory $\boldsymbol{x}(t)$ when $\|\boldsymbol{x}(t+T) - \boldsymbol{x}(t)\| \sim 0$, or equivalently when $\|\boldsymbol{x}(t+T) - \boldsymbol{x}(t)\| < \varepsilon$, where ε is a predefined threshold. That means that a recurrence $R_{t,t+\tau}^{(\varepsilon)} = 1$ can be interpreted as an increment of 2π of the phase in the time interval τ.[3]

$P^{(\varepsilon)}(\tau)$ can be viewed as a statistical measure of how often ϕ in the original phase space has increased by 2π or multiples of 2π within the time interval τ. If two systems are in PS, on the average, the phases of both systems increase by $2\pi k$, with k a natural number, within the same time interval τ. Hence, looking at the coincidence of the positions of the maxima of $P^{(\varepsilon)}(\tau)$ for both systems, we can quantitatively identify PS (from now on, we omit (ε) in $P^{(\varepsilon)}(\tau)$ to simplify the notation). The proposed algorithm then consists of two steps:

- Compute $P_{1,2}(\tau)$ of both systems based on (6.5).
- Compute the cross-correlation coefficient between $P_1(\tau)$ and $P_2(\tau)$ (Correlation between probabilities of recurrence)

$$\mathrm{CPR}_{1,2} = \frac{\langle \bar{P}_1(\tau)\bar{P}_2(\tau) \rangle_\tau}{\sigma_1 \sigma_2} , \qquad (6.6)$$

where the bar above $\bar{P}_{1,2}$ denotes that the mean value has been subtracted and σ_1 and σ_2 are the standard deviations of $P_1(\tau)$ and $P_2(\tau)$, respectively.

If both systems are in PS, the probability of recurrence is maximal simultaneously and $\mathrm{CPR}_{1,2} \approx 1$. In contrast, if the systems are not in PS, the maxima of the probability of recurrence do not occur jointly and we would expect low values of $\mathrm{CPR}_{1,2}$.

In Figs. 6.3 and 6.4, we illustrate the performance of the method with two examples of the Rössler system.

6.2.2 The Problem of Separating the Time Scales

As already mentioned, neurons can exhibit dynamics on several distinct time scales (spiking and bursting) and are also able to synchronize on both scales separately. To perform a synchronisation analysis of such a system, one has to segregate the two scales of each other. Figure 6.5 shows the RP of a Hindmarsh-Rose neuron.[4] In Fig. 6.5(a), the structures that emerged from the recurrence of the bursts can be identified quite clearly, namely the "swelling diagonal lines". In Fig. 6.5(b), one of those "swellings" is presented magnified. Here, the recurrences of the spike dynamics can be noticed as diagonal lines on a smaller scale in the RP.

Separating the scales is a non-trivial task. Filtering the time series could be one approach, but this is not recommended as the attractor of the filtered time series will be distorted, which will change the recurrence behavior.

---

[3] This can be considered as an alternative definition of the phase to (6.2) and (6.3).

[4] For the definition of the Hindmarsh-Rose neuron see Sect. 6.3 and Chap. 1, again.

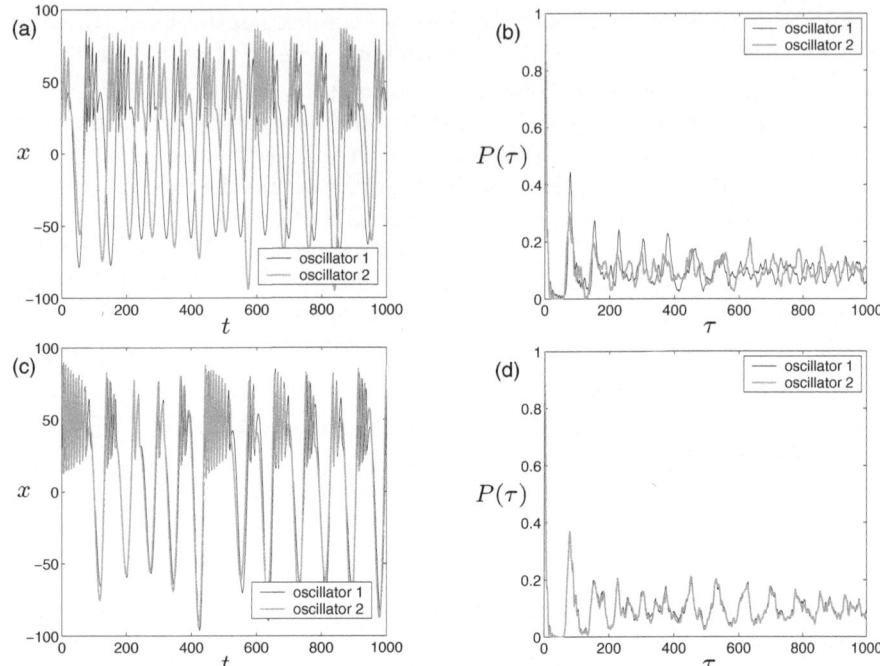

Fig. 6.4. Time series (**a** and **c**) and the probability of recurrence (**b** and **d**) of the Rössler system in a bursting regime with the parameters $a=0.38$, $b=0.4$, $c=50$, $\omega_1 = 1$ and $\omega_2 = 1.05$. The coupling strength for the non PS case (**a** and **b**) is $\mu_{\mathrm{nonPS}} = 0.005$ and $\mu_{\mathrm{PS}} = 0.23$ for the PS case (**c** and **d**), respectively. The values for CPR that have been calculated are $\mathrm{CPR}_{\mathrm{nonPS}} = 0.0258$ and $\mathrm{CPR}_{\mathrm{PS}} = 0.9684$. Clearly, the peaks do not coincide in the non-PS case and do so in the presence of PS. This example shows quite well that the algorithm is able to detect PS for systems with a very complicated flow of the phase

Therefore, separating the time scales after calculating the RP or $P(\tau)$ is a better approach. We separate the time scales in two ways: The first one requires the choice of an appropriate recurrence rate and the second one is the application of some filter to $P(\tau)$.

In Fig. 6.6(a) the recurrence probability $P(\tau)$ of a Hindmarsh-Rose neuron is presented. The large peaks correspond to the recurrence of the bursts. The arrows indicate the smaller peaks generated by the recurrence of the spikes. There are many methods for separating both scales, e.g. wavelets, etc. In this analysis, an infinite impulse response (IIR) filter has been used, which can be implemented easily by simple difference equations.

Figure 6.6(b) shows the highpass filtered $P(\tau)$. The cutoff has been chosen to be $0.2 \times$ sampling rate. The broad peaks, originated by the burst, are filtered out and the smaller peaks corresponding to the spike recurrence become clearer. Note that the filtered $P(\tau)$ cannot be interpreted as a probability of

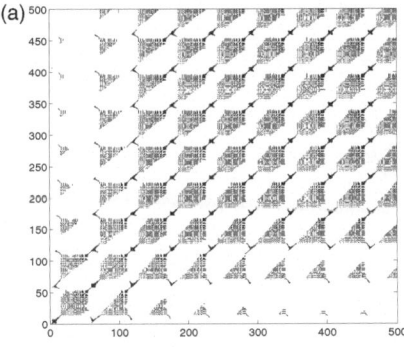 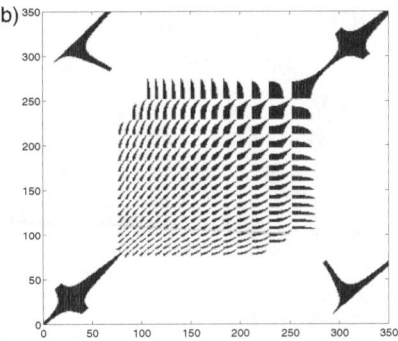

Fig. 6.5. Recurrence plot of the time series of a Hindmarsh-Rose neuron on a large scale (**a**) and zoomed in to show a small scale features (**b**)

recurrence any more, since it also assumes negative values. However, it still captures all the relevant information about the recurrence of the spiking dynamics. Thus, a separate synchronization analysis of the spike scale can now be accomplished by computing the index CPR of the filtered functions.

The recurrence rate is the parameter that specifies the number of black points in the RP and determines the threshold ε in (6.4). This parameter also influences the patterns obtained in the recurrence plot. Hence, by varying the recurrence rate, we can enhance or suppress certain information.

Figure 6.7(a) shows the RP of a Hindmarsh-Rose neuron time series, computed for a high recurrence rate of 0.5. Comparing this plot with the one in Fig. 6.5(a), it can be observed that the shorter lines originating from the recurrence of the spikes are "smeared out" The corresponding probability of recurrence $P(\tau)$ in Fig. 6.7(b) shows only the oscillations that are caused by

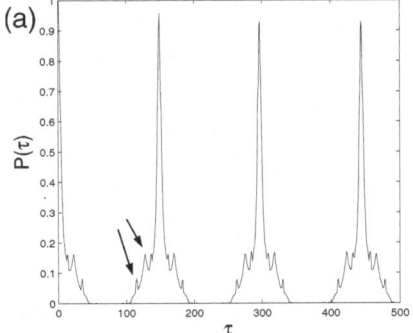 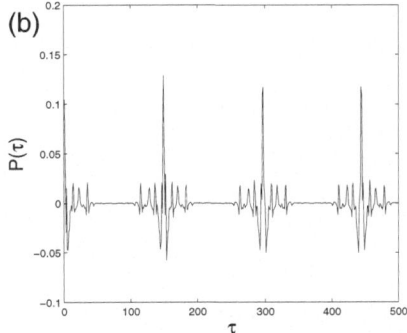

Fig. 6.6. Probability of recurrence $P(\tau)$ for the Hindmarsh-Rose neuron: (**a**) The original and; (**b**) highpass filtered. The arrows indicate the features created by the recurrence of the spikes

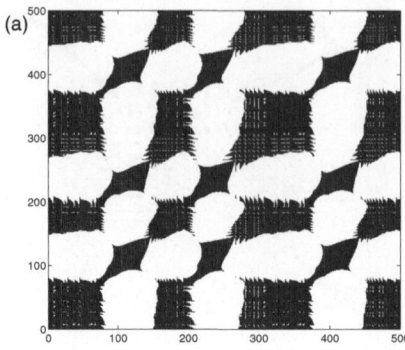 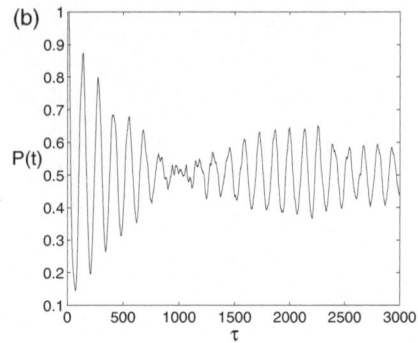

Fig. 6.7. RP with $RR = 0.5$; (**a**) and corresponding $P(\tau)$; (**b**) for an exemplary Hindmarsh-Rose neuron

the recurrence of the bursts. Consequently, the recurrence rate can be used to analyze the synchronization of the slow time scale (bursts), since the influence of the fast scale is automatically removed.

Analogously, choosing a rather low value for the recurrence rate causes the fine structures of the spike recurrences to appear more clearly. Therefore, it is advisable to use a rather low recurrence rate to analyze the synchronization of the spikes. In Fig. 6.8, the RP and the corresponding high-pass filtered $P(\tau)$ are presented. This example demonstrates quite well, how the large peaks, which are usually created by the recurrence of the bursts, are suppressed, so that the recurrence of the spikes is clearer than for higher values of the recurrence rate.

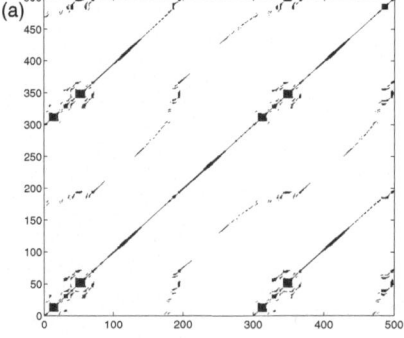 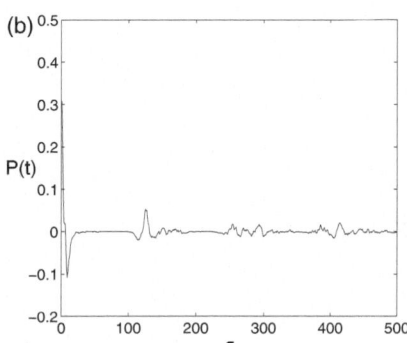

Fig. 6.8. RP with $RR = 0.05$: (**a**) and corresponding highpass filtered $P(\tau)$; (**b**) for an exemplary Hindmarsh-Rose neuron

A Few Notes on the Parameters

The RP based method has several parameters that need to be chosen in an appropriate way. These parameters are the already discussed recurrence rate and cutoff frequency of the filter, the averaging length N in (6.5), and the maximum recurrence time $\tau_{\max}$ when calculating CPR.

On the one hand, small values of N and $\tau_{\max}$ are desirable, such that the analysis can operate as locally as possible and with as small as possible computational cost. On the other hand, the values cannot be too small, since the analyses requires averaging and thus needs a large number of points for a correct calculation. Therefore, one has to determine the minimum values of N and $\tau_{\max}$ to serve both requirements. This can be done by calculating CPR for different values of these parameters. For large values, one can expect some kind of asymptotic behavior.

6.3 Application of the Algorithm

In this section, we present a few results that have been obtained by applying the proposed algorithm to networks of coupled neurons with different topologies. The neuron model that has been used is a (modified) four-dimensional Hindmarsh-Rose system (for details, see [19] and [20]),

$$
\begin{aligned}
\dot{x}_n &= \omega_{\text{fast},n}(y_n + 3x_n^2 - x_n^3 - 0.99z_n + I_n) + \mu \sum_{m=1}^{N} A_{nm}(x_m - x_n) \\
\dot{y}_n &= \omega_{\text{fast},n}(1.01 - y_n - 5.0128x_n^2 - 0.0278w_n) \\
\dot{z}_n &= \omega_{\text{slow1},n}(-z_n + 3.966(x_n + 1.605)) \\
\dot{w}_n &= \omega_{\text{slow2},n}(-0.9573w_n + 3(y_n + 1.619)),
\end{aligned}
\tag{6.7}
$$

where x_n is the membrane potential, and y_n, z_n, and w_n represent inner degrees of freedom of neuron n, with $n = 1, \ldots, N$. Whereas y_n is responsible for the fast dynamics of the spikes, z_n and w_n represent the slow dynamics of the bursts. I_n is the external input current of neuron n, $\omega_{\text{fast},n}$ determines the firing rate, and $\omega_{\text{slow1},n}$ and $\omega_{\text{slow2},n}$ determine the duration of the bursts. The neurons are electrically coupled, while the coupling topology of the neurons is given by the adjacency matrix A_{nm} (see Chap. 3). The parameter μ is the coupling strength of the whole network.

6.3.1 Analysis of Two Coupled Neurons

First, we apply the algorithm to a pair of coupled Hindmarsh-Rose neurons. We consider different parameter sets for the two neurons (see Table 6.1), so that we have three possibilities for the dynamical regime of the neurons: (i) both neurons in regular bursting regime with different frequencies, (ii) both neurons in chaotic bursting regime with different frequencies, and (iii) one neuron in spiking regime and one in regular bursting regime, both neurons with the same frequencies.

Table 6.1. A list of parameters in the examined pair of Hindmarsh-Rose neurons

	$w_{\text{slow}1,1}$	$w_{\text{slow}2,1}$	$w_{\text{fast}1}$	I_1	$w_{\text{slow}1,2}$	$w_{\text{slow}2,2}$	$w_{\text{fast}2}$	I_2
regular bursting	0.0015	0.019	1.1	3.0	0.0018	0.0012	0.9	2.9
chaotic bursting	0.0050	0.0010	1.1	3.1	0.0022	0.0007	0.9	3.1
one bursting, one spiking	0.0015	0.0009	1.0	5.0	0.0015	0.0009	1.0	2.5

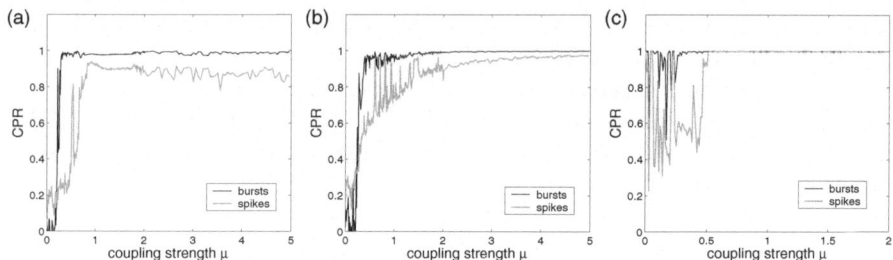

Fig. 6.9. $\text{CPR}^{\text{bursts}}$ and $\text{CPR}^{\text{spikes}}$ vs. coupling strength μ for a pair of Hindmarsh-Rose neurons with parameters according to Table 6.1: (**a**) regular bursting (**b**) chaotic bursting; (**c**) one spiking, one regular bursting

Then, we compute the synchronisation indices $\text{CPR}^{\text{bursts}}$ and $\text{CPR}^{\text{spikes}}$ for each case dependent on the coupling strength (see Fig. 6.9). For all three cases the spikes need higher coupling strengths to become phase synchronized than the bursts. This result is in good accordance with [21].

6.3.2 Analysis of Networks of Neurons

Different network topologies (random, small-world and scale-free) with Hindmarsh-Rose neurons at each node have been analyzed. Each network had $N = 200$ nodes and an average degree $\langle d \rangle$ of 10. The parameters of the neurons have been chosen as follows: $I_n \in \mathcal{N}(3.1, 0.05)$ (chaotic bursting regime), $\omega_{\text{fast},n} \in \mathcal{N}(1, 0.05)$, $\omega_{\text{slow}1,n} \in \mathcal{N}(0.002, 0.0005)$, and $\omega_{\text{slow},n} = 0.001$, where $\mathcal{N}(\tilde{\mu}, \sigma)$ denotes a Gaussian normal distribution with mean $\tilde{\mu}$ and variance σ. The coupling strength has been chosen as $\mu = g / \langle d \rangle$. The synchronization indices $\text{CPR}^{\text{bursts}}$ and $\text{CPR}^{\text{spikes}}$ have been calculated for each pair of nodes from the networks for increasing values of the coupling parameter g. Thus, we obtain two matrices $(\text{CPR}^{\text{bursts}}_{\text{nm}})$ and $(\text{CPR}^{\text{spikes}}_{\text{nm}})$, where $n, m = 1, \ldots, 200$ indicate the nodes.

In Fig. 6.10, we present a few snapshots of those CPR-matrices for different values of the coupling strength g for the scale free network. We have found that with an increasing coupling strength, the hubs (nodes with largest degree, see Chap. 3 for details) will synchronize first, while the rest of the nodes need a higher coupling strength to become synchronized. This is in good accordance with [22], where this has been shown for a scale free network of

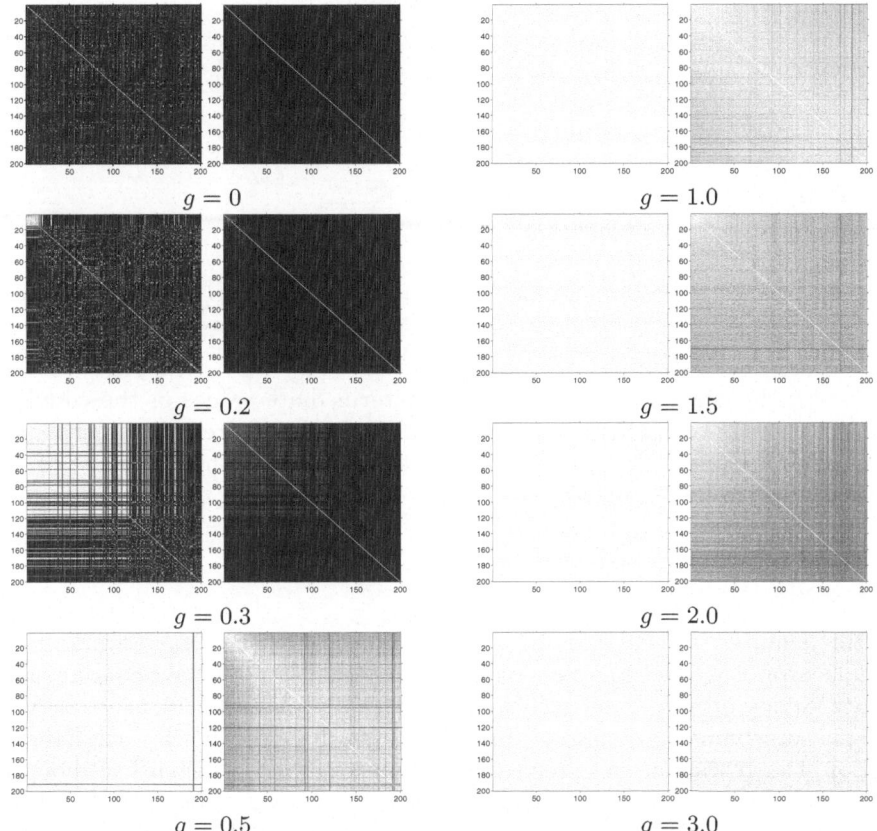

$g = 0$ $g = 1.0$

$g = 0.2$ $g = 1.5$

$g = 0.3$ $g = 2.0$

$g = 0.5$ $g = 3.0$

Fig. 6.10. Several snapshots of the CPR matrix of a network of 200 Hindmarsh-Rose neurons are presented for different coupling strengths. The left hand plot of each pair corresponds to the bursts, the right hand one to the spikes, respectively. Several phenomena stand out: 1. the hubs synchronize first, "attracting" the remaining nodes when the coupling increases further; 2. the spikes synchronize for a higher coupling strength than the bursts and; 3. there is a collapse of the spike synchronization in a certain domain of the coupling strength

Rössler oscillators. Furthermore, we have found for all three networks, as in the case of two coupled neurons, that the synchronization of the spikes sets in for higher values of the coupling strength than for the bursts.

To quantify the degree of phase synchronization of the whole network, we count the number of values in (CPR_{nm}) that are above a certain threshold and we call this number "area of synchronization". The threshold has been chosen as 0.8. In Fig. 6.11, those areas of synchronization are plotted versus the coupling strength.

An interesting result can be observed in the plot of the area of synchronization, as well in the snapshots of the CPR-matrices: there is a collapse

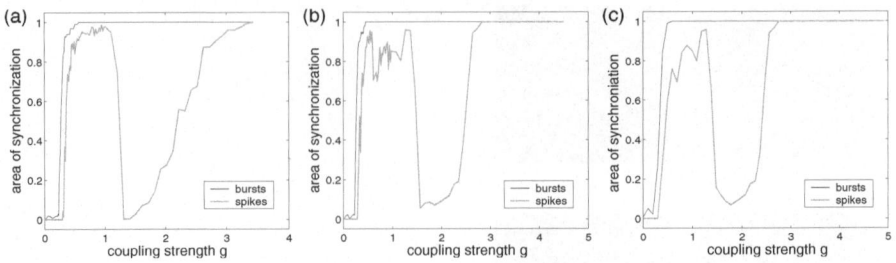

Fig. 6.11. Area of synchronization for; (**a**) random network; (**b**) small-world network; and (**c**) scale-free network for bursts and spikes, respectively

of the synchronization of the spikes for intermediate values of the coupling strength. In contrast, the synchronization of the bursts remains unchanged. This could be due to a change of the dynamics, namely the coherence of the oscillators with increasing values of the coupling strength g.

6.4 Conclusions

In this chapter, we have analyzed phase synchronization in networks with complex topology and complex dynamics. In particular, we have concentrated on dynamics on two time scales, as is typically observed in neurons with spiking and bursting dynamics. In order to analyze the synchronization behavior of such systems we extended an existing method, which is based on the concept of recurrence [11], to treat the two time scales separately. We have applied the proposed method to complex networks of Hindmarsh-Rose neurons. Our results are in accordance with [21], where it has been shown that the spikes need higher values of the coupling strength than the bursts in order to phase synchronize. Moreover, we have found that in a scale-free network of Hindmarsh-Rose neurons, the hubs synchronize first with increasing coupling strength, while the rest of the nodes need a higher coupling to synchronize, as has been reported in [22] for a scale-free network of Rössler oscillators. In addition, the most interesting result of our analysis is that we have found a collapse in the synchronization of the spikes in those complex networks for an intermediate coupling strength. This effect will be discussed in detail in a forthcoming paper.

References

1. S. H. Strogatz, Nature **410**, 268, 2001; M. E. J. Newman, SIAM Rev. **45**, 167, 2003; S. Boccaletti et al., Phys. Rep. **424**, 175, 2006; R. Albert, A.-L. Barabási, *Statistical mechanics of complex networks*, Rev. Mod. Phys., **74**, 47–97, 2002.
2. P. Erdös, and A. Rényi, Publ. Math. Inst. Hung. Acad. Sci. **5**, 17, 1960; D. J. Watts, and S. H. Strogatz, Nature **393**, 440,1998;

L. Barabási, and R. Albert, Science **286**, 509, 1999;
M. Molloy, and B. Reed, Random Struct. Algorithms **6**, 161, 1995.
3. L. Donetti et al., Phys. Rev. Lett. **95**, 188701, 2005.
4. Y. Moreno, and A. F. Pacheco, Europhys. Lett. **68** (**4**), 603, 2004;
J. G. Restrepo et al., Phys. Rev. E **71**, 036151, 2005.
5. F. M. Atay et al., Phys. Rev. Lett. **92** (**14**), 144101, 2004; W. Lu, and T. Chen,
Physica D **198**, 148, 2004;
Y. Jiang et al., Phys. Rev. E **68**, 065201(R), 2003.
6. C. Zhou, and J. Kurths, Chaos **16**, 015104 2006.
7. N. F. Rulkov et al., Phys. Rev. E, **51** (2), 980, 1995;
L. Kocarev, and U. Parlitz, Phys. Rev. Lett. **76** (11), 1816, 1996;
S. Boccaletti et al., Phys. Rep. **366**, 1, 2002.
8. B. Blasius et al., Nature **399**, 354, 1999; P. Tass et al., Phys. Rev. Lett. **81**
(**15**), 3291, 1998;
M. Rosenblum et al., Phys. Rev. E. **65**, 041909, 2002;
D. J. DeShazer et al., Phys. Rev. Lett. **87** (**4**), 044101, 2001.
9. A. Pikovsky, M. Rosenblum, J. Kurths, *Synchronization - A universal concept
in nonlinear science*, Cambridge University Press, 2001.
10. G. V. Osipov, B. Hu, C. Zhou, M. V. Ivanchenko, and J. Kurths, Phys. Rev.
Lett. **91**, 024101, 2003.
11. N. Marwan, M. C. Romano, M. Thiel, and J. Kurths, Phys. Rep. **438**, 237,
2007.
12. M. C. Romano, M. Thiel, J. Kurths, I. Z. Kiss, J. L. Hudson, *Detection of synchronization for non-phase-coherent and non-stationary data*, Europhys. Lett.,
71 (3), 466, 2005.
13. M. G. Rosenblum, A. S. Pikovsky, J. Kurths, G. V. Osipov, I. Z. Kiss, and
J. L. Hudson, Phys. Rev. Lett. **89**, 264102, 2002.
14. C. Sparrow, *The Lorenz equations: Bifurcations, chaos, and strange attractors*,
Springer-Verlag, Berlin, 1982.
15. R. N. Madan, *Chua circuit: A paradigm for chaos*, World Scientific, Singapore,
1993.
16. W. Lauterborn, T. Kurz, and M. Wiesenfeldt, *Coherent Optics. Fundamentals
and Applications*, Springer-Verlag, Berlin, Heidelberg, New York, 1993.
17. C. L. Weber Jr., and J. P. Zbilut, J. Appl. Physiology **76** (**2**) 965, 1994;
N. Marwan, N. Wessel, U. Meyerfeldt, A. Schirdewan, and J. Kurths, Phys.
Rev. E **66** (**2**), 026702, 2002;
N. Marwan, and J. Kurths, Phys. Lett. A **302** (**5–6**), 299, 2002; M. Thiel et al.,
Physica D **171**, 138, 2002.
18. M. Thiel, M. C. Romano, P. Read, J. Kurths, *Estimation of dynamical invariants without embedding by recurrence plots*, Chaos, **14** (2), 234–243, 2004.
19. J. L. Hindmarsh, R. M. Rose, *A model of neuronal bursting using three coupled
first order differential equations*, Proc. Roy. Soc. Lond. B **221**, 87–102, 1984.
20. R. D. Pinto, P. Varona, A. R. Volkovskii, A. Szücs, H. D. I. Abarbanel, M. I.
Rabinovich *Synchronous behavior of two coupled electronic neurons*, Phys. Rev.
Lett. E **62**, nr. 2, 2000.
21. M. Dhamala, V. K. Jirsa, M. Ding *Transitions to synchrony in coupled bursting
neurons*, Phys. Rev. Lett. **92**, nr. 2, p. 028101, 2004.
22. C. Zhou, J. Kurths, *Hierarchical synchronization in complex networks with het-
erogenous degrees* Chaos **16**, 015104, 2006.

Part III

Cognition and Higher Perception

Neural and Cognitive Modeling with Networks of Leaky Integrator Units

Peter beim Graben[1,3], Thomas Liebscher[2] and Jürgen Kurths[3]

[1] School of Psychology and Clinical Language Sciences,
   University of Reading, United Kingdom
   p.r.beimgraben@reading.ac.uk
[2] Bundeskriminalamt, Wiesbaden, Germany
[3] Institute of Physics, Nonlinear Dynamics Group, Universität Potsdam, Germany

Summary. After reviewing several physiological findings on oscillations in the electroencephalogram (EEG) and their possible explanations by dynamical modeling, we present neural networks consisting of leaky integrator units as a universal paradigm for neural and cognitive modeling. In contrast to standard recurrent neural networks, leaky integrator units are described by ordinary differential equations living in continuous time. We present an algorithm to train the temporal behavior of leaky integrator networks by generalized back-propagation and discuss their physiological relevance. Eventually, we show how leaky integrator units can be used to build oscillators that may serve as models of brain oscillations and cognitive processes.

7.1 Introduction

The electroencephalogram (EEG) measures the electric fields of the brain generated by large formations of certain neurons, the *pyramidal cells*. These nerve cells roughly possess an axial symmetry and they are aligned in parallel perpendicular to the surface of the cortex [1–4]. They receive excitatory input at the superficial apical dendrites from thalamic relay neurons and inhibitory input at the basal dendrites and at their somata from local interneurons [1,3–5]. Excitatory and inhibitory synapses cause different ion currents through the cell membranes thus leading to either depolarization or hyperpolarization, respectively. When these synapses are activated, a single pyramidal cell behaves as a microscopic electric dipole surrounded by its characteristic electric field [1,6].

According to the inhomogeneity of the cortical gray matter, a mass of approximately 10,000 synchronized pyramidal cells form a dipole layer whose fields sum up to the *local field potentials* that polarize the outer tissues of the scalp, which acts thereby as a low pass filter [1,3,5,6]. These filtered sum potentials are macroscopically measurable as the EEG at the surface of a subject's head (cf. Chap. 1).

Some of the most obvious features of the EEG are oscillations in certain frequency bands. The *alpha waves* are sinusoidal-like oscillations between 8–14 Hz, strongly pronounced over parietal and occipital recording sites that reflect a state of relaxation during wakefulness, with no or only low visual attention. Figure 7.1 shows a characteristic power spectrum for the alpha rhythm: There is one distinguished peak superimposed to the $1/f$ background EEG. When a subject starts paying attention, the powerful slow alpha waves disappear, while smaller oscillations with higher frequencies around 14–30 Hz (the *beta waves*) arise [2,7,8]. We will refer to this finding, sometimes called *desynchronization* of the EEG, as to the *alpha blocking* [7].[4] Alpha waves are assumed to be related to awareness and cognitive processes [11–14]. Experimental findings suggest that thalamocortical feed-back loops are involved in the origin of the alpha EEG [1,2,4,8,15,16].

The $1/f$-behavior and the existence of distinguished oscillations in the EEG such as the alpha waves are cornerstones in the evaluation of computational models of the EEG. Indeed, modeling these brain rhythms has a long tradition. Wilson and Cowan [17] were the first to use populations of excitatory and inhibitory neurons innervating each other (see Sect. 7.3.2). They introduced a two-dimensional state vector whose components describe the proportion of firing McCulloch-Pitts neurons [18] within a unit volume of neural tissue at an instance of time. This kind of ensemble statistics leads to the

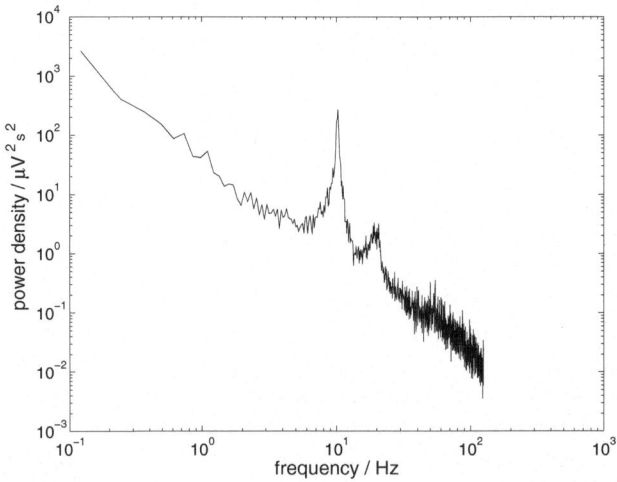

Fig. 7.1. Power spectrum of the alpha EEG at one parietal electrode site (PZ)

---

[4] The term "desynchronization" is misleading since it has no direct relation to synchronization in the sense of, for example, [9,10]. From the viewpoint of data analysis it simply means: decreasing power in the alpha band of the spectrum. However, biophysical theories of the EEG explain the loss of spectral power by a loss of coherence of neuron activity, i.e. a reduction of synchronization [7,8,11].

well-known sigmoidal activation functions for neural networks [19] through the probability distributions of either synapses or activation thresholds (see also [5]). The model further includes the refractory time in which a neuron that has been activated just before cannot be activated again, and the time-course of postsynaptic potentials as *impulse response functions*. This model has been strongly simplified by Wilson [20], leading to a network of only two recurrently coupled *leaky integrator units* (see Sect. 7.2). Wilson reported limit cycle dynamics of this system for a certain range of the excitatory input, playing the role of a control parameter. However, this network does not exhibit an equivalent to the alpha blocking, because the frequency of the oscillations becomes slower for increasing input.

Lopez da Silva et al. [21] pursued two different approaches: a distributed model of the thalamus where relay cells and interneurons are considered individually, and a "lumped" model analogous to the one of Wilson and Cowan [17] but without refractory time and with even more complicated postsynaptic potentials. In order to determine the sum membrane potential of each population as a model EEG, one has to compute the convolution integral of the postsynaptic impulse response functions with the spike rate per unit of volume. Linearizing the activation functions allows a system-theoretic treatment of the model by means of the Laplace transform, thus allowing the analytical computation of the power spectrum. Lopez da Silva et al. [21, 22] have shown that their model of thalamical or cortical feedback loops actually exhibits a peak around 10 Hz, i.e. alpha waves, in the spectrum, although they were not able to demonstrate alpha blocking. This population model [21] has been further developed by Freeman [23], Jansen et al. [24,25], Wendling et al. [26,27], and researchers from the Friston group [28–30] in order to model the EEG of the olfactory system, epileptic EEGs, and event-related potentials (ERP), respectively.

A further generalization of the Lopez da Silva et al. model [21] led Rotterdam et al. [31] to a description of spatio-temporal dynamics by considering a chain of coupled cortical oscillators. A similar approach has been pursued by Wright and Liley [32, 33] who discussed a spatial lattice of coupled unit volumes of excitatory and inhibitory elements obeying cortical connectivity statistics. The convolution integrals of the postsynaptic potentials with the spike rates were substituted by convolution sums over discrete time. The most important result for us is that the power spectrum shows the alpha peak, and, that there is a "shift to the right" (towards the beta band) of this peak with increasing input describing arousal, i.e. actually alpha blocking.

Additionally, Liley et al. [34] also suggested a distributed model of cortical alpha activity using a compartmental description of membrane potentials [35]. In such an approach, nerve cells are thought to be built up of cylindrical compartments that are governed by generalized Hodgkin-Huxley equations [36] (see also Chap. 1). Liley et al. [34] reported two oscillatory regimes of this dynamics: one having a broad-band spectrum with a peak in the beta range and the other narrowly banded with a peak around the alpha frequency.

There are also field theoretic models of neural activity [37–41] (see Chap. 8). In these theories, the unit volumes of cortical tissue are considered to be infinitesimally small. Thus, the systems of coupled ordinary differential equations are substituted by nonlinear partial differential equations. Robinson et al. [41] have proposed such a theory in order to describe thalamocortical interactions and hence the alpha EEG.

Another approach that could lead to the explanation of the EEG is Hebb's concept of a *cell assembly* [42], where *reverberatory circles* form neural oscillators. We shall see in Sect. 7.4.3 how such circles may emerge in an evolving neural network.

On the other hand, Kaplan et al. [43], van der Velde and de Kamps [44], Wennekers et al. [45], and Smolensky and Legendre [46] argue how neural networks could bridge the gap between the sub-symbolic representation of single neurons and "a symbol-like unit of thought" in models of cognitive processes. Kaplan et al. proposed that the cell assembly be an assembly of neural units that are recurrently connected to exhibit reverberatory circles, in which information needs to cycle around until the symbolic meaning is fully established. They presented a series of experiments in which they made use of physiological principles that should be present in the functioning of cell assemblies: temporally structured input, dependency on prior experience, competition between assemblies and control of its activation. A main result is that after a cell assembly is provided with input, its activation gradually increases until an asymptotic activation is reached or the input is removed. After removal of the input, the activation gradually decreases until it comes back to its resting level.

7.2 Leaky Integrator Networks

7.2.1 Description of Leaky Integrator Units

When neural signals are exchanged between different cell assemblies that are typically involved in brain functions, oscillations caused by recurrent connections between the neurons should become visible. A possible way to model this behavior is by describing each cell assembly by a leaky integrator unit [47], which integrates input over time while the internal activation is continuously decreased by a dampening leakage term. We shall present the relationship between cell assemblies and leaky integrator units in Sect. 7.3.2. However, also single neurons can be described by a leaky integrator unit, though with quite different leakage constants, as we shall see in Sect. 7.3.1. In terms of standard units (as e.g. used by Rumelhard et al. [48]), a leaky integrator unit looks like the one depicted in Fig. 7.2.

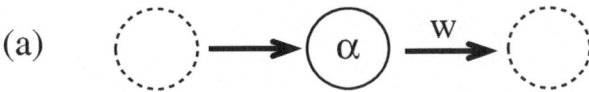

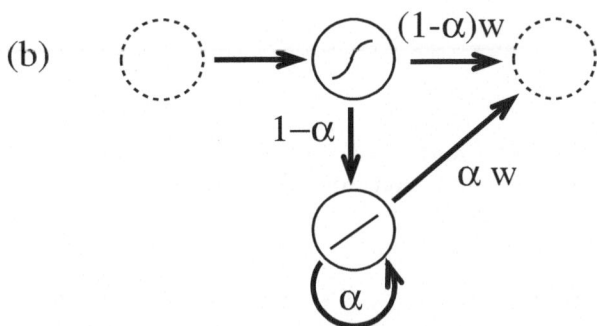

Fig. 7.2. Simulation of a leaky integrator unit (**a**) and a recurrent combination of two standard units (**b**). The function of the leakage rate α is mimicked by two parallel standard units with a logistic and a linear activation function, respectively. The synaptic weights to subsequent units are denoted by w (cf. (7.3))

The activation of this leaky integrator unit is described by

$$\frac{\mathrm{d}x_i(t)}{\mathrm{d}t} = -x_i(t) + (1 - \alpha_i)\, x_i(t) + \alpha_i f(y_i(t))$$
$$= -\alpha_i x_i(t) + \alpha_i f(y_i(t))\,, \tag{7.1}$$

or, in another form:

$$\tau_i \frac{\mathrm{d}x_i(t)}{\mathrm{d}t} + x_i(t) = f(y_i(t))\,. \tag{7.2}$$

The symbols have the following meanings:

$\frac{\mathrm{d}x_i(t)}{\mathrm{d}t}$	change of activation of unit i at time t
$x_i(t)$	activation of unit i at time t
$y_i(t)$	net input of unit i at time t
α_i	leakage rate of unit i
$\tau_i = \alpha_i^{-1}$	time constant of unit i
f	activation function of each unit; usually sigmoidal (e.g. logistic as in (7.5)) or linear.

The leakage rate α tells how much a unit depends on the actual net input. Its value is between 0 and 1. The lower the value of α, the stronger the influence of the previous level of activation and the less the influence of the actual net input. If $\alpha = 1$, the previous activation does not have any influence and the new activation is only determined by the net input (this is the case e.g. for the

standard units used by the PDP group [48]). By contrast, $\alpha = 0$ means that
the actual net input does not have any influence and the activation remains
constant. ($1 - \alpha$ could also be regarded as the strength of its *memory* with
respect to earlier activations.)

The net input of unit i is given as the sum of all incoming signals:

$$y_i(t) = \sum_j w_{ij} x_j(t) + b_i + I_i^{\text{ext}}(t) \,. \tag{7.3}$$

With

$y_i(t)$	net input of unit i at time t
w_{ij}	weight of connection from unit j to unit i
b_i	bias of unit i
I_i^{ext}	external input to unit i

Equation (7.1) is very similar to the general form of neural networks equations for continuous-valued units (described, for example, in [19]). The difference lies in the presence of the leakage term α that makes the current
activation dependent on its previous activation. We motivate (7.1) by the
equivalent recurrent network of Fig. 7.2 and we shall use it in Sect. 7.2.2 subsequently to derive a generalized back-propagation algorithm as a learning
rule for temporal patterns. On the other hand, (7.2) is well-known from the
theory of ordinary differential equations. Its associated homogeneous form

$$\tau_i \frac{\mathrm{d}x_i}{\mathrm{d}t} + x_i = 0$$

simply describes an exponential decay process. Therefore, the inhomogeneous
(7.2) can be seen as a forced decay process integrating its input on the right
hand side.

Hertz et al. [19, p. 54] discuss a Hopfield network of leaky integrator units
which is characterized by (7.2) with symmetric synaptic weights w_{ij}. Such a
network is a dynamical system whose attractors are the patterns which are
to be learned. Moreover, Hertz et al. [19, p. 55] consider another dynamical
system

$$\tau_i \frac{\mathrm{d}x_i(t)}{\mathrm{d}t} + x_i(t) = \sum_j w_{ij} f(x_j(t)) + b_i + I_i^{\text{ext}}(t) \tag{7.4}$$

having the same equilibrium solutions as (7.2). As we shall see in Sect. 7.3.1,
(7.4) appropriately models small networks of single neurons. The time-course
of activation for a leaky integrator unit using a logistic activation function

$$f(x) = \frac{1}{1 + \mathrm{e}^{-\beta x}} \tag{7.5}$$

with respect to input and leakage rate is shown in Figs. 7.3(a) and (b).

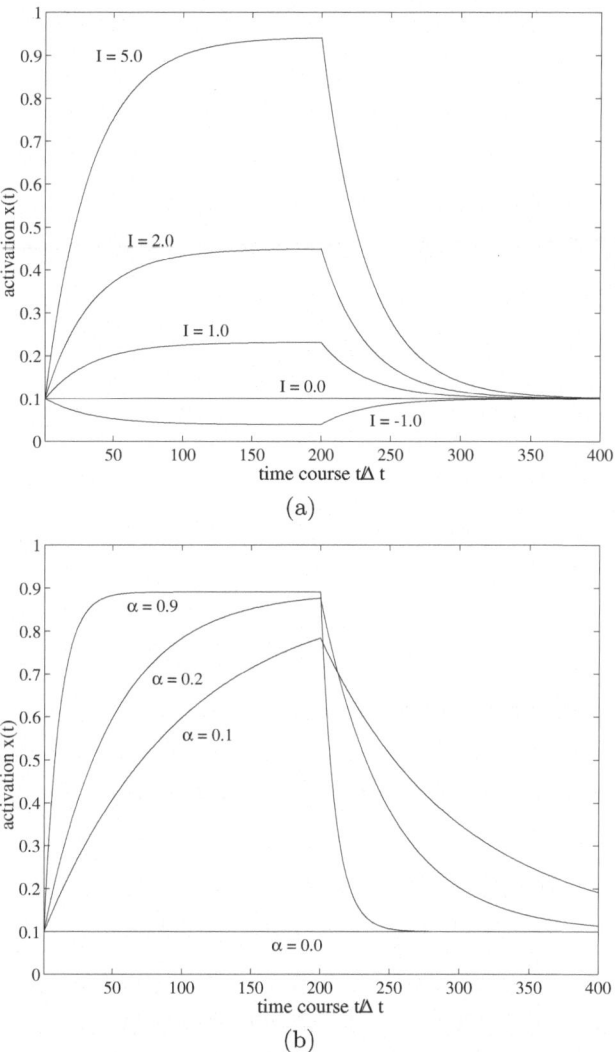

Fig. 7.3. Time-course of activation (7.1); (**a**) for different input with $\Delta t = 0.1$, $\alpha = 0.3$ and $b = -2.2$; (**b**) Time-course of activation (7.1) for different leakage rates with $\Delta t = 0.1$, $I^{\text{ext}} = 4.3$ and $b \approx -2.2$

7.2.2 Training Leaky Integrator Networks

In order to use leaky integrator units to create network models for simulation experiments, a learning rule that works in continuous time is needed. The following formulation is motivated by [49, 50] and describes how a backpropagation algorithm for leaky integrator units can be derived.

In a first step, Euler's algorithm is used to change the differential equations into difference equations:[5]

$$x_i(t + \Delta t) \approx x_i(t) + \frac{\mathrm{d}x_i(t)}{\mathrm{d}t}\Delta t$$

$$\Rightarrow \qquad \frac{\mathrm{d}x_i(t)}{\mathrm{d}t} \approx \frac{x_i(t + \Delta t) - x_i(t)}{\Delta t}. \qquad (7.6)$$

Combining (7.1) and (7.6) yields

$$\begin{aligned}
\tilde{x}_i(t + \Delta t) &= (1 - \Delta t)\tilde{x}_i(t) + \Delta t\left\{(1 - \alpha_i)\tilde{x}_i(t) + \alpha_i f(\tilde{y}_i(t))\right\} \\
&= (1 - \Delta t\alpha_i)\,\tilde{x}_i(t) + \Delta t\alpha_i f(\tilde{y}_i(t)), \qquad (7.7)
\end{aligned}$$

where tildes above variables (e.g. $\tilde{x}$) denote continuous functions that have been discretized.

Figures 7.3(a) and (b) show the time-course of activation for a leaky integrator unit with different values of external input I and leakage parameters α with $I \neq 0$ for $t \in [0, 20]$. In order to train a network, one needs to define an error function

$$E = \int_{t_0}^{t_1} f_{\mathrm{err}}\left[\boldsymbol{x}(t), t\right] \mathrm{d}t. \qquad (7.8)$$

Here, we choose the least mean square function

$$E = \frac{1}{2}\sum_i \int_{t_0}^{t_1} s_i\left[x_i(t) - d_i(t)\right]^2 \mathrm{d}t, \qquad (7.9)$$

where $d_i(t)$ is the desired activation of unit i at time t and s_i is the relative importance of this activation: $s = 0$ means unimportant and $s = 1$ means most important.

If one changes the activation of unit i at time t for a small amount, one gets a measure of how much this change influences the error function:

$$e_i(t) = \frac{\partial f_{\mathrm{err}}\left[\boldsymbol{x}(t), t\right]}{\partial x_i(t)} \qquad (7.10)$$

with

$$f_{\mathrm{err}} = \frac{1}{2}\sum_i s_i\left[x_i(t) - d_i(t)\right]^2.$$

With (7.9) as error function, we get

$$e_i(t) = s_i\left[x_i(t) - d_i(t)\right]. \qquad (7.11)$$

---

[5] Note that the following derivation could also be achieved using the variational calculus well-known from Hamilton's principle in analytical mechanics [49]. We leave this as an exercise for the reader.

Equations (7.10) and (7.11) describe the influence of a change of activation only for t. In a neural net that is described by (7.1) and (7.3), each change of activation at t also influences the activation at later times t' $(t < t')$. The amount of this influence can be described by using time-ordered derivatives [51, 52]:

$$\tilde{z}_i(t) = \frac{\partial^+ E}{\partial \tilde{x}_i(t)}$$

$$:= \frac{\partial E}{\partial \tilde{x}_i(t)} + \sum_{t' > t} \sum_j \frac{\partial^+ E}{\partial \tilde{x}_j(t')} \frac{\partial \tilde{x}_j(t')}{\partial \tilde{x}_i(t)} \qquad (7.12)$$

with $j = 1, 2, \ldots, n$ n number of units
            $t' = t + \Delta t, t + 2\Delta t, \ldots, t_1$ t_1 last defined time

$\tilde{z}_i(t)$ measures how much a change of activation of unit i at time t influences the error function at all times.

Performing the derivations in (7.12) with (7.9), (7.11), (7.7) and (7.3) and setting $t' = t + \Delta t$ gives:

$$\frac{\partial E}{\partial \tilde{x}_j(t)} = \Delta t e_i \qquad (7.13)$$

$$\frac{\partial \tilde{x}_i(t + \Delta t)}{\partial \tilde{x}_i(t)} = (1 - \Delta t \alpha_i) + \Delta t \alpha_i w_{ii} f'(\tilde{y}_i(t)) \qquad (7.14)$$

$$\frac{\partial \tilde{x}_j(t + \Delta t)}{\partial \tilde{x}_i(t)} = \Delta t \alpha_j w_{ji} f'(\tilde{y}_j(t)) \qquad (7.15)$$

for all units j that are connected with unit i.

All other derivatives are zero. With this, one gets

$$\tilde{z}_i(t) = \Delta t e_i + (1 - \Delta t \alpha_i) \tilde{z}_i(t + \Delta t)$$
$$+ \sum_j \Delta t \alpha_j w_{ji} f'(\tilde{y}_j(t)) \tilde{z}_j(t + \Delta t) . \qquad (7.16)$$

The back-propagated error signal $z(t)$ is equivalent to the δ in standard back-propagation. After the last defined activation $d_i(t_1)$, there is no further change of E, so $z_i(t_1 + \Delta t) = 0$.

Making use of Euler's method in the opposite direction, we find that the back-propagated error signal can be described by the following differential equation:

$$\frac{\mathrm{d} z_i(t)}{\mathrm{d} t} = \alpha_i z_i(t) - e_i - \sum_j \alpha_j w_{ji} f'(y_j(t)) z_j(t) . \qquad (7.17)$$

With (7.16), it is possible to calculate how the error function changes if one changes the parameters α_i and w_{ij}. Each variation also changes the activation x_i. The influence of this activation on E can be calculated using the chain rule of derivatives.

If w_{ij} changes over Δt by ∂w_{ij}, then the influence of this change on the error function can be described by

$$
\begin{aligned}
\left.\frac{\partial E}{\partial w_{ij}}\right|_t^{t+\Delta t} &:= \frac{\partial^+ E}{\partial x_i(t+\Delta t)}\frac{\partial x_i(t+\Delta t)}{\partial w_{ij}} \\
&= z_i(t+\Delta t)\alpha_i x_j(t)f'(y_i(t))\Delta t\,.
\end{aligned}
\tag{7.18}
$$

A change of ∂w_{ij} during the *whole* time $t_0 \le t \le t_1$ produces:

$$
\frac{\partial E}{\partial w_{ij}} = \alpha_i \int_{t_0}^{t_1} z_i(t)x_j(t)f'(y_i(t))\mathrm{d}t\,.
\tag{7.19}
$$

For the influence of a change in α_i on E one finds

$$
\begin{aligned}
\left.\frac{\partial^+ E}{\partial \alpha_i}\right|_t^{t+\Delta t} &= \frac{\partial E}{\partial x_i(t+\Delta t)}\frac{\partial x_i(t+\Delta t)}{\partial \alpha_i} \\
&= z_i(t+\Delta t)\left\{f(y_i(t)) - x_i(t)\right\}\Delta t\,.
\end{aligned}
\tag{7.20}
$$

For the whole time:

$$
\frac{\partial E}{\partial \alpha_i} = \int_{t_0}^{t_1} z_i(t)\left\{f(y_i(t)) - x_i(t)\right\}\mathrm{d}t\,.
\tag{7.21}
$$

Now, we have nearly all the equations that are needed to train a neural network of leaky integrator units. Finally, we must keep in mind the fact that the leakage term α must be between 0 and 1. This can be done by using

$$
\alpha = \frac{1}{1 + \mathrm{e}^{-\bar{\alpha}}}
\tag{7.22}
$$

and learning $\bar{\alpha}$ instead of α. With this replacement we set

$$
\begin{aligned}
\frac{\partial E}{\partial \bar{\alpha}_i} &= \frac{1}{1 + \mathrm{e}^{-\bar{\alpha}_i}}\left(1 - \frac{1}{1 + \mathrm{e}^{-\bar{\alpha}_i}}\right) \times \\
&\qquad \times \int_{t_0}^{t_1} z_i(t)\left\{f(y_i(t)) - x_i(t)\right\}\mathrm{d}t\,.
\end{aligned}
\tag{7.23}
$$

7.2.3 Overview of the Learning Procedure

To start the training, one needs to have the following information:

  (i) topology of the net with number of units (n) and connections
 (ii) values of the parameters $\boldsymbol{W}(0) = (w_{ij}(0))$ and $\bar{\boldsymbol{\alpha}}(0)$ at $t = 0$
(iii) activations $\boldsymbol{x}(t_0)$ at $t = 0$

(iv) time-course of the input $\boldsymbol{I}^{\text{ext}}(t), t_0 \leq t \leq t_1$
(v) time-course of the desired output $\boldsymbol{d}(t)$
(vi) activation function f for each unit
(vii) error function E
(viii) time-step size Δt that resembles the required resolution of the time-course ($\Delta t = 0.1$ turned out to be a good default value).

After having fixed these parameters according to the desired learning schedule, the goal is then to find a combination of $\boldsymbol{W}$ and $\bar{\boldsymbol{\alpha}}(0)$ that gives a minimum for E. This can be achieved by the following algorithm:

(i) At first one has to calculate the net input (7.3) for each unit successively and for each time-step *forward* in time. Simultaneously, the activations are calculated with (7.7).
(ii) With (7.9), one calculates the main error E and the error vector $\boldsymbol{e}(t)$ using (7.11).
(iii) Then, the error signals are propagated *backwards* through time with (7.16), making use of the condition $\tilde{z}_i(t_1 + \Delta t) = 0$.
(iv) Now, one calculates the gradient of each free parameter with respect to the error function E with the discrete versions of (7.19) and (7.23):

$$\frac{\partial E}{\partial w_{ij}} = \frac{1}{1 + e^{-\bar{\alpha}_i}} \sum_{t=t_0}^{t_1} \tilde{z}_i(t + \Delta t)\tilde{x}_j(t)f'(\tilde{y}_i(t))\Delta t \tag{7.24}$$

$$\frac{\partial E}{\partial \bar{\alpha}_i} = \frac{1}{1 + e^{-\bar{\alpha}_i}} \left(1 - \frac{1}{1 + e^{-\bar{\alpha}_i}}\right)$$

$$\sum_{t=t_0}^{t_1} \tilde{z}_i(t) \left\{f(\tilde{y}_i(t)) - \tilde{x}_i(t)\right\} \Delta t. \tag{7.25}$$

(v) After this, the parameters are changed along the negative gradient (*gradient descent*):

$$w_{ij} = w_{ij} - \eta_w \frac{\partial E}{\partial w_{ij}} \tag{7.26}$$

$$\bar{\alpha}_i = \bar{\alpha}_i - \eta_{\bar{\alpha}} \frac{\partial E}{\partial \bar{\alpha}_i}, \tag{7.27}$$

with η_w and $\eta_{\bar{\alpha}}$ as learning rates. ($\eta = 0.1$ is commonly a suitable starting value.) The gradient can be used for *steepest descent, conjugate gradient* or other numeric approximations (see e.g. [53]).
(vi) Having obtained the new values $\boldsymbol{W}$ and $\bar{\boldsymbol{\alpha}}$, the procedure goes back to step (i) and is followed until the main error falls below a certain value in step (ii) or this criterion is not reached after a maximal number of iterations.

(For a model that uses this type of learning algorithm with leaky integrator units, see [54]). In the context of modeling oscillating brain activity, recurrent networks of leaky integrator units become interesting. Section 7.4 will describe three typical examples.

7.3 From Physiology to Leaky Integrator Models

7.3.1 Leaky Integrator Model of Single Neurons

Let us consider the somatic membrane of a neuron i in the vicinity of its trigger zone. For the sake of simplicity, we shall assume that the membrane behaves only passively at this site. For further simplification, we do not describe the trigger zone by the complete Hodgkin-Huxley equations [36], but instead as a McCulloch-Pitts neuron [18], i.e. as a threshold device: the neuron fires if its membrane potential $U_i(t)$ exceeds the activation threshold $\theta \approx -50\,\mathrm{mV}$ from below due to the law of "all-or-nothing" [55, 56]. Because of this, the membrane potential $U_i(t)$ becomes translated into a spike train which can be modeled by a sum of delta functions

$$R_i(t) = \sum_{\substack{k:U_i(t_k)=\theta \\ \dot{U}_i(t_k)>0}} \delta(t - t_k)\,. \tag{7.28}$$

Now, we can determine the number of spikes in a time interval $[0, t]$ [35], which is given by

$$N_i(t) = \int_0^t R_i(t')\mathrm{d}t'\,.$$

Thus, from the *spike rate* per unit time, we regain the original signal

$$\frac{\mathrm{d}}{\mathrm{d}t}N_i(t) = R_i(t)\,. \tag{7.29}$$

In the next step, we consider the membrane potential U_i in the vicinity of the trigger zone which obeys Kirchhoff's First Law (see Fig. 7.4), i.e.

$$\sum_j I_{ij} = \frac{U_i - E_m}{r_m} + c_i \frac{\mathrm{d}U_i}{\mathrm{d}t}\,, \tag{7.30}$$

here, E_m is the Nernst equilibrium potential of the leakage channels with resistance r_m. c_i is the capacitance of the membrane of neuron i and I_{ij} is the current through the membrane at the chosen site coming from the synapse formed by the jth neuron with neuron i.

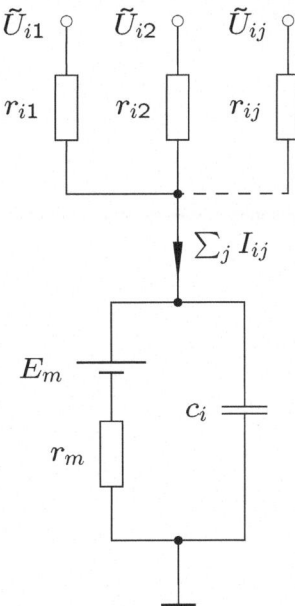

Fig. 7.4. Equivalent circuit for the leaky integrator neuron

These synaptic currents depend upon both the potential difference $\tilde{U}_{ij} - U_i$ between the postsynaptic potential $\tilde{U}_{ij}$ at the synapse connecting neuron j to i and the potential U_i at the trigger zone of i, and the intracellular resistance along the current's path r_{ij}. Therefore

$$I_{ij} = \frac{\tilde{U}_{ij} - U_i}{r_{ij}} \tag{7.31}$$

applies. Inserting (7.31) into (7.30) yields

$$\sum_j \frac{\tilde{U}_{ij} - U_i}{r_{ij}} = \frac{U_i - E_m}{r_m} + c_i \frac{dU_i}{dt} ,$$

and after some rearrangements

$$r_m c_i \frac{dU_i}{dt} + U_i \left(1 + \sum_j \frac{r_m}{r_{ij}} \right) = E_m + \sum_j \frac{r_m}{r_{ij}} \tilde{U}_{ij} . \tag{7.32}$$

After letting $E_m = 0$, without loss of generality and introducing the *time constants*

$$\tau_i = \frac{r_m c_i}{1 + \sum_j \frac{r_m}{r_{ij}}} \tag{7.33}$$

and provisory *synaptic weights*

$$\tilde{w}_{ij} = \frac{\frac{r_m}{r_{ij}}}{1 + \sum_j \frac{r_m}{r_{ij}}}, \tag{7.34}$$

we eventually obtain

$$\tau_i \frac{dU_i}{dt} + U_i = \sum_j \tilde{w}_{ij} \tilde{U}_{ij}. \tag{7.35}$$

Next, the postsynaptic potentials $\tilde{U}_{ij}$ require our attention. We assume that an action potential arriving at the presynaptic terminal of the neuron j releases, on average, one transmitter vesicle.[6] The content of the vesicle diffuses through the synaptic cleft and reacts with receptor molecules embedded in the postsynaptic membrane. After chemical reactions described by kinetic differential equations (cf. Chap. 1, [35]), opened ion channels give rise to a postsynaptic impulse response potential $G_{ij}(t)$. Because we characterize the dendro-somatic membranes as linear systems here, the postsynaptic potential elicited by a spike train $R_j(t)$ is given by the convolution

$$\tilde{U}_{ij}(t) = G_{ij}(t) * R_j(t). \tag{7.36}$$

Let us make a rather crude approximation here by setting the postsynaptic impulse response function proportional to a delta function:

$$G_{ij}(t) = g_{ij}\delta(t), \tag{7.37}$$

where g_{ij} is the *gain* of the synapse $j \to i$. Then, the postsynaptic potential is given by the product of the gain with the spike rate of the presynaptic neuron j.

Finally, we must take the stochasticity of the neuron into account as thoroughly described in Chap. 1. This is achieved by replacing the membrane potential U_j at the trigger zone by its average obtained from the distribution function, which leads to the characteristic sigmoidal activation functions [57], e.g. the logistic function (see (7.5))

$$R_j(t) = f(U_j(t)) = \frac{1}{1 + e^{-\beta[U_j(t) - \theta]}}. \tag{7.38}$$

Collecting (7.35, 7.36) and (7.38) together and introducing the proper *synaptic weights*

$$w_{ij} = g_{ij}\tilde{w}_{ij} \tag{7.39}$$

yields the leaky integrator model of a network of distributed single neurons

$$\tau_i \frac{dU_i}{dt} + U_i = \sum_j w_{ij} f(U_j(t)) \tag{7.40}$$

which is analogous to (7.4).

---

[6] The release of transmitter is a stochastic process that can be approximately described by a binomial distribution [55], and hence, due to the limit theorem of de Moivre and Laplace, is normally distributed (see Chap. 1).

7.3.2 Leaky Integrator Model of Neural Populations

According to Freeman [58] (see also [59]), a neuronal *population* ("KI" set) consists of many reciprocally connected neurons of one kind, either excitatory or inhibitory. Let us consider such a set of McCulloch-Pitts neurons [18] distributed over a unit volume i of neural tissue. We introduce the proportions of firing cells (either excitatory or inhibitory, in contrast to [17]) in volume i at the instance of time t, $Q_i(t)$, as the state variables [17, 32].

A neuron belonging to volume i will fire if its net input U_i (analogous to the membrane potential at the trigger zone, see Sect. 7.3.1) crosses the threshold θ. But now, we have to deal with an ensemble of neurons possessing randomly distributed thresholds within the unit volume i. We therefore obtain an ensemble activation function [5] (cf. Chap. 14) by integrating the corresponding probability distribution density $D(\theta)$ of thresholds [17],

$$f(U_i) = \int_0^{U_i} D(\theta)\mathrm{d}\theta\,. \tag{7.41}$$

Depending upon the modality of the distribution $D(\theta)$, the activation function could be sigmoidal or even more complicated. For unimodal distributions such as Gaussian or Poissonian distributions, $f(U_i)$ might be approximated by the logistic function (7.38). As for the single neuron model, the net input is obtained by a convolution

$$U_i(t) = \int_{-\infty}^{t} G(t - t') \sum_j w_{ij}\, Q_j(t')\mathrm{d}t'\,, \tag{7.42}$$

with "synaptic weights" w_{ij} characterizing the neural connectivity and whether the population is excitatory or inhibitory.

In the following, we shall simplify the model of Wilson and Cowan [17] by neglecting the refractory time. The model equations are then

$$Q_i(t + \tau_i) = f\left(\int_{-\infty}^{t+\tau_i} G(t - t') \sum_j w_{ij}\, Q_j(t')\mathrm{d}t'\right)\,, \tag{7.43}$$

such that $Q_i(t + \tau_i)$ is the proportion of cells being above threshold in the time interval $[t, t + \tau_i]$. Expanding the left hand side into a Taylor series at t and assuming again that $G(t - t') = \delta(t - t')$, we obtain

$$\tau_i \frac{\mathrm{d}Q_i(t)}{\mathrm{d}t} + Q_i(t) = f\left(\sum_j w_{ij}\, Q_j(t)\right)\,, \tag{7.44}$$

a leaky integrator model again, yet characterized by (7.2).

7.4 Oscillators from Leaky Integrator Units

7.4.1 Linear Model

In this section, we demonstrate that a damped harmonic oscillator can be obtained from a simple model of two recurrently coupled leaky integrator units with linear activation functions [49]. Figure 7.5 shows the architecture of this model.

The network of Fig. 7.5 is governed by (7.4):

$$\tau_1 \frac{\mathrm{d}x_1}{\mathrm{d}t} + x_1 = w_{11}x_1 + w_{12}x_2 + p \tag{7.45}$$

$$\tau_2 \frac{\mathrm{d}x_2}{\mathrm{d}t} + x_2 = w_{21}x_1 + w_{22}x_2 \,, \tag{7.46}$$

where x_1 denotes the activity of unit 1 and x_2 that of unit 2. Correspondingly, τ_1 and τ_2 are the time constants of the units 1 and 2, respectively. The synaptic weights w_{ij} are indicated in Fig. 7.5. Note that the weights w_{11} and w_{22} describe autapses [60]. The quantity p refers to excitatory synaptic input that might be a periodic forcing or any other function of time.

Equations (7.45) and (7.46) can be converted into two second-order ordinary differential equations

$$\frac{\mathrm{d}^2x_1}{\mathrm{d}t^2} + \gamma\frac{\mathrm{d}x_1}{\mathrm{d}t} + \omega_0^2 = p_1 \tag{7.47}$$

$$\frac{\mathrm{d}^2x_2}{\mathrm{d}t^2} + \gamma\frac{\mathrm{d}x_2}{\mathrm{d}t} + \omega_0^2 = p_2 \,, \tag{7.48}$$

where we have introduced the following simplifying parameters:

$$\gamma = \frac{\tau_1(1 - w_{22}) + \tau_2(1 - w_{11})}{\tau_1\tau_2}$$

$$\omega_0^2 = \frac{w_{11}w_{22} - w_{12}w_{21} - w_{11} - w_{22} + 1}{\tau_1\tau_2}$$

$$p_1 = \frac{1}{\tau_1}\frac{\mathrm{d}p}{\mathrm{d}t} + \frac{1 - w_{22}}{\tau_1\tau_2}$$

$$p_2 = \frac{w_{21}}{\tau_1\tau_2}p$$

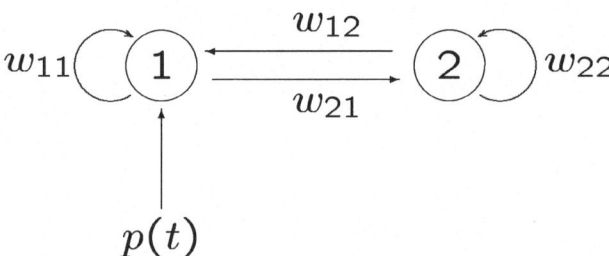

Fig. 7.5. Architecture of an oscillator formed by leaky integrator units

Now, (7.47) and (7.48) describe two damped, decoupled, harmonic oscillators with external forcing when $\gamma \geq 0$ and $\omega_0^2 > 0$, i.e. one unit must be excitatory and the other inhibitory.

7.4.2 Simple Nonlinear Model

Next, we discuss a simple nonlinear system, consisting of three coupled leaky integrator units, which provides a model of the thalamocortical loop. Figure 7.6 displays its architecture.

According to Fig. 7.6, the model (7.4) are

$$\tau_1 \frac{\mathrm{d}x_1}{\mathrm{d}t} + x_1 = -\alpha f(x_3(t)) \tag{7.49}$$

$$\tau_2 \frac{\mathrm{d}x_2}{\mathrm{d}t} + x_2 = \beta f(x_1(t)) \tag{7.50}$$

$$\tau_3 \frac{\mathrm{d}x_3}{\mathrm{d}t} + x_3 = \gamma f(x_2(t)). \tag{7.51}$$

Setting all $\tau_i = 1$ and rearranging, we get

$$\frac{\mathrm{d}x_1}{\mathrm{d}t} = -x_1 - \alpha f(x_3(t))$$

$$\frac{\mathrm{d}x_2}{\mathrm{d}t} = -x_2 + \beta f(x_1(t))$$

$$\frac{\mathrm{d}x_3}{\mathrm{d}t} = -x_3 + \gamma f(x_2(t)).$$

These equations define a vector field $\boldsymbol{F}$ with the Jacobian matrix

$$\mathrm{D}\boldsymbol{F} = \begin{pmatrix} -1 & 0 & -\alpha f'(x_3(t)) \\ \beta f'(x_1(t)) & -1 & 0 \\ 0 & \gamma f'(x_2(t)) & -1 \end{pmatrix}.$$

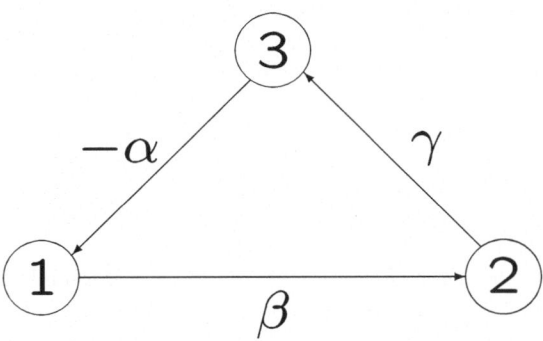

Fig. 7.6. Thalamocortical oscillator of three leaky integrator units: (*1*) pyramidal cell; (*2*) thalamus cell; (*3*) cortical interneuron (star cell)

For the activation function, we chose $f(x) = \tanh x$, which can be obtained by a coordinate transformation from the logistic function in (7.38). Therefore, $\boldsymbol{F}(x_1, x_2, x_3) = 0$ and we can look at whether the center manifold theorem [61] can be applied. The Jacobian at $(0, 0, 0)$ is

$$\mathrm{D}\boldsymbol{F}(0) = \begin{pmatrix} -1 & 0 & -\alpha \\ \beta & -1 & 0 \\ 0 & \gamma & -1 \end{pmatrix},$$

having eigenvalues

$$\lambda_1 = -1 - \sqrt[3]{\alpha\beta\gamma}$$

$$\lambda_2 = -1 + \frac{1}{2}(1 - \mathrm{i}\sqrt{3})\sqrt[3]{\alpha\beta\gamma}$$

$$\lambda_3 = -1 + \frac{1}{2}(1 + \mathrm{i}\sqrt{3})\sqrt[3]{\alpha\beta\gamma}.$$

Since $\lambda_1 < 0$ for $\alpha, \beta, \gamma \geq 0$, we seek for the weight parameters making $\mathrm{Re}(\lambda_{2|3}) = 0$. This leads to the condition

$$\alpha\beta\gamma = 8, \tag{7.52}$$

which can be easily fulfilled, for example, by setting

$$\alpha = 4, \qquad \beta = 1, \qquad \gamma = 2.$$

In this case, the center manifold theorem applies: the dynamics stabilizes along the eigenvector corresponding to λ_1, exhibiting a limit cycle in the center manifold spanned by the eigenvectors of λ_2 and λ_3. Figure 7.7 shows a numerical simulation of this oscillator. It is also possible to train a leaky integrator network using the algorithm described in Sect. 7.2.2 in order to replicate a limit cycle dynamics [49].

7.4.3 Random Neural Networks

In this last subsection, we describe a network model that is closely related to those presented in Chaps. 3, 5, 12, 13, and 14, namely a random graph carrying leaky integrator units described by (7.4) or, equivalently, (7.40), at its nodes:

$$\tau_i \frac{\mathrm{d}x_i(t)}{\mathrm{d}t} + x_i(t) = \sum_j w_{ij} f(x_j(t)).$$

We shall see that the onset of oscillatory behavior is correlated with the emergence of super-cycles in the topology of the network provided by an evolving directed and weighted Erdős-Rényi graph of N nodes where all connections between two nodes are equally likely with increasing probability [62–64].

As explained in Chap. 3, a directed Erdős-Rényi graph consists of a set of vertices V that are randomly connected by arrows taken from an edge set

$E \subset V \times V$ with equal probability q. The topology of the graph is completely described by its *adjacency matrix* $\boldsymbol{A} = (a_{ij})$ where $a_{ij} = 1$, if there is an arrow connecting the vertex j with the vertex i (i.e. $(j, i) \in E$ for $i, j \in V$) while $a_{ij} = 0$ otherwise. A directed and weighted Erdős-Rényi graph is then described by the *weight matrix* $\boldsymbol{W} = (w_{ij})$ which is obtained by element-wise multiplication of the adjacency matrix with constants g_{ij}: $w_{ij} = g_{ij} a_{ij}$. Biologically plausible models must satisfy Dale's law, which says that excitatory neurons only have excitatory synapses while inhibitory neurons only possess inhibitory synapses [56]. Therefore, the column vectors of the weight matrix are constrained to have a unique sign. We achieve this requirement by randomly choosing a proportion p of the vertices to be excitatory and the remainder to be inhibitory.

In our model, the weights become time-dependent due to the following evolution algorithm:

(i) Initialization: $\boldsymbol{W}(0) = 0$.
(ii) At evolution time t, select a random pair of nodes i, j.
(iii) If they are not connected, create a synapse with weight $w_{ij}(t + 1) = +\delta$ if j is excitatory, and $w_{ij}(t+1) = -\delta$ if j is inhibitory. If they are already connected, enhance the weight $w_{ij}(t + 1) = w_{ij}(t) + \delta$ if $w_{ij}(t) > 0$ and $w_{ij}(t+1) = w_{ij}(t) - \delta$ if $w_{ij}(t) < 0$. All other weights remain unchanged.
(iv) Repeat from (ii) for a fixed number of iterations L.

As the "learning rate", we choose $\delta = 1$, while the connectivity increases for L time steps. In order to simplify the simulations, we further set $\tau_i = 1$ for all $1 \leq i \leq N$.

Since (7.40) describes the membrane potential of the ith neuron, we can estimate its dendritic field potential by the inhomogeneity of (7.40),

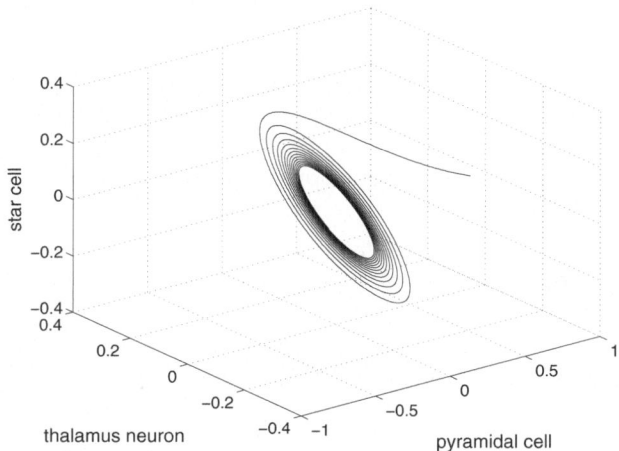

Fig. 7.7. Limit cycle of the thalamocortical oscillator in its center manifold plane

$$F_i(t) = \sum_j w_{ij} \, f(x_j) \, . \qquad (7.53)$$

Then, the model EEG[7] is given by the sum of the dendritic field potentials of all excitatory nodes

$$E(t) = \sum_{i^+} F_i(t) \, . \qquad (7.54)$$

The indices i^+ indicate that the neuron i belongs to the population of excitatory neurons, namely the EEG generating pyramidal cells.

We create such random neural networks with size $N = 100, 200, 500,$ and 1000 nodes. Since about 80% of cortical neurons are excitatory pyramidal cells, $p = 80\%$ of the network's nodes are chosen to be excitatory [66]. For each iteration of the network's evolution, the dynamics of its nodes is calculated. After preparing them with normally distributed initial conditions ($\mu = 0, \sigma = 1$), (7.40) is solved numerically with the activation functions $f_i(x) = \tanh x$ for an ensemble of $K = 10$ time series of length $T = 100$ with a step-width of $\Delta t = 0.0244$. The dendritic field potential and EEG are computed according to (7.53) and (7.54).

From the simulated EEGs, the power spectra are computed and averaged over all K realizations of the network's dynamics. In order to monitor sudden changes in the topologies of the networks, three characteristic statistics are calculated:

(1) The *mean degree* (the average number of vertices attached to the nodes) $\langle k \rangle$ of the associated undirected graphs, described by the symmetrized adjacency matrix $\boldsymbol{A}^s = \Theta(\boldsymbol{A} + \boldsymbol{A}^T)$ (Θ denotes Heaviside's step function),

(2) the *total distribution*

$$d(l) = \frac{\text{tr}(\boldsymbol{A}^l)}{l \mathcal{N}} \qquad (7.55)$$

of cycles of the exact length l [62–64, 67–70]. In (7.55), $\text{tr}(\boldsymbol{A}^l)$ provides the total number of (not necessarily self-avoiding) closed paths of length l through the network. Since any node at such a path may serve as the starting point and there are l nodes, the correct number of cycles is obtained by dividing by l. Finally, $\mathcal{N} = \sum_l \text{tr}(\boldsymbol{A}^l)/l$ is a normalization constant. From the cycle distribution (7.55), we derive

(3) an *order parameter* s for topological transitions from the averaged slopes of the envelope of $d(l)$, where the envelopes are estimated by connecting the local maxima of $d(l)$. The above mentioned procedure is repeated for each network size $M = 10$ times where we have chosen $L_{100} = 150, L_{200} = 400, L_{500} = 800,$ and $L_{1000} = 1700$ iterations of network evolution in order to ensure sufficiently dense connectivities.

---

[7] In fact, (7.54) describes better the local field potential (LFP) rather than the EEG. Considering (7.4) as a model of coupled neural populations, instead, seems to be more appropriate for describing the EEG [65].

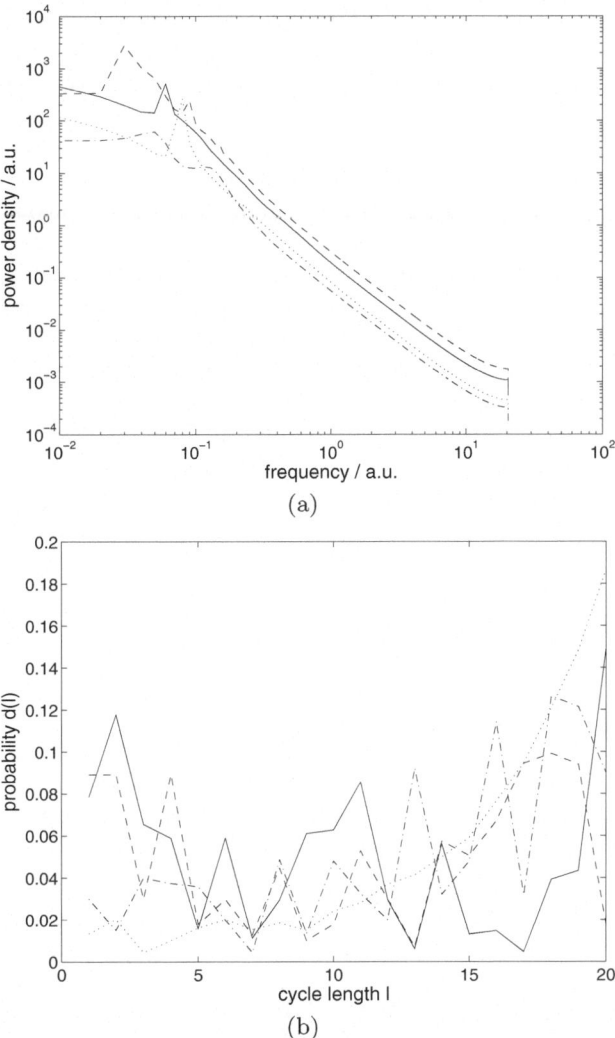

Fig. 7.8. (a) Power spectra of representative simulated time series during the oscillatory transition (critical phase) for four different network sizes: $N = 100$ (*dotted*), $N = 200$ (*dashed-dotted*), $N = 500$ (*solid*), and $N = 1000$ (*dashed*); (b) Total distributions of cycles (7.55) for the same networks

Figure 7.8 shows four representative networks in the critical phase characterized by the smallest positive value of the cycle order parameter s, averaged over the $M = 10$ network simulations, when sudden oscillations occur in the dynamics of the units, as is visible by the peaks in the power spectra [Fig. 7.8(a)]. The cycle distributions $d(l)$ [Fig. 7.8(b)] for network sizes $N = 200, 500,$ and 1000 display a transition from geometrically decaying to exponentially growing functions while this transition has already taken place for $N = 100$. As Fig. 7.8(a) reveals, the power spectra display a broad $1/f$ continuum. Superimposed to this continuum are distinguished peaks that can be regarded as the "alpha waves" of the model.

According to random graph theory, Erdős-Rényi networks exhibit a percolation transition when a giant cluster suddenly occurs for $\langle k \rangle = 1$ [62–64]. A second transition takes place for $\langle k \rangle = 2$, indicating the appearance of mainly isolated cycles in the graph. Isolated cycles are characterized by a geometrically decaying envelope of the total cycle distribution. Our simulations suggest the existence of a third transition when super-cycles are composed from merging smaller ones. This is reflected by a transition of the total cycle distribution $d(l)$ from a geometrically decaying to an exponentially growing behavior due to a "combinatorial explosion" of possible self-intersecting paths through the network (super-cycles are common in regular lattices with $\langle k \rangle \geq 3$). We detect this transition by means of a suitably chosen order parameter s derived from $d(l)$ as the averaged slope of its envelope. For decaying $d(l)$, $s < 0$ and for growing $d(l)$, $s > 0$. The appearance of super-cycles is associated with $s \approx 0$ if $d(l)$ is approximately symmetric in the range of l. In this case, sustained oscillations emerge in the network's dynamics due to the presence of reverberatory circles. For further details, see [65].

7.5 Cognitive Modeling

In this chapter we have reviewed neurophysiological findings on oscillations in the electroencephalogram as well as certain approaches to model these through coupled differential equations. We have introduced the theory of networks of *leaky integrator units* and presented a general learning rule to train these networks in such a way that they are able to reproduce temporal patterns in continuous time. This learning rule is a generalized back-propagation algorithm that has been applied for the first time to model reaction times from a psychological experiment [54]. Therefore, leaky integrator networks provide a unique and physiologically plausible paradigm for neural and cognitive modeling.

Mathematically, leaky integrator models are described by systems of coupled ordinary differential equations that become nonlinear dynamical systems by using sigmoidal activation functions. Networks of leaky integrator units may display a variety of complex behaviors: limit cycles, multistability, bifurcations and hysteresis [17]. They could therefore act as models of perceptional instability [71] or cognitive conflicts [72], as has already been demonstrated by Haken [73, 74] using *synergetic computers*. As Haken [74, p. 246] pointed out, the order parameter equations of synergetic computers are analogous to neural networks whose activation function is expanded into a power series. However, these computers are actually leaky integrator networks as we will see subsequently.

Basically, synergetic computers are time-continuous Hopfield nets [19] governed by a differential equation

$$\frac{\mathrm{d}\boldsymbol{x}}{\mathrm{d}t} - \sum_{k=1}^{K} \eta_k \boldsymbol{v}_k (\boldsymbol{v}_k^+ \boldsymbol{x}) = -B \sum_{k' \neq k}^{K} (\boldsymbol{v}_{k'}^+ \boldsymbol{x})^2 (\boldsymbol{v}_k^+ \boldsymbol{x}) \boldsymbol{x} - C(\boldsymbol{x}^+ \boldsymbol{x}) \boldsymbol{x} \qquad (7.56)$$

where $\boldsymbol{x}(t)$ denotes the activation vector of the network; the K vectors $\boldsymbol{v}_k$ are training patterns with adjoints $\boldsymbol{v}_k^+$ such that the orthonormality relations $\boldsymbol{v}_k^+ \boldsymbol{v}_l = \delta_{kl}$ hold. In this notation, $\boldsymbol{x}^+ \boldsymbol{y} = \sum_i x_i y_i$ means the *inner product* of the row vector $\boldsymbol{x}^+$ with a column vector $\boldsymbol{y}$ yielding a scalar. On the other hand, the outer product $\boldsymbol{y}\boldsymbol{x}^+$ of a column vector $\boldsymbol{y}$ with a row vector $\boldsymbol{x}^+$ is a matrix with elements $y_i x_j$.

Therefore, the second term of the left hand side of (7.56) can be rewritten as

$$\sum_{k=1}^{K} \eta_k \boldsymbol{v}_k (\boldsymbol{v}_k^+ \boldsymbol{x}) = \sum_{k=1}^{K} \eta_k (\boldsymbol{v}_k \boldsymbol{v}_k^+) \boldsymbol{x} = \left(\sum_{k=1}^{K} \eta_k \boldsymbol{v}_k \boldsymbol{v}_k^+ \right) \boldsymbol{x} = \boldsymbol{W}\boldsymbol{x}$$

where

$$\boldsymbol{W} = \sum_{k=1}^{K} \eta_k \boldsymbol{v}_k \boldsymbol{v}_k^+$$

is the synaptic weight matrix obtained by Hebbian learning of the patterns $\boldsymbol{v}_k$ with learning rates η_k.

The notion "synergetic computer" refers to the possibility of describing the network (7.56) by the evolution of *order parameters*, which are appropriately chosen as the "loads" of the training patterns $\boldsymbol{v}_k$ in a kind of principal component analysis. We therefore separate activation space and time by the ansatz

$$\boldsymbol{x}(t) = \sum_k \xi_k(t) \boldsymbol{v}_k + \boldsymbol{w}(t) \,,$$

where $\xi_k(t) = \boldsymbol{v}_k^+ \boldsymbol{x}(t)$ and $\boldsymbol{w}(t)$ is a fast decaying remainder. Multiplying (7.56) from the left with $\boldsymbol{v}_l^+$ and exploiting the orthonormality relations, we eventually obtain

$$\frac{\mathrm{d}\xi_l}{\mathrm{d}t} - \eta_l \xi_l = -B \sum_{k \neq l}^{K} \xi_k^2 \xi_l - C \left(\sum_k \xi_k^2 \right) \xi_l \,. \qquad (7.57)$$

Division by $-\eta_l = 1/\tau_l$ then yields the leaky integrator equations for the order parameters with rescaled constants B', C' and a cubic activation function

$$\tau_l \frac{\mathrm{d}\xi_l}{\mathrm{d}t} + \xi_l = B' \sum_{k \neq l}^{K} \xi_k^2 \xi_l + C' \left(\sum_k \xi_k^2 \right) \xi_l \,. \qquad (7.58)$$

The "time constants" play then the role of *attention parameters* describing the amount of attention devoted to a particular pattern. These parameters might also depend on time, e.g. for modeling habituation.

From a formal point of view, the attention model of Lourenço [75], the *cellular neural networks* (CNN) of Chua [76] (see also [77–79]) and the disease model of Huber et al. [80] can also be regarded as leaky integrator networks.

Also, higher cognitive functions such as language processing and their neural correlates such as event-related brain potentials (ERPs) [72,81,82] can be modeled with leaky integrator networks. Kawamoto [83] used a Hopfield net with exponentially decaying activation and habituating synaptic weights to modeling lexical ambiguity resolution. The activations of the units in his model are governed by the equations

$$x_i(t+1) = f\left(\delta x_i(t) + \sum_j w_{ij}x_j(t)\right). \tag{7.59}$$

Setting $\delta = 1-\alpha = 1-\tau^{-1}$ and approximating $f'(x) \approx 1$ for typical activation values yields, after a Taylor expansion of the activation function,

$$f\left(\delta x_i(t) + \sum_j w_{ij}x_j(t)\right) \approx f\left(\sum_j w_{ij}x_j(t)\right) + f'\left(\sum_j w_{ij}x_j(t)\right)\delta x_i(t),$$

the leaky integrator equation (7.2).

Smolensky and Legendre [46] consider Hopfield nets of leaky integrator units that can be described by a Lyapunov function E. They call the function $H = -E$ the *harmony* of the network and argue that cognitive computations maximize this harmony function at the sub-symbolic level. Additionally, the harmony value can also be computed at the symbolic level of linguistic representations in the framework of harmonic grammars or optimality theory [84]. By regarding the harmony as an order parameter of the network, one could also model neural correlates of cognitive processes, e.g., ERPs.

This has recently been attempted by Wennekers et al. [83], who built a six-layer model of the perisylvian language cortex by randomly connecting leaky integrator units within each layer (similar to our exposition in Sect. 7.4.3). The network was trained with a Hebbian correlation learning rule to memorize "words" (co-activated auditory and motor areas) and "pseudowords" (activation of the auditory layer only). After training, cell assemblies of synchronously oscillating units across all six layers emerged. Averaging their event-related oscillations in the recall phase then yielded a larger amplitude for the "words" than for the "pseudowords", thus emulating the mismatch negativity (MMN) ERP known from word recognition experiments [45].

Acknowledgements

We gratefully acknowledge helpful discussions with W. Ehm, P. Franaszcuk, M. Garagnani, A. Hutt, V. K. Jirsa, and J. J. Wright. This work has been supported by the Deutsche Forschungsgemeinschaft (research group "Conflicting

Rules in Cognitive Systems"), and by the Helmholtz Institute for Supercomputational Physics at the University of Potsdam.

References

1. O. Creutzfeld and J. Houchin. Neuronal basis of EEG-waves. In *Handbook of Electroencephalography and Clinical Neurophysiology*, Vol. 2, Part C, pp. 2C-5–2C-55. Elsevier, Amsterdam, 1974.
2. F. H. Lopes da Silva. Neural mechanisms underlying brain waves: from neural membranes to networks. *Electroencephalography and Clinical Neurophysiology*, 79: 81–93, 1991.
3. E.-J. Speckmann and C. E. Elger. Introduction to the neurophysiological basis of the EEG and DC potentials. In E. Niedermeyer and F. Lopez da Silva, editors, *Electroencephalography. Basic Principles, Clinical Applications, and Related Fields*, Chap. 2, pp. 15–27. Lippincott Williams and Wilkins, Baltimore, 1999.
4. S. Zschocke. *Klinische Elektroenzephalographie*. Springer, Berlin, 1995.
5. W. J. Freeman. Tutorial on neurobiology: from single neurons to brain chaos. *International Journal of Bifurcation and Chaos*, 2(3): 451–482, 1992.
6. P. L. Nunez and R. Srinivasan. *Electric Fields of the Brain: The Neurophysics of EEG*. Oxford University Press, New York, 2006.
7. E. Başar. *EEG-Brain Dynamics. Relations between EEG and Brain Evoked Potentials*. Elsevier/North Holland Biomedical Press, Amsterdam, 1980.
8. M. Steriade, P. Gloor, R. R. Llinás, F. H. Lopes da Silva, and M.-M. Mesulam. Basic mechanisms of cerebral rhythmic activities. *Electroencephalography and Clinical Neurophysiology*, 76: 481–508, 1990.
9. C. Allefeld and J. Kurths. Testing for phase synchronization. *International Journal of Bifurcation and Chaos*, 14(2): 405–416, 2004.
10. C. Allefeld and J. Kurths. An approach to multivariate phase synchronization analysis and its application to event-related potentials. *International Journal of Bifurcation and Chaos*, 14(2): 417–426, 2004.
11. R. Srinivasan. Internal and external neural synchronization during conscious perception. *International Journal of Bifurcation and Chaos*, 14(2): 825–842, 2004.
12. G. Pfurtscheller. EEG rhythms — event related desynchronization and synchronization. In H. Haken and H. P. Koepchen, editors, *Rhythms in Physiological Systems*, pp. 289–296, Springer, Berlin, 1991.
13. E. Başar, M. Özgören, S. Karakaş, and C. Başar-Eroğlu. Super-synergy in brain oscillations and the grandmother percept is manifested by multiple oscillations. *International Journal of Bifurcation and Chaos*, 14(2): 453–491, 2004.
14. W. Klimesch, M. Schabus, M. Doppelmayr, W. Gruber, and P. Sauseng. Evoked oscillations and early components of event-related potentials: an analysis. *International Journal of Bifurcation and Chaos*, 14(2): 705–718, 2004.
15. N. Birbaumer and R. F. Schmidt. *Biologische Psychologie*. Springer, Berlin, 1996.
16. M. Steriade. Cellular substrates of brain rhythms. In E. Niedermeyer and F. Lopez da Silva, editors, *Electroencephalography. Basic Principles, Clinical Applications, and Related Fields*, Chap. 3, pp. 28–75. Lippincott Williams and Wilkins, Baltimore, 1999.

17. H. R. Wilson and J. D. Cowan. Excitatory and inhibitory interactions in localized populations of model neurons. *Biophysical Journal*, 12: 1–24, 1972.
18. W. S. McCulloch and W. Pitts. A logical calculus of ideas immanent in nervous activity. *Bulletin of Mathematical Biophysics*, 5: 115–33, 1943. Reprinted in J. A. Anderson and E. Rosenfeld (1988), pp. 83ff.
19. J. Hertz, A. Krogh, and R. G. Palmer. *Introduction to the Theory of Neural Computation*. Perseus Books, Cambridge (MA), 1991.
20. H. R. Wilson. *Spikes, Decisions and Actions. Dynamical Foundations of Neuroscience*. Oxford University Press, New York (NY), 1999.
21. F. H. Lopes da Silva, A. Hoecks, H. Smits, and L. H. Zetterberg. Model of brain rhythmic activity: the alpha-rhythm of the thalamus. *Kybernetik*, 15: 27–37, 1974.
22. F. H. Lopes da Silva, A. van Rotterdam, P. Bartels, E. van Heusden, and W. Burr. Models of neuronal populations: the basic mechanisms of rhythmicity. In M. A. Corner and D. F. Swaab, editors, *Perspectives of Brain Research*, Vol. 45 of *Progressive Brain Research*, pp. 281–308. 1976.
23. W. J. Freeman. Simulation of chaotic EEG patterns with a dynamic model of the olfactory system. *Biological Cybernetics*, 56: 139–150, 1987.
24. B. H. Jansen and V. G. Rit. Electroencephalogram and visual evoked potential generation in a mathematical model of coupled cortical columns. *Biological Cybernetics*, 73: 357–366, 1995.
25. B. H. Jansen, G. Zouridakis, and M. E. Brandt. A neurophysiologically-based mathematical model of flash visual evoked potentials. *Biological Cybernetics*, 68: 275–283, 1993.
26. F. Wendling, F. Bartolomei, J. J. Bellanger, and P. Chauvel. Epileptic fast activity can be explained by a model of impaired GABAergic dendritic inhibition. *European Journal of Neuroscience*, 15: 1499–1508, 2002.
27. F. Wendling, J. J. Bellanger, F. Bartolomei, and P. Chauvel. Relevance of nonlinear lumped-parameter models in the analysis of depth-EEG epileptic signals. *Biological Cybernetics*, 83: 367–378, 2000.
28. O. David, D. Cosmelli, and K. J. Friston. Evaluation of different measures of functional connectivity using a neural mass model. *NeuroImage*, 21: 659–673, 2004.
29. O. David and K. J. Friston. A neural mass model for MEG/EEG: coupling and neuronal dynamics. *NeuroImage*, 20: 1743–1755, 2003.
30. O. David, L. Harrison, and K. J. Friston. Modelling event-related respones in the brain. *NeuroImage*, 25: 756–770, 2005.
31. A. van Rotterdam, F. H. Lopes da Silva, J. van den Ende, M. A. Viergever, and A. J. Hermans. A model of the spatial-temporal characteristics of the alpha rhythm. *Bulletin of Mathematical Biology*, 44(2): 283–305, 1982.
32. J. J. Wright and D. T. L. Liley. Simulation of electrocortical waves. *Biological Cybernetics*, 72: 347–356, 1995.
33. J. J. Wright and D. T. J. Liley. Dynamics of the brain at global and microscopic scales: neural networks and the EEG. *Behavioral and Brain Sciences*, 19: 285–320, 1996.
34. D. T. J. Liley, D. M. Alexander, J. J. Wright, and M. D. Aldous. Alpha rhythm emerges from large-scale networks of realistically coupled multicompartmental model cortical neurons. *Network: Computational. Neural Systems*, 10: 79–92, 1999.

35. C. Koch and I. Segev, editors. *Methods in Neuronal Modelling. From Ions to Networks*. Computational Neuroscience. MIT Press, Cambridge (MA), 1998.
36. A. L. Hodgkin and A. F. Huxley. A quantitative description of membrane current and its application to conduction and excitation in nerve. *Journal Physiology*, 117: 500–544, 1952.
37. V. K. Jirsa and H. Haken. Field theory of electromagnetic brain activity. *Physical Review Letters*, 77(5): 960–963, 1996.
38. V. K. Jirsa. Information processing in brain and behavior displayed in large-scale scalp topographies such as EEG and MEG. *International Journal of Bifurcation and Chaos*, 14(2): 679–692, 2004.
39. J. J. Wright, C. J. Rennie, G. J. Lees, P. A. Robinson, P. D. Bourke, C. L. Chapman, E. Gordon, and D. L. Rowe. Simulated electrocortical activity at microscopic, mesoscopic, and global scales. *Neuropsychopharmacology*, 28: S80–S93, 2003.
40. J. J. Wright, C. J. Rennie, G. J. Lees, P. A. Robinson, P. D. Bourke, C. L. Chapman, E. Gordon, and D. L. Rowe. Simulated electrocortical activity at microscopic, mesoscopic and global scales. *International Journal of Bifurcation and Chaos*, 14(2): 853–872, 2004.
41. P. A. Robinson, C. J. Rennie, J. J. Wright, H. Bahramali, E. Gordon, and D. L. Rowe. Prediction of electroencephalic spectra from neurophysiology. *Physical Reviews E*, 63, 021903, 2001.
42. D. O. Hebb. *The Organization of Behavior*. Wiley, New York (NY), 1949.
43. S. Kaplan, M. Sonntag, and E. Chown. Tracing recurrent activity in cognitive elements (TRACE): a model of temporal dynamics in a cell assembly. *Connection Science*, 3: 179–206, 1991.
44. F. van der Velde and M. de Kamps. Neural blackboard architectures of combinatorial structures in cognition. *Behavioral and Brain Sciences*, 29:37–108, 2006.
45. T. Wennekers, M. Garagnani, and F. Pulvermüller. Language models based on hebbian cell assemblies. *Journal of Physiology - Paris*, 100: 16–30, 2006.
46. P. Smolensky and G. Legendre. *The Harmonic Mind. From Neural Computation to Optimality-Theoretic Grammar*, Vol. 1: Cognitive Architecture. MIT Press, Cambridge (MA), 2006.
47. R. B. Stein, K. V. Leung, M. N. Oğuztöreli, and D. W. Williams. Properties of small neural networks. *Kybernetik*, 14: 223–230, 1974.
48. D. E. Rumelhart, J. L. McClelland, and the PDP Research Group, editors. *Parallel Distributed Processing: Explorations in the Microstructure of Cognition*, Vol. I. MIT Press, Cambridge (MA), 1986.
49. B. A. Pearlmutter. Learning state space trajectories in recurrent neural networks. *Neural Computation*, 1(2): 263–269, 1989.
50. B. A. Pearlmutter. Gradient calculations for dynamic recurrant neural networks: A survey. *IEEE Transaction on Neural Networks*, 6(5): 1212–1228, 1995.
51. P. Werbos. Back-propagation through time: What it does and how to do it. Vol. 78 of *Proc. IEEE*, pp. 1550–1560, 1990.
52. P. Werbos. Maximizing long-term gas industry profits in two minutes in Lotus using neural network models. *IEEE Transaction on Systems, Man, and Cybernetics*, 19: 315–333, 1989.
53. W. H. Press, S. A. Teukolsky, W. T. Vetterling, and B. P. Flannery. *Numerical Recipies in C*. Cambridge University Press, New York, 1996.

54. T. Liebscher. Modeling reaction times with neural networks using leaky integrator units. In K. Jokinen, D. Heylen, and A. Nijholt, editors, *Proc. 18th Twente Workshop on Language Technology*, Vol. 18 of *TWLT*, pp. 81–94, Twente (NL), 2000. Univ. Twente.

55. E. R. Kandel, J. H. Schwartz, and T. M. Jessel, editors. *Principles of Neural Science*. Appleton & Lange, East Norwalk, Connecticut, 1991.

56. P. Dayan and L. F. Abbott. *Theoretical Neuroscience*. Computational Neuroscience. MIT Press, Cambridge (MA), 2001.

57. D. J. Amit. *Modeling Brain Function. The World of Attractor Neural Networks*. Cambridge University Press, Cambridge (MA), 1989.

58. W. J. Freeman. *Mass Action in the Nervous System*. Academic Press, New York (NY), 1975.

59. W. J. Freeman. How and why brains create meaning from sensory information. *International Journal of Bifurcation and Chaos*, 14(2): 515–530, 2004.

60. C. S. Herrmann and A. Klaus. Autapse turns neuron into oscillator. *International Journal of Bifurcation and Chaos*, 14(2): 623–633, 2004.

61. J. Guckenheimer and P. Holmes. *Nonlinear Oscillations, Dynamical Systems, and Bifurcations of Vector Fields*, Vol. 42 of *Springer Series of Appl. Math. Sciences*. Springer, New York, 1983.

62. R. Albert and A.-L. Barabási. Statistical mechanics of complex networks. *Reviews of Modern Physics*, 74(1): 47–97, 2002.

63. S. Bornholdt and H. G. Schuster, editors. *Handbook of Graphs and Networks. From the Genome to the Internet*. Wiley-VCH, Weinheim, 2003.

64. B. Bollobás. *Random Graphs*. Cambridge University Press, Cambridge (UK), 2001.

65. P. beim Graben and J. Kurths. Simulating global properties of electroencephalograms with minimal random neural networks. *Neurocomputing*, doi:10.1016/j.neucom.2007.02.007, 2007.

66. A. J. Trevelyan and O. Watkinson. Does inhibition balance excitation in neocortex? *Progress in Biophysics and Molecular Biology*, 87: 109–143, 2005.

67. S. Itzkovitz, R. Milo, N. Kashtan, G. Ziv, and U. Alon. Subgraphs in random networks. *Physical Reviews E*, 68, 026127, 2003.

68. G. Bianconi and A. Capocci. Number of loops of size h in growing scale-free networks. *Physical Review Letters*, 90(7), 2003.

69. H. D. Rozenfeld, J. E. Kirk, E. M. Bollt, and D. ben Avraham. Statistics of cycles: how loopy is your network? *Journal of physics A: Mathematical General*, 38: 4589–4595, 2005.

70. O. Sporns, G. Tononi, and G. M. Edelman. Theoretical neuroanatomy: Relating anatomical and functional connectivity in graphs and cortical connection matrices. *Cerebral Cortex*, 10(2): 127–141, 2000.

71. J. Kornmeier, M. Bach, and H. Atmanspacher. Correlates of perceptive instabilities in visually evoked potentials. *International Journal of Bifurcation and Chaos*, 14(2): 727–736, 2004.

72. S. Frisch, P. beim Graben, and M. Schlesewsky. Parallelizing grammatical functions: P600 and P345 reflect different cost of reanalysis. *International Journal of Bifurcation and Chaos*, 14(2): 531–549, 2004.

73. H. Haken. *Synergetic Computers and Cognition. A top-down Approach to Neural Nets*. Springer, Berlin, 1991.

74. H. Haken. *Principles of Brain Functioning*. Springer, Berlin, 1996.

75. C. Lourenço. Attention-locked computation with chaotic neural nets. *International Journal of Bifurcation and Chaos*, 14(2): 737–760, 2004.

76. L. O. Chua. *CNN: A paradigm for complexity*. World Scientific, Singapore, 1998.

77. D. Bálya, I. Petrás, T. Roska, R. Carmona, and A. R. Vázquez. Implementing the multi-layer retinal model on the complex-cell cnn-um chip prototype. *International Journal of Bifurcation and Chaos*, 14(2): 427–451, 2004.

78. V. Gál, J. Hámori, T. Roska, D. Bálya, Zs. Borostyánköi, M. Brendel, K. Lotz, L. Négyessy, L. Orzó, I. Petrás, Cs. Rekeczky, J. Takács, P. Venetiáner, Z. Vidnyánszky, and Á Zarándy. Receptive field atlas and related CNN models. *International Journal of Bifurcation and Chaos*, 14(2): 551–584, 2004.

79. F. S. Werblin and B. M. Roska. Parallel visual processing: A tutorial of retinal function. *International Journal of Bifurcation and Chaos*, 14(2): 843–852, 2004.

80. M. T. Huber, H. A. Braun, and J.-C. Krieg. Recurrent affective disorders: nonlinear and stochastic models of disease dynamics. *International Journal of Bifurcation and Chaos*, 14(2): 635–652, 2004.

81. P. beim Graben, S. Frisch, A. Fink, D. Saddy, and J. Kurths. Topographic voltage and coherence mapping of brain potentials by means of the symbolic resonance analysis. *Physical Reviews E*, 72: 051916, 2005.

82. H. Drenhaus, P. beim Graben, D. Saddy, and S. Frisch. Diagnosis and repair of negative polarity constructions in the light of symbolic resonance analysis. *Brain and Language*, 96(3): 255–268, 2006.

83. A. H. Kawamoto. Nonlinear dynamics in the resolution of lexical ambiguity: A parallel distributed processing account. *Journal of Memory and Language*, 32: 474–516, 1993.

84. A. Prince and P. Smolensky. Optimality: from neural networks to universal grammar. *Science*, 275: 1604–1610, 1997.

A Dynamic Model of the Macrocolumn

James J. Wright[1,2]

[1] Liggins Institute, and Department of Psychological Medicine,
   University of Auckland, School of Medicine, Auckland, New Zealand
[2] Brain Dynamics Center, Westmead Hospital,
   University of Sydney, Sydney, Australia
   jj.w@xtra.co.nz

Summary. Neurons within a cortical macrocolumn can be represented in continuum state equations that include axonal and dendritic delays, synaptic densities, adaptation and distribution of AMPA, NMDA and GABA postsynaptic receptors, and back-propagation of action potentials in the dendritic tree. Parameter values are independently specified from physiological data. In numerical simulations, synchronous oscillation and gamma activity are reproduced and a mechanism for self-regulation of cortical gamma is demonstrated. Properties of synchronous fields observed in the simulations are then applied in a model of the self-organization of synapses, using a simple Hebbian learning rule with decay. The patterns of connection of maximally stable configuration are compared to real cortical synaptic connections that emerge in neurodevelopment.

8.1 Introduction

This chapter gives an account of two complementary approaches to modeling the axo-dendritic dynamics, and of the functionally related synaptic dynamics, within a small volume of cerebral cortex. Choice of the appropriate volume is somewhat arbitrary, but a useful scale is that which has been described using a variety of related criteria as the macrocolumn, or corticocortical column a volume approximately 300 microns in surface diameter [1–3]. The extension of the dendritic and intracortical proximal axonal trees of pyramidal cells within the cortex conform to this 300 micron approximation, as is indicated in Fig. 8.1, and because of the branching structure of both dendrites and local axons, the density of synaptic connections between neurons declines with distance from the cell body [4, 5]. Consequently, the strength of interaction of neurons up to 300 microns apart is comparatively high. On the other hand, the largest fraction of synapses within any volume of cortex arises from cell bodies outside the column, and the sparse connectivity of neurons inside the column makes the intermingling of adjacent "columns" inevitable [3, 5]. As will be shown, the theoretical modeling reported here may help to provide a new definition

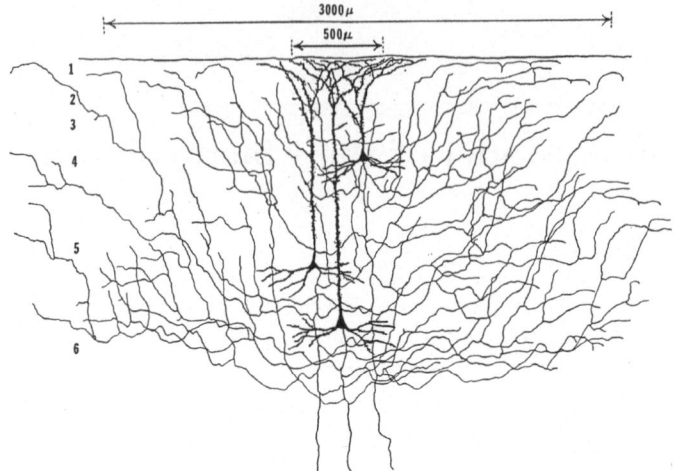

Fig. 8.1. Dendritic trees of pyramidal cells. (Braitenberg and Schüz, 1991)

of the macrocolumn, as the scale of a synaptic map of activity relayed from the wider cortex.

To simplify description, the cortex will be considered to be two-dimensional, thus largely ignoring organization in depth.

Analysis of cortical functional anatomy at the macrocolumnar scale has been found of particular utility within the visual cortex [6, 7], because this scale is also the scale of ramification in cortex of terminal axons from the direct visual pathway [2, 3, 6, 7]. Studies of responses of individual cells and of groups of cells to moving objects in the visual field, have shown that neural responses are organized so that responses to moving lines are not only selective for the orientation of the line (orientation preference, or OP), but also to the velocity, angle relative to motion, and extension of the lines [8]. Where binocular vision is present, neurons are organized into bands, each of which are about as broad as a macrocolumn, with alternating bands — the ocular dominance (OD) columns — selectively responding to one or other eye. This yields a unit system — the hypercolumn [1, 6] — with the capacity to process information from a specific small part of the visual field [9]. It can also be shown that the function of each such unit is modulated by the contextual activity of the surrounding visual field [10].

It is presently unclear how this anatomical detail is involved in the processing of visual information in neural-network terms. The modeling described here attempts to solve parts of this problem.

Synchronous oscillation is a physiological phenomenon relevant to all considerations of the processing of cortical information. When separate neurons are concurrently activated by discrete stimuli, they begin to fire synchronously, emitting action potentials with maximum cross-correlation at zero lag [11–14]. The emission of action potentials and fluctuation of the local dendritic potential is typically at around 50 Hz — although not uniquely so — and because this frequency falls within the gamma-band of electrocortical activity, it is referred to as "gamma synchrony". Gamma synchrony is believed to underlie the psychological process of perceptual "binding", allowing states of the cortex to code for many different perceptual states, by use of combinations of a smaller set of unit states per neuron [15, 16].

To provide an account of the relation of synchronous oscillation to functional anatomical organization at the scale of the macrocolumn, a continuum model of electrocortical activity is proposed. The model uses discretized integral state equations and includes effects of axonal and dendritic delays, back-propagation in the dendritic tree, reversal potentials, synaptic densities, and kinetics of AMPA, GABA and NMDA receptors, and has been reported in an earlier form in relation to global electrocortical activity [17]. To the extent practicable, all parameters are obtained from independent physiological and anatomical estimates, and all lie in the physiologically plausible range. In this and related models, gamma activity is reproduced, associated with synchronous oscillation [18–22]. As a further step toward realism, the most recent development introduces a mechanism of control of transition into autonomous gamma, which is initiated and suppressed by the level of subcortical reticular activation, and the transcortical synaptic flux originating from outside the macrocolumn.

Having described a means of generation of synchronous fields of activity in cortex, a model [23] for the self-organization of synapses can then be applied, thus extending consideration to learning-related modifications of synapses. According to this model, "local maps" are formed by self-organization of synapses during development, and each local map is analogous to a projection of the primary visual cortex (V1) onto a Möbius strip. The scale of each map is that of a macrocolumn, and they represent the most stable synaptic state in fields of neural synchrony, under conditions of uniform metabolic load and of Hebbian learning with decay. In this maximally stable state, all synapses are either saturated, or have minimum pre/postsynaptic coincidence. Each local map is arrayed as approximately a mirror-image reflection of each of its neighbors, accounting for a number of major features of local anatomical organization. Preliminary consideration is given to the impact of dynamic perturbation upon the stable synaptic configuration, and the implications for perception and cortical information processing.

8.2 A Continuum Model of Electrocortical Activity

Electrotonic and pulse activity in the cortex can be treated as activity in a wave medium, as shown by numerous workers [24–28] (cf. Chap. 1). All members of this family of models have strong resemblances, but there are differences in both the details of the state equations and the parameters applied.

The form given here has been developed to be applicable to cortex at a variety of scales, and to permit application alongside related models [17, 29].

8.2.1 State Equations

Synaptic Flux Density

The distribution of neuron cell bodies sending afferent connections to a cortical point, r, is $f(r, r')$, where $\{r'\}$ are all other points in the field. The synaptic flux density, $\varphi_p(r, t)$, the average input pulse rate per synapse at r, is given by

$$\varphi_p(r, t) = \int f(r, r') Q_p\left(r', t - \frac{|r - r'|}{v_p}\right) d^2 r', \tag{8.1}$$

where the normalized axonal spread $f(r, r')$ satisfies

$$\int f(r, r') d^2 r = \int f(r, r') d^2 r' = 1,$$

$Q_p(r', t)$ are the pulse densities of neurons in the afferent field,
v_p is the velocity of axonal conduction,
$p = e, i$ indicates whether the afferent neurons are excitatory or inhibitory.

An alternative to (8.1) is a damped wave equation, for which the implicit axonal spread is approximately a two-dimensional Gaussian [28], is:

$$\left(\frac{\partial^2}{\partial t^2} + 2\gamma_p \frac{\partial}{\partial t} + \gamma_p^2 - v_p^2 \nabla^2\right) \varphi_p(r, t) = \gamma_p^2 Q_p(r, t), \tag{8.2}$$

where γ_p is the ratio of action potential conduction velocity and the axonal range, for excitatory and inhibitory axons respectively. The wave equation is much more numerically efficient, but for the present work, the integral form has been retained.

Synapto-dendritic Transformations of Synaptic Flux

Afferent synaptic activity ultimately gives rise to a change in membrane polarization at the trigger point for the generation of action potentials. The change in membrane polarization depends upon the types of postsynaptic receptor, adaptive changes in receptor configurations, membrane reversal potentials, position of the synapses on the dendritic tree, and the state of the postsynaptic neuron — notably, whether or not it has recently discharged an action potential.

All these processes can be reduced in first approximation to steady-state equations and linear impulse responses. To compress the equations and emphasize analogies, the following conventions apply: $p = e, i$ indicates presynaptic neurons and $q = e, i$ indicates postsynaptic neurons. The superscripts

$[R] = [\text{AMPA}], [\text{NMDA}]$ when $p = e$, and $[R] = [\text{GABA}]$ when $p = i$, indicate receptor type, described further below.

Within the synapse the afferent synaptic flux is modified by changes in the conformation of ion channels [30–33]. A normalized impulse response function, $J^{[R]}(\tau)$, describes the rise and fall of receptor adaptation to a brief afferent stimulus:

$$J^{[R]}(\tau) = \left[\sum_n \frac{B_n^{[R]}}{\beta_n^{[R]}} - \sum_m \frac{A_m^{[R]}}{\alpha_m^{[R]}} \right]^{-1} \times \left[\sum_n B_n^{[R]} \exp(-\beta_n^{[R]}\tau) \right. \tag{8.3}$$

$$\left. - \sum_m A_m^{[R]} \exp(-\alpha_m^{[R]}\tau) \right],$$

where $\int_0^{\infty} J^{[R]}(\tau)\mathrm{d}\tau = 1$, $\{A_n^{[R]}, B_n^{[R]}, \alpha_n^{[R]}, \beta_n^{[R]}\}$ are constants, and $J^{[R]} = 0$ if $\tau = 0$.

The postsynaptic depolarization, $\psi_{qp}^{[R]}(\boldsymbol{r}, t)$, is the time-varying change of membrane voltage produced via synaptic receptors of a specific type, consequent on the synaptic flux density, and is defined without initial regard to the position of specific synapses on the dendritic tree. $\Psi_{qp}^{[R]}(\boldsymbol{r})$ is the steady-state value of $\psi_{qp}^{[R]}(\boldsymbol{r}, t)$:

$$\Psi_{qp}^{[R]} = g_p^{[0]} \exp[-\lambda^{[R]}\varphi_p] \left(\frac{V_p^{rev} - V_q}{V_p^{rev} - V_q^{[0]}} \varphi_p \right), \tag{8.4}$$

where

$g_p^{[0]}$ is the synaptic gain at resting membrane potential,
$\lambda^{[R]}$ is a measure of steady-state synaptic adaptation to φ_p,
V_p^{rev} is the excitatory or inhibitory reversal potential,
$V_q^{[0]}$ is the resting membrane potential, and
V_q is the average membrane potential.

Another normalized impulse response function, $H(\tau)$, describes the rise and fall of postsynaptic membrane potential

$$H(\tau) = \frac{ab}{b - a}(\mathrm{e}^{-a\tau} - \mathrm{e}^{-b\tau}), \tag{8.5}$$

where $\int_0^{\infty} H(\tau)\mathrm{d}\tau = 1$, a, b, are constants, and $H = 0$ if $\tau = 0$.

Consequently, from (8.1–8.5), and where "*" indicates convolution in time,

$$\psi_{qp}^{[R]} = H_p * (J^{[R]} * \Psi_{qp}^{[R]}). \tag{8.6}$$

Transmission of Postsynaptic Depolarization to Initiate Action Potentials

At the release of an action potential at the soma, a retrograde propagation takes place, depolarizing the dendritic membrane throughout the proximal dendritic tree [34]. This must have major implications for the weight of individual synapses in determining any subsequent action potential generation, depending on the recent history of activity in the neuron. Those synapses within the zone of back-propagation can be called "near" synapses, and those more distal in the dendritic trees "far" synapses. It is assumed that when the neuron is fully re-polarized, the greatest weight in the generation of a subsequent action potential can be ascribed to activity at the near synapses, because of their weighting by proximity to the axon hillock. On the release of an action potential, the near synapses have their efficacy reduced to zero during the absolute refractory period, and the distal synaptic trees become partially depolarized, so that whether or not a subsequent action potential is generated at the end of the refractory period is relatively weighted toward activity at the far synapses, conducted via cable properties with delay to the trigger point. Thus, in the continuum formulation, the impact of transmission of total postsynaptic flux to the trigger point depends upon cable delays in near and far dendritic trees, the fraction of neurons which have recently fired, and the relative distribution of synapses and receptor types in the near and far dendritic trees.

Following the normalized format of (8.4) and (8.5), the cable delay, L^j, is given by

$$L^j = a^j \exp(-a^j \tau) , \tag{8.7}$$

where a^j are constants, and $j = n, f$ indicate synapses positioned in the near and far dendritic trees, respectively.

Consequently, the fractions $A^f(t), A^n(t)$, of neurons responding primarily to near or far synapses, are

$$A^f(t) = \frac{Q_q}{Q_q^{\max}} \tag{8.8}$$

$$A^n(t) = 1 - \frac{Q_q}{Q_q^{\max}} , \tag{8.9}$$

where $Q_q^{\max}$ is the maximum firing rate of neurons and reflects the refractory period, while Q_q is the pulse density at r.

Since the distribution of postsynaptic receptors differs in near and far trees, fractional distributions $r^{j[R]}$ can be defined with

$$r^{n[R]} + r^{f[R]} = 1 . \tag{8.10}$$

$r^{j[R]}$ can be used to fractionally weight the synaptic numbers, N_{qp}, the number of excitatory or inhibitory synapses per neuron.

Equation (8.6) can then be aggregated over types of afferent neuron and number of synapses and types of receptor in the near and far dendritic trees, as

$$V_q(t) = V_q^{[0]} + \sum_p \sum_j \sum_{[R]} r^{j[R]} N_{qp} A^j (L^j * \psi_{qp}^{[R]}),$$ (8.11)

where $V_q(t)$ is the potential of the dendritic membrane at the trigger points, and $V_q^{[0]}$ is the resting membrane potential. The value of $V_e(t)$ — the potential in the excitatory (pyramidal) neurons — scales as the local field potential (LFP; cf. Chap. 1), and $V_q(t)$ can be applied as a surrogate for V_q in (8.4).

Generated Pulse Density

Generation of action potentials then follows the sigmoidal relation

$$Q_q(t) = \frac{Q_q^{\max}}{1 + \exp[-\pi(V_q - \theta_q)/(\sqrt{3}\sigma_q)]}.$$ (8.12)

θ_q is the mean value of V_q at which 50% of neurons are above threshold for the emission of action potentials. σ_q approximates one standard deviation of probability of emission of an action potential in a single cell, as a function of V_q.

8.2.2 Parameter Values

Parameter values for the state equations have been obtained from anatomical and physiological measurements, or inferred from direct measurements. They are presented in Tabs. 8.1–8.5. In most instances values are only known approximately, and confidence intervals are unknown or uncertain. Parameters expressed as constants or as linear processes must, in reality, be time-varying and nonlinear to some degree. However, sensitivity studies to be reported elsewhere indicate that despite reservations on the accuracy of individual parameters, the system properties to be reported are relatively robust to parameter

Table 8.1. Synaptic numbers and gain factors [3, 5, 21, 35, 36]

$N_{ee,cc}$	Excitatory to excitatory corticocortical synapses/cell	3710 dimensionless
$N_{ie,cc}$	Excitatory to inhibitory corticocortical synapses/cell	3710 dimensionless
$N_{ee,ic}$	Excitatory to excitatory intracortical synapses/cell	410 dimensionless
$N_{ei,ic}$	Inhibitory to excitatory intracortical synapses/cell	800 dimensionless
$N_{ie,ic}$	Excitatory to inhibitory intracortical synapses/cell	410 dimensionless
$N_{ii,ic}$	Inhibitory to inhibitory intracortical synapses/cell	800 dimensionless
$N_{ee,ns}$	Synapses per excitatory cell from subcortical sources	100 dimensionless
$N_{ie,ns}$	Synapses per inhibitory cell from subcortical sources	0 dimensionless
$g_e[0]$	Excitatory gain per synapse at rest potential	2.4×10^{-6} Vs
$g_i[0]$	Inhibitory gain per synapse at rest potential	-5.9×10^{-6} Vs

Table 8.2. Threshold values [21, 37]

$Q_e^{\max}$	Maximum firing rate of excitatory cells	$100\,\mathrm{s}^{-1}$
$Q_i^{\max}$	Maximum firing rate of inhibitory cells	$200\,\mathrm{s}^{-1}$
V_e^{rev}	Excitatory reversal potential	$0\,\mathrm{V}$
V_i^{rev}	Inhibitory reversal potential	$-0.070\,\mathrm{V}$
$V_{q,p}^{[0]}$	Resting membrane potential	$-0.064\,\mathrm{V}$
θ_q	Mean dendritic potential when 50% of neurones firing	$-0.035\,\mathrm{V}$
σ_q	Standard deviation of neuron firing probability, versus mean dendritic potential	$0.0145\,\mathrm{V}$

variation. Since the state equations are given in terms of scalar gains and normalized impulse responses embedded within convolutions, errors in individual parameters can partially cancel. With the following tables, sources for the values are given as numbered references with each table title, and further qualifications given in the associated text.

The parameters $\mathfrak{a}^j$ have not been specifically sourced, but are approximate physiologically realistic delays.

These parameters were obtained by deriving steady-state and impulse response functions from mass-action models of receptor/transmitter interactions.

Distribution of three receptor types were considered as representative of a much wider group of receptors. These were the principal fast excitatory glutamate receptor (AMPA), the principal fast inhibitory GABA receptor (GABA$_\mathrm{A}$) and the principal slow and voltage-dependent glutamate receptor (NMDA). NMDA is distributed predominantly in the distal dendritic tree [38] and the others more uniformly — and their distribution may be subject to dynamic functional variation. Thus, the values applied are rather arbitrary. Sensitivity analyses indicate that robust results may be obtained despite considerable variation of the values applied, and adjustment of these parameters depended upon obtaining match to the average firing rates of cells observed in cortex [39].

Table 8.3. Membrane time constants [21]

a_{ee}	EPSP decay time-constant in excitatory cells $68\,\mathrm{s}^{-1}$
b_{qp}	EPSP and IPSP rise time-constants $\quad 500\,\mathrm{s}^{-1}$
a_{ei}	IPSP decay time-constant in excitatory cells $47\,\mathrm{s}^{-1}$
a_{ie}	EPSP decay time-constant in inhibitory cells $176\,\mathrm{s}^{-1}$
a_{ii}	IPSP decay time-constant in inhibitory cells $82\,\mathrm{s}^{-1}$
$\mathfrak{a}^j$	Delay attributable to position of near and far $\mathfrak{a}^n = 1000\,\mathrm{s}^{-1}$ synapses on dendritic tree

$$\mathfrak{a}^f = 200\,\mathrm{s}^{-1}$$

Table 8.4. Receptor adaptation gains and time constants [30–33]

$\lambda^{[R]}$	Receptor adaptation pulse-efficacy decay constants	[AMPA] = 0.012 s
		[NMDA] = 0.037 s
		[GABA$_A$] = 0.005 s
$B_n^{[R]}$	Receptor onset coefficients	[AMPA]$_1$ = 1.0
		[NMDA]$_1$ = 1.0
		[GABA$_A$]$_1$ = 1.0
		dimensionless
$A_n^{[R]}$	Receptor offset coefficients	[AMPA]$_1$ = 0.0004
		[AMPA]$_2$ = 0.6339
		[AMPA]$_3$ = 0.3657
		[NMDA]$_1$ = 0.298
		[NMDA]$_2$ = 0.702
		[GABA]$_1$ = 0.0060
		[GABA]$_2$ = 0.9936
		dimensionless
$\beta_n^{[R]}$	Receptor onset time-constants	[AMPA]$_1$ = 760.0 s^{-1}
		[NMDA]$_1$ = 50.5 s^{-1}
		[GABA]$_1$ = 178.0 s^{-1}
$\alpha_n^{[R]}$	Receptor offset time-constants	[AMPA]$_1$ = 21.8 s^{-1}
		[AMPA]$_2$ = 60.3 s^{-1}
		[AMPA]$_3$ = 684.0 s^{-1}
		[NMDA]$_1$ = 0.608 s^{-1}
		[NMDA]$_2$ = 3.3 s^{-1}
		[GABA]$_1$ = 11.2 s^{-1}
		[GABA]$_2$ = 127 s^{-1}

8.2.3 Application to the Macrocolumn

The continuum model was applied numerically in discrete form, using a 20×20 matrix of "elements", each of which can be considered as situated at the position r, surrounded by other elements at positions $\{r'\}$, and each coupled to the others so as to create an approximation of the "Mexican Hat"

Table 8.5. Receptor distribution

$r^{n[R]}$	Relative weighting of receptors on near dendritic field	[AMPA] = $1 - r^{f[R]}$
		[NMDA] = $1 - r^{f[R]}$
		[GABA] = $1 - r^{f[R]}$
$r^{f[R]}$	Relative weighting of receptors on far dendritic field	[AMPA] = 0.5
		[NMDA] = 1.0
		[GABA] = 0.375

configuration [40] of excitatory and inhibitory intracortical connections within a macrocolumn, in accord with a two-dimensional Gaussian version of (8.1). Where γ_p represents the standard deviation of axonal range. The connection densities as a function of distance are thus

$$\varphi_p(\boldsymbol{r},t) = \int \frac{1}{2\pi\gamma_p^2} \exp\left[-\frac{|\boldsymbol{r}-\boldsymbol{r}'|^2}{2\gamma_p^2}\right] Q_p(t-\delta_p)\mathrm{d}^3\boldsymbol{r}', \qquad (8.13)$$

The value of γ_p was 4.9 simulation elements for the excitatory couplings, and 4.5 simulation elements for the inhibitory couplings. A wide range of plausible axonal conduction velocities were applied, and results found insensitive to variation for all small conduction lags, consistent with the size of

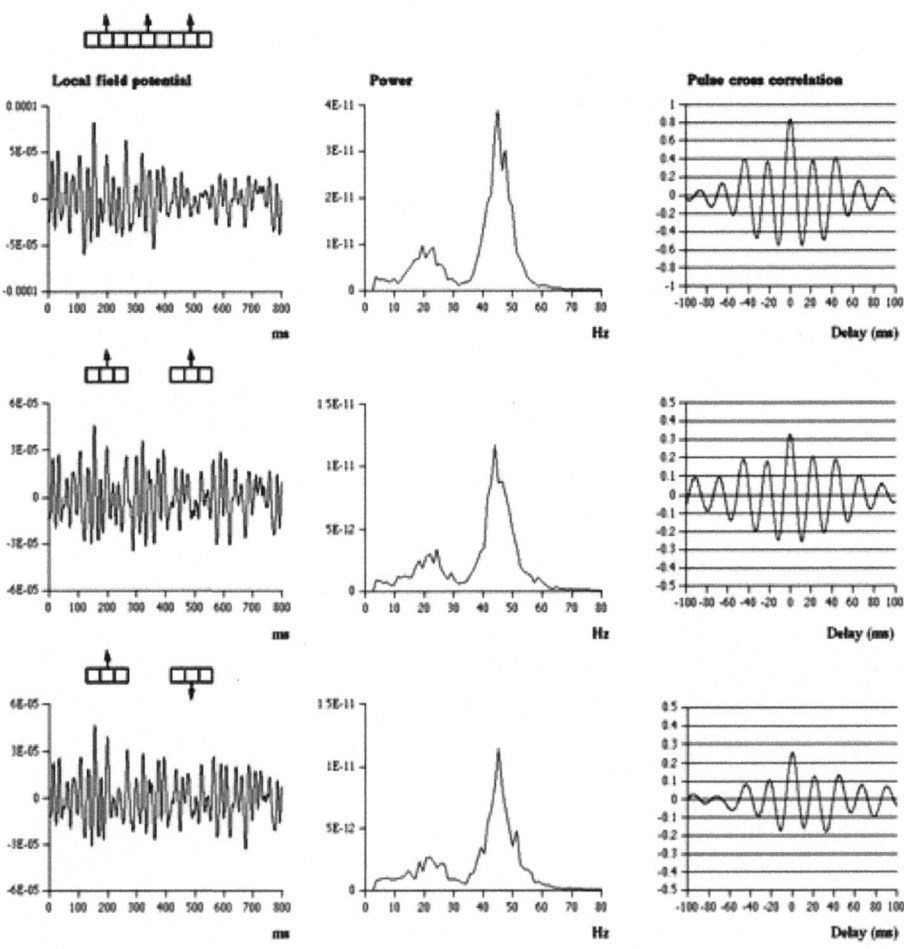

Fig. 8.2. Simulation of synchronous oscillation induced by moving bars (size and movement shown as arrowed icons) in the visual field. Plotted are local field potential time-series, power spectral content, and cross-correlations of two sites in the cortical field, when driven by simulated moving bars — with each "bar" a field of zero-mean white noise, uncorrelated in separate bars. (Wright et al., 2000)

the macrocolumn. In the results shown below, axonal delay was 0.4 ms per element. Simulation time-step was 0.1 ms.[3]

The tabulated parameters were applied distinguishing the synapses for nonspecific cortical activation from the reticular formation ($N_{ee,ns}$) and those reaching the macrocolumn from the surrounding cortex ($N_{ee,cc}$ and $N_{ie,cc}$) as the principal sites of external input to the macrocolumn.

The non-specific afferent flux was considered excitatory and terminating on excitatory cortical neurons only (consistent with its predominant input to the upper layers of the cortex, where the pyramidal cell dendritic trees predominate) [3]. The afferent flux from trans-cortical sources terminated on both excitatory and inhibitory neurons.

8.2.4 Comparison of Simulation to Experimental Data

Figure 8.2 shows that when the simulation is configured to imitate results representative of synchronous oscillation (differential response to short and long moving bars in the visual field), LFP time-signatures, LFP spectra, and pulse cross-correlations are like those seen in real data [11–15]. (These results were obtained with an earlier simulation having properties identical in the respects shown, to the present simulation.)

Figure 8.3 shows that the balance of the non-specific afferent synaptic flux and the trans-cortical synaptic flux entering the macrocolumn can act

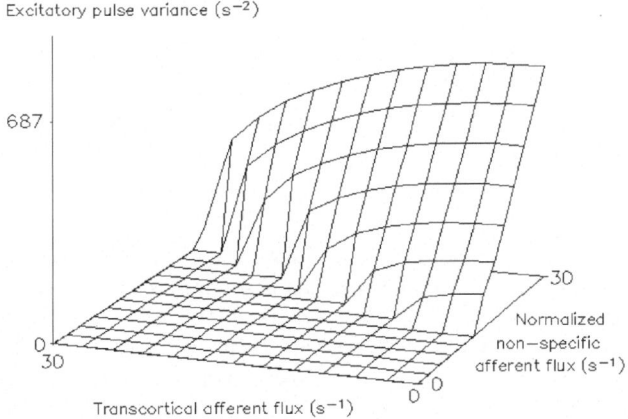

Fig. 8.3. Amplitude of simulated gamma band oscillation as a function of the excitatory synaptic flux delivered to pyramidal neurons only (the nonspecific afferent flux) versus the excitatory synaptic flux delivered to both pyramidal and inhibitory neurons (the transcortical afferent flux). (Units of nonspecific afferent flux have been "normalized" to avoid specification of synaptic efficacies of connections from subcortical sources)

---

[3] In computation, due to the serial nature of the algorithm, delay by a time step must be assumed between (8.4) and (8.11), and also between (8.12) and (8.11).

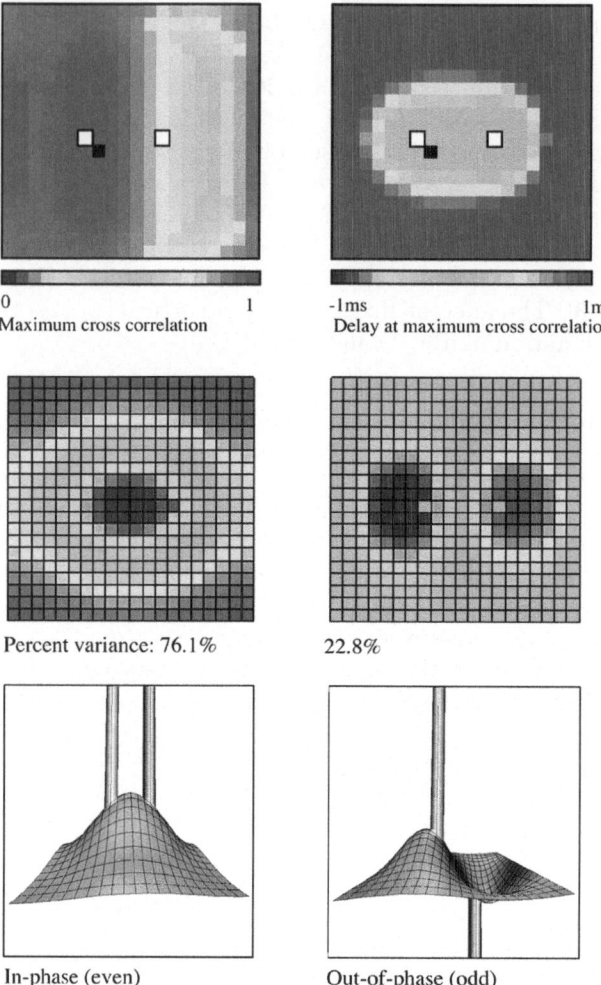

0   1   -1ms   1ms
Maximum cross correlation   Delay at maximum cross correlation

Percent variance: 76.1%   22.8%

In-phase (even)   Out-of-phase (odd)

Fig. 8.4. Essential properties of synchronous oscillation. **Top figures** A representation of the simulated cortical surface. Open squares represent the sites of input of uncorrelated white noise. The *filled square* is the reference point from which cross-correlations are calculated with respect to the rest of the field. **Top left** Maximum positive cross-correlation (over all lags). **Top right** Delay associated with maximum positive cross-correlation. **Middle figures** The first and second principle eigenmodes of spatial activity on the same simulated surface. **Bottom figures** Schematic "freeze frame" images of local field potentials (or pulse densities) on the simulated cortical surface when the twin inputs are in-phase or anti-phase signals. (Wright et al. 2003)

as a control parameter, respectively initiating and suppressing the onset of oscillation in the macrocolumn according to the balance of excitatory and inhibitory tone and providing an explanation of gamma bursts, with a pulse variance consistent with gamma oscillation [39].

Figure 8.4 shows the basis of the synchronous fields generated within simulations of this type [18–20, 22]. In brief, synchronous oscillation arises from a distinctive property of the cortical wave medium. "Odd" components in any pair of the Fourier components in signals input to dendrites are selectively dissipated — since dendrites are summing junctions. This selective elimination leaves the synchronous "even" components of activity at any two sites predominant.

The ubiquity of synchronous oscillation in the cortical field, and the mechanism of synchrony — which is not confined to the gamma band, but applies to all frequencies [41] — has implications for the self-organization of synapses in the developing brain, as we argue in the next section.

8.3 Synaptic Dynamics

The second part of the model summarized here is concerned with self-organization of synapses during antenatal and/or early postnatal visual development [23]. The emphasis here is upon the most stable configuration of synapses that can emerge under a Hebbian learning rule that incorporates "decay" (forgetting) under metabolic constraints, given an initial set of connections of random strength, density and with distribution consistent with anatomical findings, as well as the dynamical properties described above.

8.3.1 Initial Connections and the Transmission of Information in Early V1

Decline of synaptic density with distance occurs in the local intracortical connections at the scale of macrocolumns and in the longer intracortical connections spanning a fraction of the extent of the visual cortex (V1) [3–5]. Via polysynaptic transmission, information can potentially reach each macrocolumnar-sized area from the whole, or a substantial part, of V1. Thus, the distribution of terminal axonal ramifications in intracortical axons defines the scale of a *local map* of approximately macrocolumnar size, and associate the scale of V1 with a *global map* — the map of the visual field.

8.3.2 Visual Spatial Covariance, and Synchronous Fields

Because of the decline of synaptic density with distance, fields of synchronous oscillation decline in magnitude with distance [18–20, 22] as can be seen in

Fig. 8.4. Visual stimuli themselves exhibit a decline in cross-covariance with distance. Thus, cross-covariance of activity in V1 declines with distance at both the global, V1, scale, and the local, macro-columnar, scale.

8.3.3 Learning Rule

As a generic simplification of synaptic plasticity at multiple time-scales, a simple Hebbian rule with decay can be applied. These learning-related synaptic modifications fall outside the mechanisms included in the preceding account, and discussion of their physiological analogs is deferred to the conclusion.

At each synapse, the coincidence of pre- and postsynaptic activity, $r_{Q\varphi}$, is given by a relation of the form

$$r_{Q\varphi} \propto \sum_t Q_e(t) \times \varphi_e(t) , \qquad (8.14)$$

where $Q_e(t) \in \{0, 1\}$ is the postsynaptic firing state, and $\varphi_e(t) \in \{0, 1\}$ is the presynaptic firing state. A multiplication factor, H_s, operating on the gain of synapses at steady states of pre- and postsynaptic firing is approximately

$$H_s = H_{\max} \exp(-\lambda/r_{Q\varphi}) , \qquad (8.15)$$

where λ is a suitable constant. With changes in either the pre- or postsynaptic firing state, H_s can increase or fade over time, at rates differing for fast and slow forms of memory storage.

8.3.4 Individual Synaptic States of Stability

It can be shown [23] that under these learning rules, synapses can approach a stable, unchanging state only by approaching either one of two extremes — either *saturated* or *sensitive*. In the saturated state, H_s and $r_{Q\varphi}$ are at maxima, while in the sensitive state, H_s and $r_{Q\varphi}$ are at minima. Conversely, $\frac{dH_s}{dr_{Q\varphi}}$, the sensitivity to change in synaptic gain, is at a minimum for saturated synapses and a maximum for sensitive synapses — hence the choice of the names.

8.3.5 Metabolic Uniformity

Competition for metabolic resources within axons adds a constraint to Hebbian rules [42]. The metabolic energy supply of all small axonal segments can be presumed to remain approximately uniform, while the metabolic demand of saturated synapses, which have high activity, will be much greater than for sensitive synapses. Therefore, the proportion of saturated and sensitive synapses must be uniform along axons, and consequently, the densities of both saturated and sensitive synapses must decline with the distance of the presynapses from the cell bodies of origin.

8.3.6 The Impact of Distance/Density and Saturation/Sensitivity on Overall Synaptic Stability

All positions in V1, $\{P_{j,k}\}$, can be given an ordered numbering in the complex plane, $1 \ldots, j, \ldots, k, \ldots, 2n$, and all positions within a macrocolumn located at P_0, $\{p_{j,k}\}$, can be similarly numbered. The total perturbation of synaptic gains for the synapses from V1 entering the macrocolumn, $\Psi(pP)$, and the total perturbation of synaptic gains within the macrocolumn, $\Psi(pp)$, can thus be written as

$$\Psi(pP) = \sum_{j=1}^{j=n} \sum_{k=1}^{k=n} \sigma_{SAT}(p_j P_k) S_{SAT}(p_j P_k) + \qquad (8.16)$$

$$\sum_{j=1}^{j=n} \sum_{k=1}^{k=n} \sigma_{SENS}(p_j P_k) S_{SENS}(p_j P_k)$$

$$\Psi(pp) = \sum_{j=1}^{j=n} \sum_{k=1}^{k=n} \sigma_{SAT}(p_j p_k) S_{SAT}(p_j p_k) + \qquad (8.17)$$

$$\sum_{j=1}^{j=n} \sum_{k=1}^{k=n} \sigma_{SENS}(p_j p_k) S_{SENS}(p_j p_k) \, ,$$

where $\sigma_{SAT}(p_j P_k, p_j p_k)$ and $\sigma_{SENS}(p_j P_k, p_j p_k)$ are the densities of saturated and sensitive synapses respectively, and $S_{SAT}(p_j P_k, p_j p_k)$ and $S_{SENS}(p_j P_k, p_j p_k)$ are the corresponding variations of synaptic gains over a convenient short epoch.

Since the densities of synapses decline with increasing cell separation, then as a simple arithmetic property of sums of products, minimization of $\Psi(pp)$ requires neurons separated by short distances to most closely approach maximum saturation, or maximum sensitivity. Yet, metabolic uniformity requires that both sensitive and saturated synaptic densities must decline with distance from the cell bodies of origin, and remain in equal ratio. An apparent paradox arises, since sensitive synapses must link pre- and postsynaptic neurons with minimal pre- and postsynaptic pulse coincidence, yet the reverse is true for saturated synapses. Also apparently paradoxically, minimization of $\Psi(pP)$ requires that saturated connections afferent to any p_j arise from highly covariant, and therefore closely situated, sites in V1, while sensitive connections afferent to p_j must arise from well-separated sites. Yet, metabolic uniformity requires that both sensitive and saturated presynapses arise from cells at the same site. The paradoxes exist only in the Euclidean plane, and can be resolved as in the next subsection.

8.3.7 Möbius Projection, and the Local Map

By re-numbering $\{P_{j,k}\}$ as $\{P_{j1,j2,k1,k2}\}$, and $\{p_{j,k}\}$ as $\{p_{j1,j2,k1,k2}\}$, the subscript numbers $1, \ldots, j1, \ldots, j2, \ldots, n, (n+1), \ldots, k1, \ldots, k2, \ldots, 2n$ can be assigned in the global map so that $j1$ and $j2$ are located diametrically opposite and equidistant from P_0, while in the local map $j1$ and $j2$ have positions analogous to superimposed points located on opposite surfaces of a Möbius strip. This generates a *Möbius projection* (the *input map*) from global to local, and a *Möbius ordering* within the local map. That is,

$$\frac{P_{jm}^2}{|P_{jm}|} \to p_{km}, \qquad m \in \{1, 2\} \tag{8.18}$$

and

$$p_{jm} \to p_{km} \qquad m \in \{1, 2\}. \tag{8.19}$$

In (8.18), the mapping of widely separated points in the global map converge to coincident points on opposite surfaces of the local map's Möbius representation. In (8.19), the density of saturated synaptic connections now decreases as $|j1 - k1|$ and $|j2 - k2|$, while the density of sensitive couplings decreases as $|j2 - k1|$ and $|j1 - k2|$.

The anatomical parallel requires $j1$ and $j2$ in the local map to represent two distinct groups of neurons. To attain maximum synaptic stability within the local map, an intertwined mesh of saturated couplings forms, closed after passing twice around the local map's center, with sensitive synapses locally linking the two turns of the mesh together. In this fashion, both saturated and sensitive synapses decline in density with distance as required. The input map is of corresponding form, conveying an image of the activity in V1 analogous to projection onto a Möbius strip.

Evolution of these patterns of synaptic connections is shown in Figs. 8.5 and 8.6.

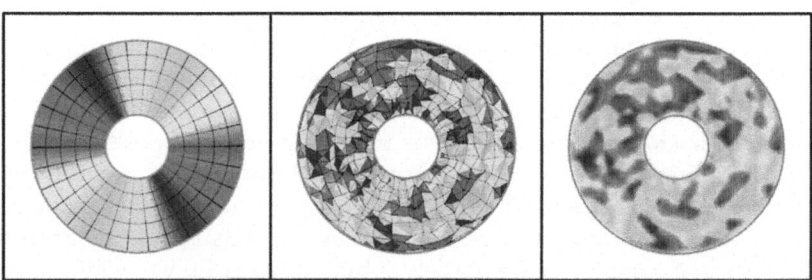

Fig. 8.5. Initial conditions for local evolution of synaptic strength. **Left.** The global field (V1) in polar co-ordinates. Central defect indicates the position of a local area of macro-columnar size. Polar angle is shown by the color spectrum, twice repeated. **Middle.** Zones of random termination (shown by color) of lateral axonal projections from global V1 in the local area. Central defect is an arbitrary zero reference. **Right.** Transient patterns of synchronous oscillation generated in the local area, mediated by local axonal connections

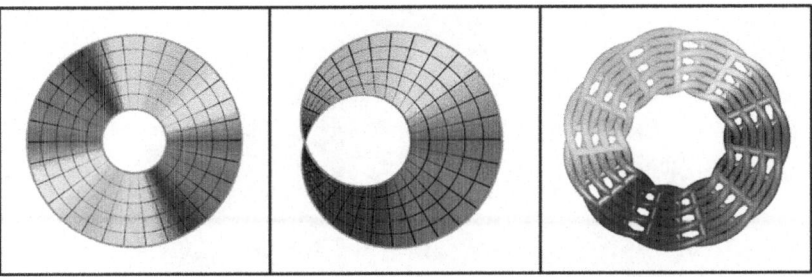

Fig. 8.6. Evolution of synaptic strengths to their maximally stable configuration. **Left**. The global field (V1), as represented in Fig. 8.1. **Middle**. Saturated synaptic connections input from the global field now form a Möbius projection of the global field, afferent to the local neuronal field, and forming a local map. **Right**. Saturated local synapses, within the local map, form a mesh of connections closed over $0 - 4\pi$ radians. The central defect now corresponds to the position within the local map, of the local map within the global map. Sensitive synapses (not shown) link adjacent neurons as bridges between the $0 - 2\pi$ and $2\pi - 4\pi$ limbs of the mesh of saturated connections. (Wright et al. 2006)

8.3.8 Monosynaptic Interactions between Adjacent Local Maps

The input and local maps can, in principle, emerge with any orientation, and with either left or right-handed chirality. However, chirality and orientation of adjacent local maps is also constrained by a requirement for overall stability. Adjacent local maps should form an approximately mirror image relation, as shown in Fig. 8.7, because in that configuration, homologous points within the local maps have the densest saturated and sensitive synaptic connections, thus meeting minimization requirements analogous to those of (8.16) and (8.17).

8.3.9 Projection of Object Motion to the Local Map and Dendritic Integration

Since the emergent input and local maps form a 1:1 representation of points in the global map, they enable the relay of information delivered to V1 by the visual pathway to every local map. This pattern is relayed to the local map according to

$$O\left(\frac{P_{jm}^2}{|P_{jm}|}, t - \frac{P_{jm}}{\nu}\right) \Rightarrow O(p_{jm}, t), \qquad (8.20)$$

where $O(P, t)$ is the pattern of neuronal firing generated in V1 by a visual object and ν is the axonal conduction velocity.

When signals from global V1 are received in the local map, they are subject to integration over time in local dendrites. If we represent local dendritic potentials as $V(p, t)$, average synaptic gain (incorporating the Hebbian gain factor) as g, dendritic rise and fall time-constants as a, b, then

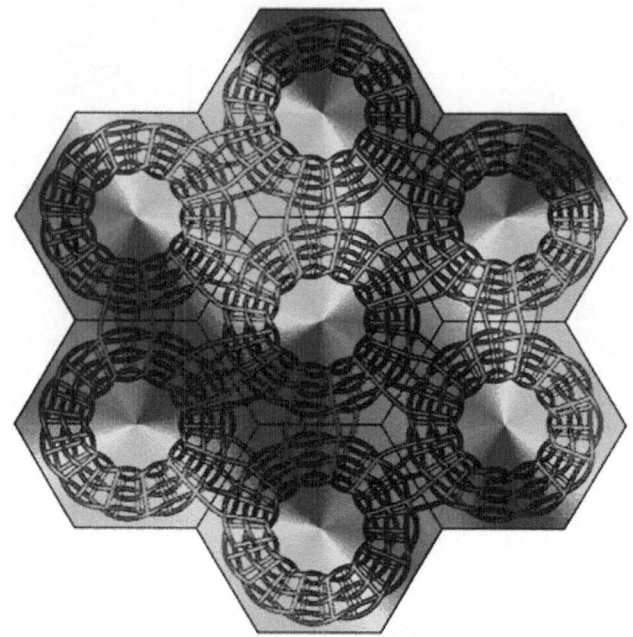

Fig. 8.7. Mutual organization of saturated coupling within and between local maps. Orientation and chirality of OP in macrocolumns. (Wright et al. 2006)

$$V(p_{jm}, t) = [g(e^{-a\tau} - e^{-b\tau})] * \left[O(P_{jm}, t - \frac{P_{jm}}{\nu} - \tau)\right], \qquad t, \tau \geq 0 \quad (8.21)$$

expresses the way moving objects in the visual field exert threshold or sub-threshold effects on action potential generation within the local map. The impact of delayed conduction from the global to the local maps can account for the recently discovered [8] dependence of OP on stimulus velocity, angle of stimulus orientation to direction of motion, and extension of the stimulus [23].

8.3.10 Effects of Perturbation

Because all activity in global V1 is projected to each local map, visual stimuli must act to perturb synaptic gains away from the stable configuration. Further, cells at any two corresponding positions on the mesh of saturated connections positioned on the opposite $0-2\pi$ and $2\pi-4\pi$ limbs of the mesh and connected with sensitive synapses of high density are maximally sensitive to perturbation when concurrently stimulated by some visual object. Figure 8.8 shows the effect of a stimulus such as a moving line in the visual field, which will give rise to a strong perturbation, followed by a relaxation back toward the stable configuration as the stimulus is withdrawn. Perturbation interactions within and between local maps on short time scales may account for phenomena of perceptual closure.

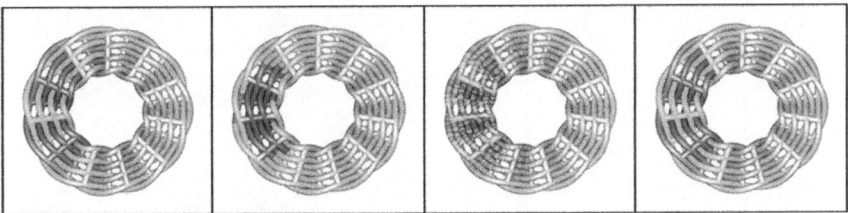

Fig. 8.8. Perturbation of synaptic saturation and sensitivity by extended stimuli.
Left. A representation of the connections formed by a small group of neurons with
cell bodies located at 3 o'clock in the local map. Saturated connections (*red*) and
sensitive connections (*green*) are arrayed at their maximally stable configuration.
Second from left. An afferent volley is delivered to neurons at 3 o'clock in the
local map, arising from sites on both sides of the position of the local map, so
that neurons in the $0 - 2\pi$ and $2\pi - 4\pi$ limbs of the mesh are forced into highly
correlated firing. **Second from right.** On withdrawal of the perturbing afferent
volley, the synaptic configuration generated by the perturbation begins to decay.
Right. The maximally stable configuration is again attained

Decay to the maximally stable configuration may occur on multiple time-
scales, and with continuing perturbation may be retarded indefinitely if
growth mechanisms overcome the prior requirements of metabolic uniformity.

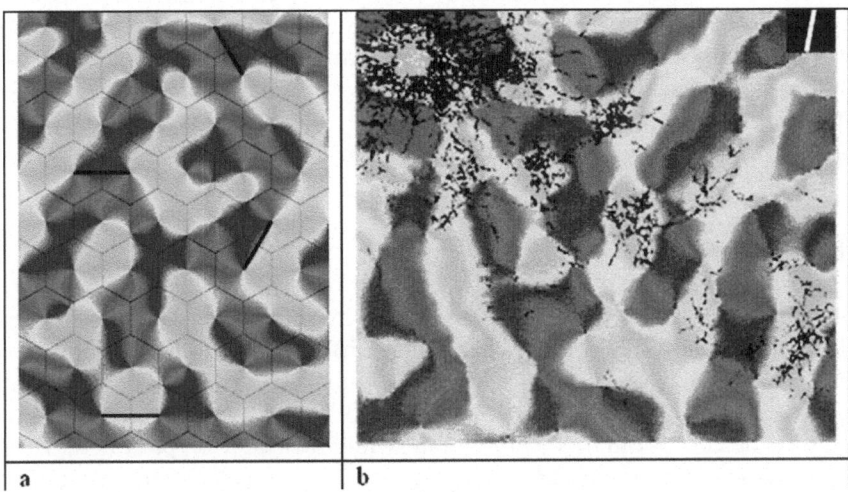

Fig. 8.9. Simulated and real maps of orientation preference: (**a**) Final configuration
of OP consequent to seeding the development of fields of OP with the local map
mirror-image pairs shown joined by solid lines. (Wright et al. 2006); (**b**) Real OP
as visualized in the tree shrew by Bosking et al. (1997). Intracortical connections
superimposed in black connect zones of like OP

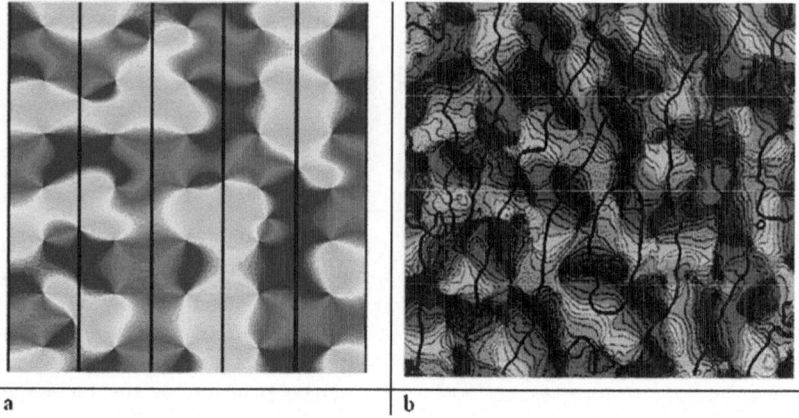

Fig. 8.10. (a) Simulated OD columns (Wright et al. 2006); (b) Real OD columns, as visualized by Obermayer & Blasdel (1993)

8.3.11 Comparison to Standard Anatomical Findings

Figures 8.9 and 8.10 show the results of simulations [23] based on the topological principles described, matched to experimental data [43, 44]. Other experimental data accounted for include direction preference fractures and the occurrence of OD columns. The OD columns arise as an "exception to the rule". Representation of visual input from each eye separately is required, since images seen with binocular disparity by the two eyes are spatially lag-correlated. This violates the requirement that information be mapped from global to local map with preservation of cross-covariance with distance. Separate representation of the images of each eye is a required compromise to achieve overall maximum stability.

8.4 Conclusion

The two pieces of research work described in this chapter offer a contrast and a convergence. The model of axo-dendritic dynamics depends upon the quantitative choice of parameter values obtained from physiological data to the extent possible. The properties of this model are relatively robust to perturbation of the parameter values, but wide variation of parameters leads to dynamics wholly unmatched to real electrocortical activity, the parameters that result in realism needing to be in particular proportions to each other [45]. Moreover, limited aspects of real dynamics can be reproduced by partial models, which utilize apparently different mechanisms [17, 29] (cf. Chaps. 7 and 5). More complete models will require confrontation of simulations with many separate, but concurrent, classes of data.

Conversely, the model of synaptic organization is essentially independent of parameter values, and depends upon topological effects, which emerge in between-scale dynamic interactions in cortex. It appears that dynamic interactions may contribute to the organization of realistic anatomical connections, thus supplementing the actions of the many growth factors, and chemical gradients contributing at biochemical level to the formation and dissolution of connections (e.g. [46]).

The basis for learning-related modifications of synaptic gains was earlier left undiscussed. The mechanisms are likely to be multiple, and occurring on many time scales. A likely major candidate for inclusion is long term potentiation (LTP) and depression (LTD). Recent work [47, 48] on learning-rule modeling and experiments in hippocampus have led to the proposal of a learning rule and a link to LTP/D consistent with the requirements for synaptic stability and sensitivity in the stable state and during perturbation (cf. Chap. 2). At a more abstract level, comparison with the information-theoretic coherent-infomax principle is apparent [38]. In a unified version, the two models may provide a framework for more detailed comparisons with experimental data, while also enabling analysis of their information storage and processing properties.

The author presumes, but has not proved, that the two models are mutually compatible and could be combined in a single numerical simulation.

References

1. P. L. Nunez. *Neocortical Dynamics and Human EEG Rhythms*. Oxford University Press, New York, Oxford. pp. 99–114, 1995.
2. J. Szentagothai. Local neuron circuits of the neocortex. In: *The Neurosciences 4th Study Program*. (F. O. Schmitt and F. G. Worden, Eds.). MIT Press, Cambridge, Mass. pp. 399–415, 1979.
3. V. Braitenberg and A. Schüz. *Anatomy of the Cortex: Statistics and Geometry*. Springer-Verlag, New York. 1991.
4. D. A. Scholl. *The Organization of the Cerebral Cortex*. Wiley, New York. 1956.
5. D. T. J. Liley and J. J. Wright. Intracortical connectivity of pyramidal and stellate cells: estimates of synaptic densities and coupling symmetry. *Network*, 5: 175–189, 1994.
6. V. B. Mountcastle. An organizing principle for cerebral function: the unit module and the distributed system. In: *The Neurosciences 4th Study Program*. (F. O. Schmitt and F. G. Worden, Eds.). MIT Press, Cambridge, Mass., 1979.
7. E. R. Kandel, J. H. Schwartz and T. M. Jessell. *Principles of Neural Science*. 3rd Edition. Prentice-Hall International (UK), London. pp. 421–439, 1991.
8. A. Basole, L. E. White and D. Fitzpatrick. Mapping multiple features of the population response of visual cortex. *Nature*, 423: 986–990, 2003.
9. D. H. Hubel and T. N. Wiesel. Receptive fields, binocular interaction, and functional architecture of the in the cat's visual cortex. *J. Physiol.*, 160: 106–154, 1962.

10. M. Fiorani, M. G. P. Rosa, R. Gattass and C. E. Rocha-Miranda. Dynamic surrounds of receptive fields in primate striate cortex: a physiological basis for perceptual completion? *Proc. Natl. Acad. Sci. USA*, 89: 8547–8551, 1992.

11. R. Eckhorn, R. Bauer, W. Jordon, M. Brosch, W. Kruse, M. Monk, H. J. Reitböck. Coherent oscillations: a mechanism of feature linking in the in the visual cortex? *Biological Cybernetics*, 60: 121–130, 1988.

12. C. M. Gray, P. König, A. K. Engel and W. Singer. Oscillatory responses in cat visual cortex exhibit intercolumnar synchronization which reflects global stimulus properties. *Nature*, 388: 334–337, 1989.

13. C. M. Gray, A. K. Engel, P. König and W. Singer. Synchronization of oscillatory neuronal responses in cat striate cortex: temporal properties. *Visual Neuroscience*, 8: 337–347, 1992.

14. C. M. Gray and W. Singer. Stimulus-specific neuronal oscillations in orientation columns of cat visual cortex. *Proc. Natl. Acad. Sci. USA*, 86: 1698–1702, 1989.

15. W. Singer and C. M. Gray. Visual feature integration and the temporal correlation hypothesis. *Annual Rev. Neuroscience*, 18: 555–586, 1995.

16. M. P. Stryker. Is grandmother an oscillation? *Nature*, 388: 297–298, 1989.

17. J. J. Wright, C. J. Rennie, G. J. Lees, P. A. Robinson, P. D. Bourke, C. L. Chapman, E. Gordon, and D. L. Rowe. Simulated electrocortical activity at microscopic, mesoscopic, and global scales. *Neuropsychopharmacology*, 28: S80–S93, 2003.

18. J. J. Wright. EEG simulation: variation of spectral envelope, pulse synchrony and 40 Hz oscillation. *Biological Cybernetics*, 76: 181–184, 1997.

19. P. A. Robinson, C. J. Rennie and J. J. Wright. Synchronous oscillations in the cerebral cortex. *Physical Review*, E 57: 4578–4588, 1998.

20. J. J. Wright, P. D. Bourke and C. L. Chapman. Synchronous oscillation in the cerebral cortex and object coherence: simulation of basic electrophysiological findings. *Biological Cybernetics*, 83: 341–353, 2000.

21. C. J. Rennie, J. J. Wright and P. A. Robinson. Mechanisms of cortical electrical activity and emergence of gamma rhythmn. *J. Theoretical Biol.*, 205(1): 17–35, 2000.

22. C. L. Chapman, P. D. Bourke and J. J. Wright. Spatial eigenmodes and synchronous oscillation: coincidence detection in simulated cerebral cortex. *J. Math. Biol.*, 45: 57–78, 2002.

23. J. J. Wright, D. M. Alexander and P. D. Bourke. Contribution of lateral interactions in V1 to organization of response properties. *Vision Research*, 46: 2703–2720, 2006.

24. W. J. Freeman. *Mass Action in the Nervous System*. Academic Press, New York. 1975.

25. H. Haken. *Principles of Brain Functioning*. Springer, Berlin. 1996.

26. P. L. Nunez. *Neocortical Dynamics and Human EEG Rhythms*. Oxford University Press, New York, Oxford. 1995.

27. V. K. Jirsa and H. Haken. Field theory of electromagnetic brain activity. *Phys. Rev. Lett.*, 77: 960–963, 1996.

28. P. A. Robinson, C. J. Rennie and J. J. Wright. Propagation and stability of waves of electrical activity in the cortex. *Phys. Rev. E*, 55: 826–840, 1997.

29. P. A. Robinson, C. J. Rennie, D. L. Rowe, S. C. O'Connor, J. J. Wright, E. Gordon et al. Neurophysical modeling of brain dynamics. *Neuropsychopharmacology*, 28: S74 – S79, 2003.

30. R. A. Lester and C. E. Jahr. NMDA channel behavior depends on agonist affinity. *J. Neuroscience*, 12: 635–643, 1992.

31. C. Dominguez-Perrot, P. Feltz and M. O. Poulter. Recombinant GABAA receptor desensitization: the role of the gamma2 subunit and its physiological significance. *J. Physiol.*, 497: 145–159, 1996.

32. W. Hausser and A. Roth. Dendritic and somatic glutamate receptor channels in rat cerebellar Purkinje cells. *J. Physiol.*, 501.1: 77–95, 1997.

33. K. M. Partin, M. W. Fleck and M. L. Mayer. AMPA receptor flip/flop mutants affecting deactivation, desensitization and modulation of cyclothiazide, aniracetam and thiocyanate. *J. Neuroscience*, 16: 6634–6647, 1996.

34. G. J. Stuart and B. Sakmann. Active propagation of somatic action potentials into neocortical cell pyramidal dendrites. *Nature*, 367: 69–72, 1994.

35. A. M. Thompson, D. C. West, J. Hahn and J. Deuchars. Single axon IPSPs elicited in pyramidal cells by three classes of interneurones in slices of rat cortex. *Journal of Physiology* (London) 496.1: 81–102, 1997.

36. A. M. Thompson. Activity-dependent properties of synaptic transmission at two classes of connections made by rat neocortical pyramidal axons *in vitro*. *Journal of Physiology* (London) 502.1: 131–147, 1997.

37. E. R. Kandel, J. H. Schwartz and T. M. Jessell. *Principles of Neural Science*. 3rd Edition. Prentice-Hall International (UK), London. pp. 81–118, 1991.

38. W. A. Phillips and W. Singer. In search of common foundations for cortical computations. *Behavioral and Brain Sciences*, 20: 657–722, 1997.

39. M. Steriade, I. Timofeev and F. Grenier. Natural waking and sleep states: a view from inside neocortical neurons. *J. Neurophysiol.*, 85: 1969–1985, 2001.

40. K. Kang, M. Shelley and H. Sompolinsky. Mexican Hats and pinwheels in visual cortex. *Proc. Natl. Acad. Sci. USA.*, 100: 2848–2853, 2003.

41. S. L. Bressler, R. Coppola and R. Nakamura. Episodic multiregional cortical coherence at multiple frequencies during visual task performance. *Nature*, 366: 153–156, 1993.

42. S. Grossberg and J. R. Williamson. A neural model of how horizontal and interlaminar connections of visual cortex develop into adult circuits that carry out perceptual grouping and learning. *Cerebral Cortex*, 11: 37–58, 2001.

43. W. H. Bosking, Y. Zhang, B. Schofield and D. Fitzpatrick. Orientation selectivity and the arrangement of horizontal connections in tree shrew striate cortex. *J. Neuroscience*, 17(6): 2112–2127, 1997.

44. K. Obermayer and G. G. Blasdel. Geometry of orientation and ocular dominance columns in monkey striate cortex. *J. Neuroscience*, 13(10): 4114–4129, 1993.

45. J. J. Wright. Simulation of EEG: dynamic changes in synaptic efficacy, cerebral rhythms and dissipative and generative activity in cortex. *Biological Cybernetics*, 81: 131–147, 1999.

46. Y. Yin, M. T. Henzl, B. Lorber, T. Nakazawa, T. T. Thomas, F. Jiang, R. Langer and L. Benowitz. Oncomodulin is a macrophage-derived signal for axon regeneration in retinal ganglion cells. *Nature Neuroscience*, 9(6): 843–852, 2006.

47. M. Tsukada and X. Pan. The spatio-temporal learning rule and its efficiency in separating spatio-temporal patterns. *Biological Cybernetics*, 92: 139–146, 2005.

48. T. Aihara, Y. Kobayashi and M. Tsukada. Spatiotemporal visualization of long-term potentiation and depression in the hippocampal CA1 area. *Hippocampus*, 15: 68–78, 2005.

Part IV

Implementations

Building a Large-Scale Computational Model of a Cortical Neuronal Network

Lucia Zemanová, Changsong Zhou and Jürgen Kurths

Institute of Physics, University of Potsdam, Germany
zemanova@agnld.uni-potsdam.de

Summary. We introduce the general framework of the large-scale neuronal model used in the 5th Helmholtz Summer School — Complex Brain Networks. The main aim is to build a universal large-scale model of a cortical neuronal network, structured as a network of networks, which is flexible enough to implement different kinds of topology and neuronal models and which exhibits behavior in various dynamical regimes. First, we describe important biological aspects of brain topology and use them in the construction of a large-scale cortical network. Second, the general dynamical model is presented together with explanations of the major dynamical properties of neurons. Finally, we discuss the implementation of the model into parallel code and its possible modifications and improvements.

9.1 Introduction

In the last few decades, an innumerable amount of information about the mammalian brain has been collected [1,2]. The anatomical properties of the cortices of different animal species have been explored in detail with modern imaging techniques revealing the functions of various brain regions and giving insight into the processes of perception and cognition.

Neural modeling represents a powerful and effective tool for the investigation and understanding of the development and organization of the brain, and of the dynamical processes. The wide spectrum of neuronal models captures and describes processes ranging from the behavior of a single cell at the microscopic level to large-scale neuronal population activity. 'Bottom-up' modeling is a common strategy used to design large cortical networks [3–6]. In this approach, the basic dynamical and topological unit of the system is a single neuron. The specific pattern of interconnections between the simple units can be represented as a complicated network. Depending on the network structure, the model can stand for a local neuronal ensemble of a cortical area or for the hierarchically organized architecture of the brain. The selection of the concrete neuronal model should take the main dynamical behaviors, such as spiking or bursting, into account.

Table 9.1. Parameters of the network — structure and connections

Parameter	Description
m	Number of areas
n	Number of neurons per area
z	Number of connections per neuron within an area
p_{ring}	Density of connections inside one area
p_{rew}	Probability of rewiring
p_{inh}	Ratio of inhibitory neurons
p_3	Ratio of neurons receiving synapses from a connected area
p_4	Ratio of neurons with synapses towards a connected area
$g_{1,\text{exc}}$	Non-normalized strength of intra-areal excitatory synapses
$g_{1,\text{inh}}$	Non-normalized strength of intra-areal inhibitory synapses
$g_{2,\text{exc}}$	Non-normalized strength of inter-areal excitatory synapses

The main idea of this chapter is to introduce a general framework for building a complex large-scale brain network that can be used to study the relationship between network topology and spreading of activity (see Chaps. 14 and 13) and present a large-scale cortical model using the 'bottom-up' approach. We discuss the neuronal properties of a single unit and the structure of the network connecting these neurons. Our aim is to build a general neuronal model able to capture and mimic various dynamical processes, as well as the wide spectrum of possible neuronal topologies. Furthermore, we would like to use this complex model to investigate the relationship between the structure and the function of the system.

In Sect. 9.2, we introduce the concept of the connectome. Subsequently, the model of the network topology and structural details are presented. All network parameters are summarized in Table 9.1. In Sect. 9.3, we deal with the

Table 9.2. Parameters of the neuronal dynamics

Parameter	Description
I_{base}	Constant base current
V_{exc}	Reversal potential for excitatory synapses
V_{inh}	Reversal potential for inhibitory synapses
D	Intensity of the Gaussian white noise
G_{ex}	Strength of Poissonian current (Pc)
N_p	Number of Pc
λ	Frequency of Pc
$\tau_{1,\text{exc}}, \tau_{2,\text{exc}}$	Rise and decay times of excitatory synaptic current
$\tau_{1,\text{inh}}, \tau_{2,inh}$	Rise and decay times of inhibitory synaptic current
A_+, A_-	Magnitude of the LTP, LTD
τ_+, τ_-	Rise and decay rate of the LTP, LTD
$t_{\text{del},1,\text{exc}}$	Delay of intra-areal excitatory synapses
$t_{\text{del},1,\text{inh}}$	Delay of intra-areal inhibitory synapses
$t_{\text{del},2,\text{exc}}$	Delay of inter-areal excitatory synapses

dynamical characteristics of neurons. The basic neuronal properties are listed and their specific role in the neuronal dynamics is explained. We again present an overview of all dynamical variables used in the model (cf. Table 9.2). Furthermore, the general framework of the large-scale neuronal model is summarized and its implementation into parallel code is described. At the end, we discuss possible improvements and extensions of the model.

9.2 Topology

9.2.1 Connectome

Mammalian brains consist of a vast number of neurons that are interconnected in complex ways [2]. In recent years, the network of anatomical links connecting neuronal elements, the connectome, has been the subject of intensive investigation. From numerous neurohistological studies, information about the morphology, location and connections of different types of neuronal cells, microcircuits and anatomical areas has been collected and sorted. These data play an important role in creating a global image of the brain. The implementation of such topological information in a large-scale neuronal model might help us to understand the mechanisms of temporal and spatial spreading of the cortical activity.

Although the details of the neuronal network architecture are not fully known, several levels of cortical connectivity can be defined [2].

Microscopic Connectivity

In the human brain, approximately 10^{11} neurons are linked together by 10^{14} to 10^{15} connections, which correspond to 10^4 synapses per neuron. The network is rather sparsely connected, with mainly local connectivity. Neurohistological studies of animal cortical tissue have pointed out that each neuron makes contact to its closest neighbors only by one synapse or not at all [7, 8]. Generally, individual neuronal interconnections are partially predetermined by genetic constraints and later modified by adaptation rules and processes like 'spike-timing-dependent plasticity' (STDP) [9], nutrition, and learning, often happening on the daily base.

For many reasons — the high number of neurons, the complex topology, frequent changes in the connectivity, the rapid decrease of living neurons in the dead tissue and invasive histological techniques (staining, neurotracers, etc.) — it is not possible to extract the complete realistic connectivity of neuronal ensembles either for animals or for humans [7].

Thus, the connections, especially at this microscopic scale, have to be modelled as a graph, whose structure ranges from simple networks such as random [5, 10–12], small-world [13, 14], or globally coupled networks [15–17] to more realistic networks reflecting spatial growth of the cortex [18] (cf. Chap. 4).

Mesoscopic Connectivity

A cortical minicolumn, an ensemble of neurons organized in the vertical direction, is considered to be a basic functional unit processing information in the brain of mammals. Such local circuit consists of only approx. 80–100 neurons, but the exact anatomical details of its structure are still not fully described [2,8]. It is assumed that the minicolumnar architecture is more complex than just random or distance dependent connections patterns. A set of these functionally specialized and precisely rewired small neuronal populations gives rise to the cortical column. Therefore, the minicolumn is deemed to be a basic building block of the complete connectome [2,19].

Macroscopic Connectivity

In the cerebral cortex, neurons are organized into numerous regions (areas) that differ in cytoarchitecture and function. These areas, originally defined and listed by Brodmann, may be assumed to be basic elements at the macroscale. Several studies have examined the topology of the neuronal fiber connections linking different areas in the animal brain [20–22]. For various species, like rat, cat and monkey, cortical maps were extracted that capture the presence and the strength of cortical connections between the areas. Unfortunately, the current histological techniques using mainly tracer injections have toxic effects on the neuronal tissue and thus it is not possible to perform similar studies on humans. Other imaging methods like Diffusion Tensor Imaging [23] are still under development and do not bring sufficiently satisfactory results.

The detailed knowledge of the anatomical connectivity at the systems level offers a good starting point to explore the undergoing dynamical processes.

Databases

Even though the human connectome still remains unrevealed, a large amount of information concerning animal anatomy has been already summarized and presented in various databases on several web sites. At the mesoscopic scale the database Microcircuit [24] or Wormatlas [25] offers insights into local circuit connectivity. The database Cocomac [26] contains connectivity maps of macroscopic brain networks of macaque monkey and BrainMaps [27] maps the anatomical details of different animal species like domestic mouse, rat, cat, and several types of monkeys.

9.2.2 Topology of Network Model

Due to the modular and hierarchical organization of the human connectome (brain), simple models of individual levels do not offer an appropriate insight into the complex dynamics occurring in such a complex topology. Thus, our model combines the microscopic and macroscopic levels into one framework.

The higher level copies the known connectivity of real neuroanatomical data, especially the interconnectivity between 53 cat cortical areas [20, 21]. At the lower level, single cortical areas of the cat brain are modeled by large neuronal ensembles. Implementation of these two layers gives rise to a specific topology — a *network of networks*. Recent analysis has confirmed a crucial role of this type of hierarchical network structure in the uncovering of dynamical properties of the system (see Chaps. 4 and 5). In the following section, we will describe details of the topology of the model and discuss possible modifications.

Global Cortical Network

As a representation of the large scale connectivity in our model, we chose the cat cortical map, see Fig. 9.1. The cat cortex, together with the cerebral cortex of the macaque monkey, are the most completely described brain systems among the mammals. The first collation of the cat corticocortical connections, including 65 areas and 1139 reported links, was presented by Scannell et al. [20]. The results of the study were later completed and reorganized which led to the origin of a corticocortical network of 53 cortical areas and additional thalamocortical network of 42 thalamic areas [21]. We will consider only corticocortical connections in our modeling.

The corticocortical network of the cat is composed of 53 highly reciprocally interconnected brain areas, see Fig. 9.2. The density of afferent and efferent axonal fibers is expressed in three levels — 3 for the strongest bundles of fibers, 2 for intermediate or unknown density and 1 for the weakest connections. The value 0 characterizes absent or unknown connections. These values convey more the ranks of the links than the absolute density of the fibers, in the sense that a '2' is stronger than a '1' but weaker than a '3'. All together, there are around 830 connections in the corticocortical network with an average of 15 links per area [20, 21]. (For more details, see Chaps. 3 and 4).

Generally, in the network of cat cortical connections, four distinct subsystems can be identified. The three sensory or sensorimotor subsystems — visual

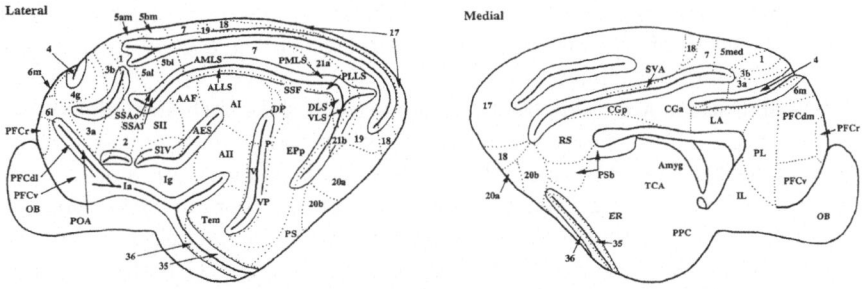

Fig. 9.1. Topographical map of cat cerebral cortex (from [20])

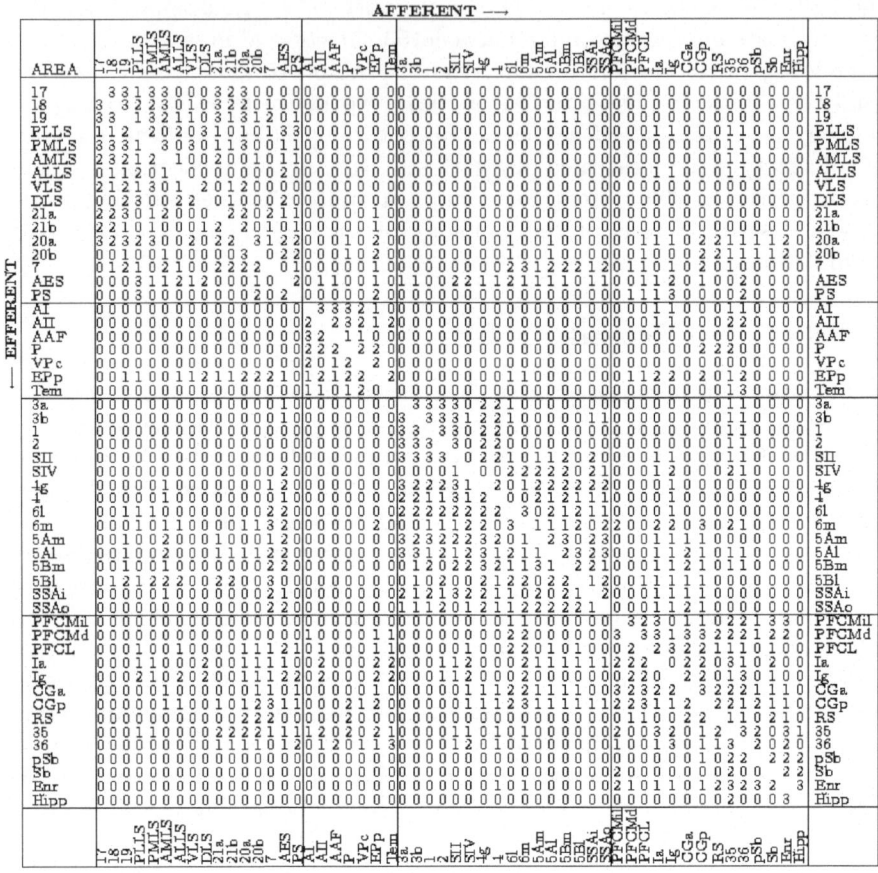

Fig. 9.2. Connectivity matrix representing connections between 53 cortical areas of the cat brain

(V, 16 areas), auditory (A, 7 areas), somatosensory-motor (SM, 16 areas) — involve regions participating in the processing of sensory information and execution of motoric function. The fourth subsystem — frontolimbic (FL, 14 areas) — consists of various cortical areas related to higher brain functions, like cognition and consciousness.

The subsystems are defined as sets of cortical areas with specialized function. To obtain the optimal arrangement of the areas into clusters, several methods based on network connectivity were applied [21,28,29]. For example, in the evolutionary optimization algorithm, the number of connections between units of the cluster should be maximized while inter-cluster connections are minimized. The four resulting clusters largely agree with the functional subsystems.

The corticocortical network has been a subject of much detailed analysis based on graph theory (clustering coefficient, average pathlength, matching index and many other statistical properties) [30] and theoretical neuroanatomy (e.g. segregation and integration) [31]. (See more in Chaps. 3 and 4). The knowledge of the topological properties provides a good starting point for our investigation of the relationship of the structure and dynamics. Our model, however, is flexible to allow for the inclusion of any known cortical connectivity or artificially created network of long-range cortical connections. The cat cortical map is, in evolutionary terms, not so closely related to the structure of the human cortex. To minimize this difference, one can replace the cat matrix by the cortical map extracted from macaque monkey or possibly by a map of the human connectome in the future.

Local Neuronal Network

As we have already mentioned, the individual areas differ in cytoarchitecture and function. Due to these natural distinctions, we model each area as a local network, i.e. a population of neurons having its own topology. Considering the fact that local connections are more frequent than long range ones (although the exact neuronal topology is unknown), we have chosen a small-world architecture as a minimal model [13]. This type of network, originally proposed by Watts and Strogatz [32], represents a transition between random and regular connectivity. At the beginning, each unit of the network connects to a number z of the nearest neighbors, specified by a connection density parameter p_{ring} as $z = p_{ring} \times n$. Later, links are rewired with a probability p_{rew} to a randomly selected node, which introduces the long-range connections. So, the parameters p_{ring} and p_{rew} are crucial for the selection of specific network character (regular, small-world or random), see [32] or Chap. 3.

The small-world topology disposes of improved structural properties like short average pathlength and large clustering coefficient. From the dynamical point of view, it is known that synchronization is enhanced on such networks because of these two characteristics. Such an improvement in the ability to achieve synchronization plays an important role in neural signaling. Many studies also confirmed the presence of the small-world properties in various biological networks, including cortical networks [30, 33].

Previous Chaps. (3, 5) presented a general overview of different kind of networks, their network properties and the influence of these properties on the network dynamics.

We distinguished two types of neurons — excitatory and inhibitory. It is known that approximately 75–80% of the neurons are excitatory (pyramidal type) and the remaining 20–25% are inhibitory neurons (interneurons) [1,5]. In our simulations, we randomly select the inhibitory neurons with a probability $p_{inh} = 0.25$. Since only pyramidal neurons are involved in the long-range inter-areal connections, we consider all inter-areal links to be excitatory.

In the following part, we are mainly interested in the specification of the strength of the different types of synapses. Generally, due to the smaller amount of inhibitory neurons (and thus inhibitory synapses), these connections are usually stronger than the excitatory ones. We assume different coupling strengths for the excitatory ($g_{1,ex}$) and inhibitory ($g_{1,in}$) synapses within a cortical area. The modification of $g_{1,ex}$ and $g_{1,in}$ allows us to balance the excitatory and inhibitory inputs to the neurons within a single cortical area and achieve the 'natural' firing rate of neurons in the range of 1–3 Hz. To exclude the dependence of the neuron firing rate on the network size, we additionally normalize the coupling strength by the square root of the number of connections per area (z). In Chap. 14, the students present an efficient description of the search for the optimal coupling parameters.

Additionally, we also have to consider signals coming from other cortical areas (inter-areal links). If two areas are connected, only 5% of neurons within each area will receive or send signals to the other area. On average, up to 30–40% of neurons of one area can be involved in communication with other areas [34]. The coupling strength $g_{2,ex}$ of the inter-areal connections is scaled

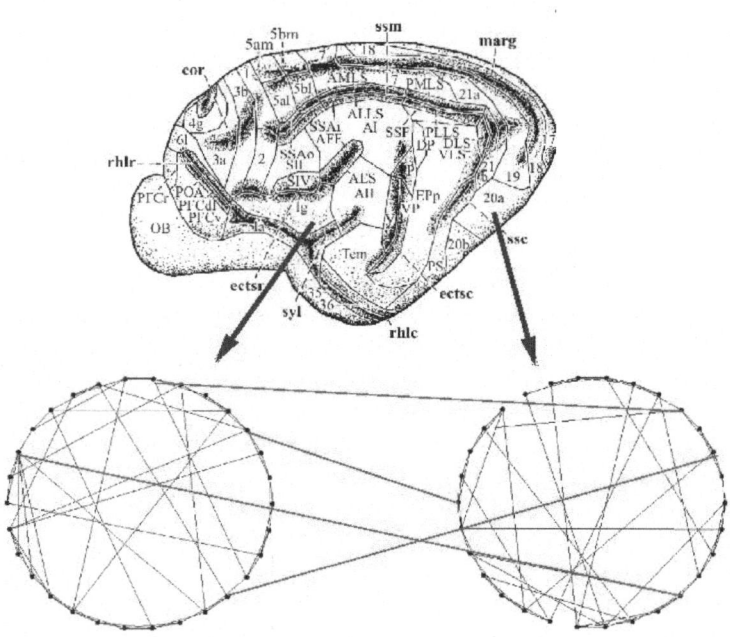

Fig. 9.3. The modeled system — a network of networks. Note that local subnetworks have small-world structure

here by the square of the total number of neurons from the distant area sending signals to the specific neuron.

Table 9.1 offers an overview of all network parameters presented in the model of the network topology.

Let us briefly summarize the network structure of the model: The system represents a *network of networks*, see Fig. 9.3. The macroscopic level corresponds to the known anatomical connectivity map of 53 cat cortical areas. At the microscopic level, a single cortical area is modeled by a large neuronal population of excitatory and inhibitory neurons. The topology and the size of the local network can be adjusted by changing the network parameters. Four possible patterns of the local connectivity structure are available — global (all-to-all), regular, small-world and random. Omitting the layered structure of the cortical area and corresponding topological details, we randomly choose 5% of the neurons to receive an input from, and 5% of neurons to send an output to the various connected cortical areas. The coupling strengths are tuned to reproduce the 'natural' firing rate of neurons in their resting state.

9.3 Dynamics

Whether the model system will plausibly reflect the biological behavior or not does not only depend on the network structure but also on the presence of other necessary properties of neurons. We are aware that this book and chapter resulting from the Summer School are not able to capture all these properties, so here, we will discuss only the most relevant ones.

Neurons are highly specialized cells of the nervous system responsible for the processing and transmission of information encoded in the form of electrical activity. To handle such a peculiar task, neurons possess complex morphology, including a wide dendritic tree with branches contacting many neighboring cells, and an axon with its special myelin sheath conveying action potential effectively and quickly. Additionally, several types of ion channels incorporated in the cell membrane moderate the ionic currents and flexibly respond to incoming signals (see Chaps. 1 and 2).

An implementation of all these dynamical and spatial properties would lead to a complex structural model of the neuron, computationally expensive and thus unsuitable for large-scale neuronal simulations. (However, this approach is used in the Blue Brain Project, which uses supercomputers to simulate the neuronal dynamics of the brain on different levels [6].) Rather than representing neurons as a spatial unit with a complex geometry, they should be modeled as dynamical systems with emphasis on the various ionic currents. These ionic currents determine the neural response to the stimuli and its excitability, which are of the main importance for the neural dynamics.

The general answer of a neuron to the stimuli is an 'all-or-nothing' activity. Neurons only fire when the total synaptic or external input reaches some threshold. If the inputs are weak, only temporal and spatial summation of such inputs will cause the neuron to fire. After the emission of a spike, the upcoming short refractory period prevents the neuron from firing for a certain time interval, even under the application of strong stimuli. Such a neuronal response can be captured by a simple threshold or excitable model.

Thus, excitable neurons can be modeled by a variety of point spike models, e.g. Integrate-and-fire model, Hindmarsh-Rose model, Izhikevich model (see Chap. 1 and several overviews [35, 36]). In several simulations during the Summer School, we chose the Morris-Lecar model, which is able to mimic different types of the behavior categorized according to the neuron excitability and is computationally efficient [37, 38].

Here we introduce the general global dynamics of the neurons described by two variables V and W, see (9.1, 9.2).

$$\dot{V}_i = f(V_i, I^{\text{base}}) + I_i^{\text{syn}}(t) + I_i^{\text{ext}}(t) \tag{9.1}$$
$$\dot{W}_i = h(W_i) + D\xi_i(t) \tag{9.2}$$

The dynamics of the fast variable V imitating the membrane potential are predetermined by a function f of two arguments: the membrane potential V and the basic current I^{base}, which flows into the neuron and sets up the neuronal excitability. Moreover, the membrane potential V is modified by the total synaptic current I^{syn} coming from other connected neurons and external current I^{ext}, representing perturbations from lower brain parts. The slow recovery variable W, modeled by a function h, accounts for the activity of various ion channels. Neurons are additionally stimulated by Gaussian white noise ξ of intensity D, which simulates inherent neuronal stochastic disturbance.

Now we will describe each individual input and its properties.

- Noise
  In the living brain, neurons in the normal state usually do not exhibit strong activity. According to some estimates, they are silent 99% of the time, just sitting below the critical threshold and being ready to fire [39]. In our model, the neurons are also initially set in the excitable state.
  To mimic the intrinsic stochastic character of neuronal dynamics, caused by stochastic processes like synaptic transmission, Gaussian white noise is included:

$$\langle \xi_i(t)\xi_j(t-\tau)\rangle = \delta_{ij}\delta(\tau).$$

The tunable parameter D in (9.2) scales the intensity of this random input. We would like to emphasize that due to the stochastic term in the system, the Euler method is more appropriate for numerical integration. Furthermore, the excitatory neurons are stimulated by multiple inputs of Poissonian noise I^{Poiss} (9.3), where N_p is a number of Poissonian inputs of frequency $\lambda = 3$ Hz. Poissonian input simulates external influences, e.g.

from subcortical areas [10, 11, 39, 40]. In order to obtain the natural firing rate of individual neurons (1–3 Hz), we can also vary the strength of the Poissonian current G_{ex} (normalized to the square root of number of Poissonian processes) until the expected firing rate is reached.

$$I_i^{\text{ext}}(t) = \frac{G_{\text{ex}}}{\sqrt{N_p}} \sum_l^{N_p} I_i^{\text{Poiss,l}}(t) \tag{9.3}$$

- Synapses
  Neurons in brain tissue are connected together in a sparse network through two types of synapses: electrical and chemical [1].
  Electrical coupling (linear) appears only locally through the close contact of the membrane of the neurons. The information about the change of membrane potential of one neuron is transmitted directly as a current flowing through ion channels, called gap junctions.
  Chemical synaptic connections (nonlinear) represent the majority of connections between neurons in the neocortex. The principle of signal transmission is based on the release of chemical messengers from the depolarized presynaptic neuron, which consequently bind to the receptors of the postsynaptic neuron and cause the flow of ions in the cytoplasm. Depending on the type of the receptor, we can distinguish excitatory or inhibitory neurons, which occur in the ratio about 3:1 ($p_{\text{inh}} = 0.25$). Several models of chemical coupling have been proposed, varying from simple [41] to more complex ones [1, 14, 15]. In our model, we consider only the chemical type of neuronal coupling (cf. Chap. 1).

The term I_i^{syn} in (9.1) represents a total synaptic current to the ith cell, i.e. the sum of signals (spikes) $k = 1, \ldots, m$ from all pre-synaptic neurons, $j = 1, \ldots, n$, as shown in (9.4).

The response from all synapses is modeled by (i) the nonlinear function $\alpha(t)$ describing the neuronal response and (ii) the difference between the membrane potential of the postsynaptic neuron V_i and the reversal potential V_s (see a similar approach in [11]). V_s stands for V_{exc} or V_{inh} depending on whether the neuron is excitatory or inhibitory.

$$I_i^{\text{syn}}(t) = \sum_k^m g_{ij}\alpha_j(t - t_{j,\text{spike}}^k - t_{\text{del}})[V_i(t) - V_s] \tag{9.4}$$

The parameter g_{ij} determines the connectivity and coupling strength between the postsynaptic i and presynaptic j neurons. In the case of disconnected neurons, we have $g_{ij} = 0$; $g_{ij} > 0$ indicates the presence of excitatory links and $g_{ij} < 0$ the presence of inhibitory ones.

The gain function $\alpha(t)$ (9.5) expresses the dynamics of the neural response with $\tau_{1,s}$ and $\tau_{2,s}$ as parameters of rise and decay times, where s symbolizes whether the neurons are excitatory or inhibitory.

Time $t^k_{j,\text{spike}}$ is the spiking time of the k-th input spike of neuron i from presynaptic neuron j. The variable t_{del} represents the time delay in the signal transmission typical for the neuronal connection of i and j. For more details, see Chaps. 1, 2 and 8.

$$\alpha_j(t) = \frac{1}{\tau_{1,s} - \tau_{2,s}} [e^{(-t/\tau_{1,s})} - e^{(-t/\tau_{2,s})}] \qquad (9.5)$$

- Plasticity

  Neurons and neuronal connections in the brain evolve throughout life. These changes are characterized by a decrease of the number of the neurons and an increase of the density of the neural connections. According to Hebb's postulate, the most often used connections are strengthened, while the weakest ones atrophy.

  Recently, several researchers described a mechanism of spike-timing-dependent plasticity (STDP) [9,42] (cf. Chap. 2). We used it to modify the weight of the coupling between pairs of neurons. The amount of synaptic modification depends on the exact time difference Δt between postsynaptic t_i and presynaptic t_j spike arrival (see (9.6, 9.7)). If the presynaptic neuron j fires first ($\Delta t > 0$), long term potentiation (LTP) is induced and the synapse is strengthened. In the opposite case, the synapse is weakened (long term depression, LTD).

$$\Delta t = t_i - t_j \qquad (9.6)$$

$$g_{ij}(\Delta t) = \begin{cases} A_+ \exp(-\Delta t/\tau_+) & \text{if } \Delta t > 0 \, ; \\ -A_- \exp(-\Delta t/\tau_-) & \text{if } \Delta t < 0 \, . \end{cases} \qquad (9.7)$$

  The parameters $A_\pm$ correspond to the maximum number of synaptic changes (when $\Delta t \to 0$). The time parameters $\tau_\pm$ determine the range of the temporal window for synaptic strengthening and weakening. Here, we set the values: $A_+ = 0.01$, $A_- = -0.012$ and $\tau_+ = 20.0$, $\tau_- = 20.0$. For more details, see [40, 42].

- Time Delay

  Spikes require some time to propagate within the network; this time can be determined from the axonal conduction velocities, which depend on the length and diameter of axonal fiber [43]. In our model, we have omitted all spatial properties of the neuron but transduction delays t_{del} have been considered to capture the time scale of neuronal communication. They play a crucial role in the neuronal dynamics [40, 44], e.g. neuronal synchrony can be enhanced. The typical time delay between neurons varies between 0.1–20 ms corresponding to axon conduction velocities around 1–20 m/s [45].

  The time delay $t_{\text{del},1}$ of the range of 1–10 ms was initially set up for local neurons within one area. For the inter-areal delay, we considered values $t_{\text{del},2}$ of 10–30 ms.

Chapter 2 offers more information and details of the synaptic properties of neuronal connections, their models and the mechanism of spike-timing-dependent plasticity. Here, we finally summarize the parameters of the neurons defined in this section, whose alteration provides freedom in the exploration of the neuronal dynamics.

9.4 Parallel Implementation of the Code

We have presented a large-scale network model of the cortex that accounts for several biological features at different scales. From our previous experience, we know that even simple neural dynamics, omitting properties like time delay, synaptic plasticity, neural response etc., demand large computational power, see Chap. 5 and [46, 47]. The inclusion of these omitted components causes the simulation of the system to be computationally infeasible to run on single CPUs. To reduce computational time and to improve the efficacy of the code, the parallelization of the code was the only possibility to perform simulations on a reasonable time scale [48] (cf. Chap. 10). For the parallel communication, we chose the message-passing interface (MPI) [49]. The main idea is based on the exchange of packages between different CPUs, i.e. sending and receiving messages. All details of the code and process of parallelization are described in Chap. 11.

The flexibility of the program allows one to replace various parts with new ones or redefined properties. The groups of students had free access to this parallel code and used it for their own simulations. Students chose and implemented the neural models, in some cases including their own modules (Chaps. 13 and 14).

9.5 Summary

In this chapter, we introduced the general concepts of neuronal modeling, especially the construction of a large-scale computational model of cortical neuronal network. First, we reviewed the main neurophysiological properties that should be included in such a complicated model. The general idea of the connectome was introduced and the structural properties of neural networks discussed. Second, we described the general dynamical features of single neurons and interactions between them. All these individual structural and dynamical properties are explained in more depth and summarized in the following chapters, which give basic information about the different biological and physical phenomena occurring in the brain, e.g. dynamics of individual neurons and populations (Chaps. 1, 2, 7, and 8), structure and its relation to the dynamics (Chaps. 3, 5, and 7). Additionally, we discussed the need of the parallelization of the code. More details and parallel implementation are given in Chaps. 10 and 11. This general framework we created was later used by the groups of students during the summer school.

Although, we have previously described our approach as bottom-up modeling (where we go from single neuron dynamics to the dynamics of cortical areas by averaging), the model also exhibits features of a top-down modeling scheme. We started from the systems level, cortical areas, connected according to the cat map. The internal structure of each area was later expressed as the local network of the neurons. From the structural point of the view, we have omitted the complex character of the connections on the cellular level, e.g. layered structure, morphology of the different types of neurons etc. Future possible improvements could include hierarchical organization of neurons into different layers together with substructures like cortical minicolumns and columns. But such a detailed approach would increase the number of parameters and demand even higher computational power. The improved model follows the same goal as an ambitious project, the Blue Brain Project (BBP), attempting to create a computational model of the mammalian brain [6]. The current effort of BBP concentrates on an accurate computational replica of the neocortical column using one of the fastest supercomputers in the world. Later, simulations of the whole brain with detailed anatomical structure and dynamical properties are planned to discover the secrets of dynamical processes in the brain.

Acknowledgments

We would like to thank G. Zamora and J. Ong for discussion and helpful suggestions.

References

1. E. R. Kandel, J. H. Schwartz, and T. M. Jessell, Principles of Neural Science, 2000, 4th ed., New York, McGraw Hill.
2. O. Sporns, G. Tononi, and R. Kötter, The human connectome: a structural description of the human brain, PLoS Comput. Biol. 1(4) (2005) 0245–0251.
3. J. J. Wright, C. J. Rennie, G. J. Lees, P. A. Robinson, P. D. Bourke, C. L. Chapman, E. Gordon, and D. L. Rowe, Simulated electrocortical activity at microscopic, macroscopic and global scales, Neuropsychopharmacology 28 (2003) S80–S93.
4. N. Rulkov, I. Timofeev, and M. Bazhenov, Oscillations in large-scale cortical networks: map-based model, J. Comput. Neurosci. 17 (2004) 203–223.
5. E. M. Izhikevich, J. A. Gally, and G. M. Edelman, Spike-timing dynamics of neuronal groups, Cereb. Cortex 14 (2004) 933–944.
6. H. Markram, The blue brain project, Nat. Rev. Neurosci. 7 (2006) 153–160.
7. V. Braitenberg and A. Schüz, Anatomy of the Cortex: Statistics and Geometry, 1991, Springer, Berlin.

8. D. Rodney and M. Kevan, Neocortex, 459–509, in Synaptical Organisation of the Brain, G. Shepherd, 1991, Springer, New York.
9. G.-q. Bi and M.-m. Poo, Synaptic modification by correlated activity: Hebb's postulate revisited, Annu. Rev. Neurosci. 24 (2001) 139–166. ·
10. N. Brunel, Dynamics of sparsely connected networks of excitatory and inhibitory spiking neurons, J. Comput. Neurosci. 8 (2000) 183–208.
11. P. Kudela, P. J. Franaszczuk, and G. K. Bergey, Changing excitation and inhibition in simulated neural networks: effects on induced bursting behavior, Biol. Cybern. 88 (2003) 276–285.
12. V. P. Zhigulin, Dynamical motifs: building blocks of complex dynamics in sparsely connected random networks, Phys. Rev. Lett. 92, 23 (2004) 238701.
13. N. Masuda and K. Aihara, Global and local synchrony of coupled neurons in small-world networks, Biol. Cybern. 90 (2004) 302–309.
14. L. Lago-Fernández, R. Huerta, F. Corbacho, and J. A. Sigüenza, Fast response and temporal coherent oscillations in small-world networks, Phys. Rev. Lett. 84 (2000) 2758–2761.
15. D. Hansel and G. Mato, Existence and stability of persistent states in large neuronal networks, Phys. Rev. Lett. 86 (2001) 4175–4178.
16. J. Ito and K. Kaneko, Spontaneous structure formation in a network of chaotic units with variable connection strengths, Phys. Rev. Lett. 88 (2001) 028701.
17. I. Belykh, E. de Lange, and M. Hasler, Synchronization of bursting neurons: what matters in the network topology, Phys. Rev. Lett. 94 (2005) 188101.
18. M. Kaiser and C. C. Hilgetag, Spatial growth of real-world networks, Phys. Rev. E 69 (2004) 036103.
19. D. Buxhoeveden and M. F. Casanova, The minicolumn hypothesis in neuroscience, Brain 125 (2002) 935–951.
20. J. W. Scannell, C. Blakemore, and M. P. Young, Analysis of connectivity in the cat cerebral cortex, J. Neurosci. 15 (1995) 1463–1483.
21. J. W. Scannell, G. A. P. C. Burns, C. C. Hilgetag, M. A. O'Neill, and M. P. Young, The connectional organization of the cortico-thalamic system of the cat, Cereb. Cortex 9 (1999) 277–299.
22. D. J. Felleman and D. C. Van Essen, Distributed hierarchical processing in the primate cerebral cortex, Cereb. Cortex 1 (1991) 1–47.
23. Z. Ding, J. C. Gore, and A. W. Anderson, Classification and quantification of neuronal fiber pathways using diffusion tensor MRI, Magn. Reson. Med. 49 (2003) 716–721.
24. Neocortical Microcircuit Database, Copyright 2003 Brain & Mind Institute, EPFL, Lausanne, Switzerland, http://microcircuit.epfl.ch/, (2006)
25. WormAtlas. Z. F. Altun and D. H. Hall (eds.), 2002–2006, http://www.wormatlas.org/, (2006)
26. Cortical Connectivity in Macaque, http://cocomac.org/ (2006)
27. Brain Maps, copyright UC Regents Davis campus, 2005–2006, http://brainmaps.org/, (2006)
28. C. C. Hilgetag, G. A. P. C. Burns, M. A. O'Neill, J. W. Scannell, and M. P. Young, Anatomical connectivity defines the organization of clusters of cortical areas in macaque monkey and cat, Phil. Trans. R. Soc. Lond. B 355 (2000) 91–110.
29. C. C. Hilgetag and M. Kaiser, Clustered organization of cortical connectivity, Neuroinformatics 2 (2004) 353–360.

30. O. Sporns, D. R. Chialvo, M. Kaiser, and C. C. Hilgetag, Organization, development and function of complex brain networks, Trends Cogn. Sci 8 (2004) 418–425.

31. O. Sporns, Network analysis, complexity and brain function, Complexity 8 (2003) 56–60.

32. D. J. Watts and S. H. Strogatz, Collective dynamics of 'small-world' networks, Nature 393 (1998) 440–442.

33. V. M. Eguíluz, D. R. Chialvo, G. Cecchi, M. Baliki, and A. Vania Apkarian, Scale-free brain functional networks, Phys. Rev. Lett. 94 (2005) 018102.

34. M. P. Young, The architecture of visual cortex and inferential processes in vision, Spatial Vis. 13 (2000) 137–146.

35. M. I. Rabinovich, P. Varona, A. I. Selverston, and H. D. I. Abarbanel, Dynamical principles in neuroscience, Rev. Mod. Phys. 78 (2006) 1213–1265.

36. E. M. Izhikevich, Which model to use for cortical spiking neurons? IEEE Trans. Neural Netw. 15 (2004) 1063–1070.

37. J. Rinzel and G. B. Ermentrout, Analysis of neural excitability and oscillations, methods in neuronal modeling: From synapses to networks, 135–169, C. Koch and I. Segev, 1989, The MIT Press, Cambridge, MA.

38. M. St-Hilaire and A. Longtin, Comparison of coding capabilities of type I and type II neurons, J. Comput. Neurosci 16 (2004) 299–313.

39. W. J. Freeman, Characteristics of the synchronization of brain activity imposed by finite conduction velocities of axons, Int. J. Bifurcation Chaos Appl. Sci. Eng. 10 (2000) 2307–2322.

40. E. M. Izhikevich, Polychronization: computation with spikes, Neural Comput. 18 (2006) 245–282.

41. M. V. Ivanchenko, G. V. Osipov, V. D. Shalfeev, and J. Kurths, Synchronized bursts following instability of synchronous spiking in chaotic neural networks, arXiv.org:nlin/0601023 (2006).

42. S. Song, K. D. Miller, and L. F. Abbot, Competitive Hebbian learning through spike-timing-dependent synaptic plasticity, Nat. Neurosc. 3, 9 (2000) 919–926.

43. C. W. Eurich, K. Pawelzik, U. Ernst, J. D. Cowan, and J. G. Milton, Dynamics of self-organized delay adaptation, Phys. Rev. Lett. 82 (1999) 1594–1597.

44. M. Dhamala, V. K. Jirsa, and M. Ding, Enhancement of neural synchrony by time delay, Phys. Rev. Lett. 92 (2004) 074104.

45. N. Kopell, G. B. Ermentrout, M. A. Whittington, and R. D. Traub, Gamma rhythms and beta rhythms have different synchronization properties, Proc. Natl. Acad. Sci. USA 97 (2000) 1867–1872.

46. L. Zemanová, C. Zhou, and J. Kurths, Structural and functional clusters of complex brain networks, Physica D 224 (2006) 202–212.

47. C. Zhou, L. Zemanová, G. Zamora, C. C. Hilgetag, and J. Kurths, Hierarchical organization unveiled by functional connectivity in complex brain networks, Phys. Rev. Lett. 97 (2006) 238103.

48. A. Morrison, C. Mehring, T. Geisel, A. Aertsen, and M. Diesmann, Advancing the boundaries of high-connectivity network simulation with distributed computing, Neural Comput. 17 (2005) 1776–1801.

49. W. Gropp, E. Lusk, and A. Skjellum, Using MPI: Portable parallel programming with the message-passing interface, 1999, 2nd ed., The MIT Press, Cambridge, MA.

Maintaining Causality in Discrete Time Neuronal Network Simulations

Abigail Morrison and Markus Diesmann

Brain Science Institute, RIKEN, Wako, Japan

Summary. When designing a discrete time simulation tool for neuronal networks, conceptual difficulties are often encountered in defining the interaction between the continuous dynamics of the neurons and the point events (spikes) they exchange. These problems increase significantly when the tool is designed to be distributed over many computers. In this chapter, we bring together the methods that have been developed over the last years to handle these difficulties. We describe a framework in which the temporal order of events within a simulation remains consistent. It is applicable to networks of neurons with arbitrary subthreshold dynamics, both with and without delays, exchanging point events either constrained to a discrete time grid or in continuous time, and is compatible with distributed computing.

10.1 Introduction

Neural network simulations are crucial for the advancement of computational neuroscience, as the nonlinear activity dynamics is only partially accessible by purely analytical methods and experimental techniques are still severely limited in their ability to observe and manipulate large numbers of neurons. The brain is an unusual physical system, as it consists of elements (neurons) which can best be described by a set of differential equations, yet the interaction between these elements is mediated by point-like events (action potentials or spikes). It is, moreover, a very complex system — for example, each neuron in the cortex receives in the order of 10^4 connections from other neurons, both within its immediate area and from more remote parts of the brain. Simulating networks with this degree of complexity naturally suggests the use of distributed computing techniques. However, the meshing of continuous-time dynamics and discrete-time communication makes it notoriously difficult to define a consistent and sufficiently general framework for the integration of the dynamics.

There are two classical approaches to simulation: time-driven and event-driven. In the former, a computational time step h is defined. One iteration of a simulation involves each neuron advancing its dynamics over one time step. If its conditions for generating an action potential are met, a spike is delivered to each of the neurons to which it projects. After all neurons have been updated, the next iteration begins. In the latter approach, an event queue manages the order in which spikes are delivered. Each neuron is only updated when it receives an event. If its conditions for generating an action potential are met, the new event is inserted into the queue. This algorithm can be defined very simply and can be very efficient if the neuronal dynamics is invertible — for example, if the arrival of a spike causes an immediate jump in the membrane potential which then decays exponentially. In this case, the neuron can only fire at the arrival of an incoming event, so the behavior of the neuron between the arrival of one event and the next is not relevant for the correct integration of the network. For neuron models with non-invertible dynamics, such as those where the maximum excursion of the membrane potential occurs some time after the arrival of a spike, it is much harder to define an event-driven algorithm. More sophisticated mechanisms are needed: for example, neurons might place provisional events in the queue if they are close to their firing conditions, but may have to revise their predicted spike times upon the arrival of further events [1, 2]. In the following, we concentrate on the time-driven approach as defined above, which can incorporate any kind of subthreshold dynamics without changes being made to the updating and spike delivery algorithm, and has been shown to have good performance in simulating large-scale neuronal networks and to scale excellently when distributed [3].

Here, we present a framework which defines the interactions between the neurons without damaging causality, i.e. such that the order in which neurons are updated does not affect the outcome of a simulation. The framework is suitable for distributed computing. In Sect. 10.2, we cover the basics of point event interaction between continuous-time neuronal elements. We first discuss the historically important concept of neuronal networks with no propagation delay and describe an updating scheme ensuring that the simulation results are independent of the order in which neurons are updated (Sect. 10.2.1). We then demonstrate how this scheme needs to be adapted to incorporate delays that are multiples of the computational time step h (Sect. 10.2.2). Such networks have traditionally constrained spike times to the discrete time grid. However, for networks with propagation delays greater or equal to the computational time step, this constraint can be relaxed. In Sect. 10.3 we show how the scheme can be extended to permit neurons to generate and receive off-grid point events. Finally, we discuss how the propagation delays between neurons can be exploited to optimize communication efficiency between machines in a distributed environment (Sect. 10.4).

10.2 Networks with Discrete Spike Times

In the following, we will assume that communication between neurons is mediated by synapses. When a neuron spikes, all of its outgoing synapses send a discrete event to their respective postsynaptic neurons. The event is parameterized with a weight w, which is interpreted by the postsynaptic neuron with respect to the postsynaptic dynamics it implements. For example, the postsynaptic neuron may interpret w as the size of an instantaneous jump in its membrane potential, or as the maximum amplitude of a postsynaptic current implemented as an alpha function (see Chap. 1). In Sect. 10.2.2, the event is further parameterized by an integer delay k, which expresses the propagation delay d between the neurons in units of the computational time step h, i.e. $d = k \cdot h$.

10.2.1 Networks without Propagation Delays

Consider the following situation: neuron i and neuron j have a strong reciprocal inhibitory connection, such that a spike causes an instantaneous reduction in the membrane potential of the postsynaptic neuron. Each neuron is receiving enough input to drive it to spike at time t. If neuron i is updated to time t first, the spike is instantaneously delivered to neuron j. When j is updated, the strong inhibition prevents the membrane potential from passing the threshold, and so it does not itself generate a spike. Conversely, if neuron j is updated first, neuron j spikes at time t and neuron i is inhibited.

The order dependence in the above example is extremely undesirable. However, with a small conceptual adjustment, the simulation can be made internally consistent. The convention is to define that the generation of a spike may only be influenced by spikes which precede it. This is depicted in the flowchart in Fig. 10.1(a). When the neuron is updated from t to $t + h$, it first modifies its state according to the new spikes from its upstream neurons which fired at time t (operator G), for example incrementing the membrane potential or postsynaptic current. Then the subthreshold dynamics is carried out to propagate the modified neuron state, including the new events, to $t + h$ (operator F_h). At this point, the spiking criteria are applied; if they are fulfilled, the neuron emits a spike. Thus, the effects of the spike on the neuron state are consistent with receiving a spike at t, as can be seen in the membrane potential of neuron *post* in Fig. 10.1(b), but the earliest time the neuron can emit a spike as a result of receiving that spike is $t + h$. This is the equivalent of considering spikes to have an infinitesimal ϵ-delay, and has the effect of making simulations consistent, in that the order of updates does not affect the outcome. In our previous example of two neurons with mutual inhibition, both neurons would fire at time t. Assuming no refractory period, the effect of the mutual inhibition would result in a hyperpolarization of both neurons at time $t + h$.

10.2.2 Networks with Propagation Delays

Minimal Delay

It is particularly simple to alter the algorithm described in Sect. 10.2.1 to one in which all propagation delays in the network are equal to the computation

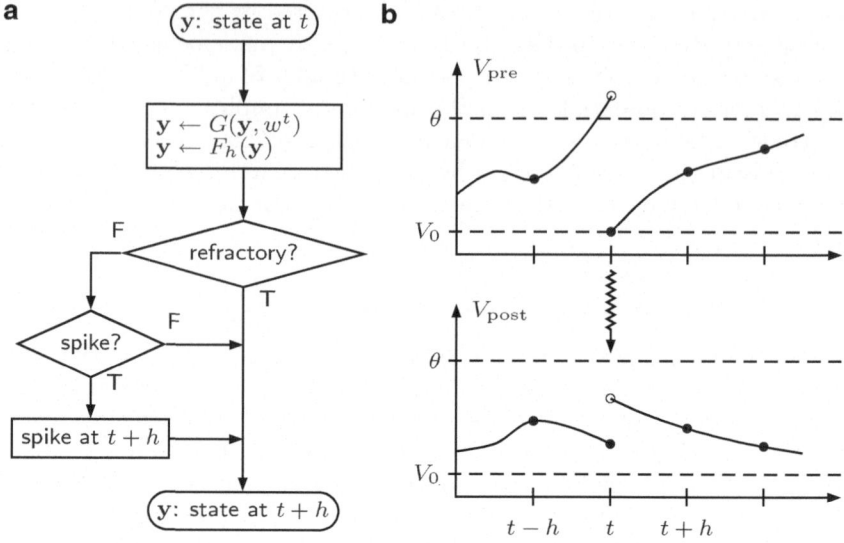

Fig. 10.1. Schematic of a discrete time simulation with no propagation delay: (**a**) Neuron update algorithm. The flowchart depicts the order of operations required to propagate the state $\mathbf{y}$ of an individual neuron by one time step h. Operator G modifies the state according to the incoming events and operator F_h performs the subthreshold dynamics; (**b**) Spike transmission and its effect on the postsynaptic neuron. The membrane potential V_{pre} of neuron *pre* crosses the threshold θ in the time step $(t-h, t]$, so a spike is emitted at time t and the membrane potential is reset to V_0. The spike arrives at neuron *post* with no delay (*zig-zag arrow*). Filled circles denote the values of the membrane potential that can be reported by the neuron at the end of a time step. Intermediate (non-observable) values of the membrane potential are shown as unfilled circles

time step h. In fact, all it amounts to is changing the order of the two operators F_h and G, see Fig. 10.1 and Fig. 10.2. If neuron *pre* spikes at time t, this spike is delivered immediately to neuron *post* (Fig. 10.2(b)). When neuron *post* is being updated from t to $t+h$, first the subthreshold dynamics are performed to propagate the neuron by a step of h (operator F_h), then the neuron state is modified to include the new events visible at $t+h$, including the spike sent by neuron *pre*. Note that for this case and the case where no propagation delay is assumed, the infrastructure of the simulation is the same. A spike produced at time t is delivered immediately to its target, but due to the different order of operations, the effect of the spike is instantaneous in the first case, but delayed by h in the second.

The data structure used to store the pending events can be very simple. If all the synapses have the same dynamics, varying only in amplitude, then each neuron only requires one buffer to store incoming events. Examples of neuron models that only require one buffer are those in which synaptic interactions

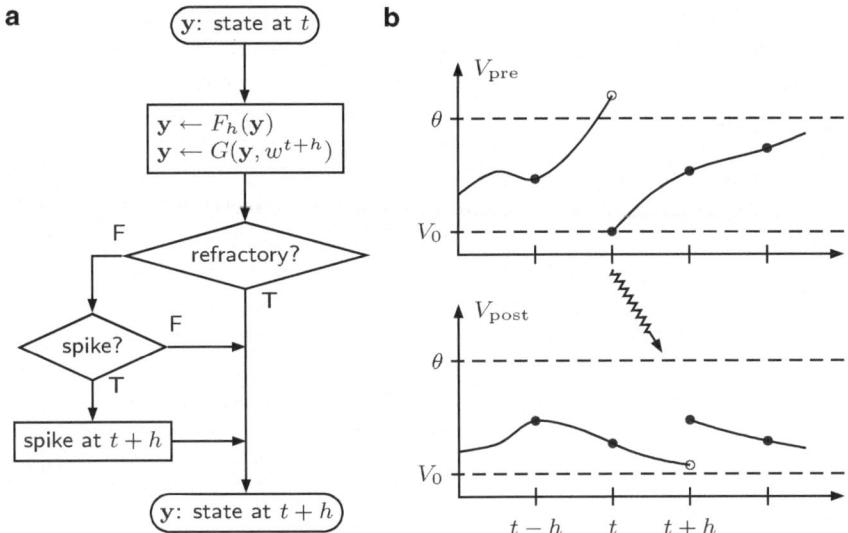

Fig. 10.2. Schematic of a discrete time simulation with propagation delays: (**a**) Neuron update algorithm. The flowchart depicts the order of operations required to propagate the state y of an individual neuron by one time step h. Operator F_h applies the subthreshold dynamics and operator G modifies the state according to the incoming events. Note that the order of these two operations is the reverse of the order shown in Fig. 10.1; (**b**) Spike transmission and its effect on the postsynaptic neuron. As in Fig. 10.1(b), neuron *pre* emits a spike at time t. The spike arrives at neuron *post* with a minimal delay of h (*zig-zag arrow*). Filled circles denote the values of the membrane potential that can be reported by the neuron at the end of a time step. Intermediate (non-observable) values of the membrane potential are shown as unfilled circles

cause an instantaneous increment to the membrane potential, or induce exponential postsynaptic currents. Other neuron models may have more than one set of synaptic dynamics, such as a longer time constant for inhibitory than for excitatory interactions. Clearly, in this case, one buffer per time constant would be required. However, for the sake of simplicity, we will focus on neuron models with only one set of synaptic dynamics.

Depending on the implementation, either a one-element or a two-element buffer is sufficient to maintain causality in the system. If the global scheduling algorithm iterates through all the neurons twice — once to advance the dynamics, the second time to apply the spiking criteria and deliver any generated events — then a one-element buffer is sufficient, as the 'read' and 'write' phases are cleanly separated. However, for reasons of cache effectiveness it may be preferable to iterate through all the neurons only once — i.e. for each neuron, advance its dynamics, apply its spiking criteria and deliver the new events if it spikes. In this case, the 'read' and 'write' phases are no longer cleanly separated, and a two-element buffer is required.

A two-element buffer is depicted in Fig. 10.3. One side of the buffer can be considered as the 'read' side, the other as the 'write' side. When neuron i is modifying its state to incorporate the new events (operator G in Fig. 10.2), it collects the summed weights becoming visible at $t + h$ from the 'read' side. The act of reading clears that side of the buffer. If any neurons that project to i emit a spike at $t + h$, the weight of this event is added to the 'write' side. After all the neurons have been updated, all their buffers are toggled so that the empty 'read' sides are now 'write' sides, and the 'write' sides, containing those events which become visible at $t + 2h$, are now 'read' sides. Thus, the order in which the neurons are updated does not affect the outcome, as events generated in one time step are always cleanly separated from those generated in the next.

Exactly the same structure can be used for networks with no propagation delay, except the assignation of times to buffer elements is shifted by h: in Fig. 10.3(a), the left side receives events for the time step $t + h$ while the summed weight of events becoming visible in time step t is read out of the right side.

General Delay

A system to simulate a network with a minimal propagation delay h can be converted into one encompassing many different delays, as long as they are all integer multiples of h, by replacing the simple two-element buffer with a ring buffer. A traditional ring buffer is an implementation of a queue. The data is represented in a contiguous series of segments. New elements are appended to one end of the series, and the oldest elements are popped off the other end.

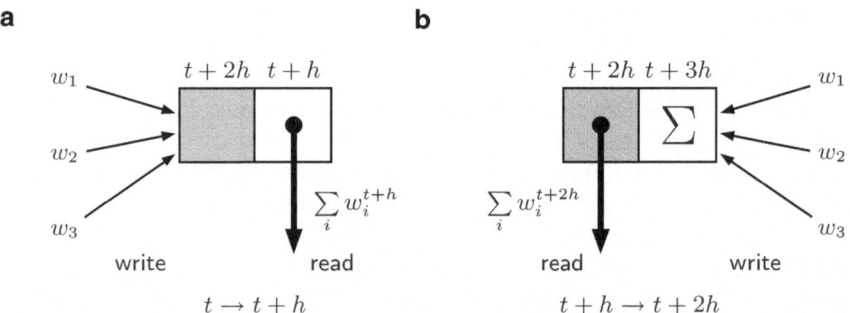

Fig. 10.3. A two-element buffer, suitable for use in networks with a delay of h: (a) In the time step $(t, t + h]$, the gray side of the buffer is the 'write' side, and it sums the weights of events generated in this time step that are to become visible in the next step. The white side of the buffer is the 'read' side, containing the summed weights of all the events received in the previous time step, $\Sigma_i w_i^{t+h}$. Once the neuron has read out the buffer, the 'read' side is emptied; (b) After all neurons have been updated, the neuron buffers are toggled. Now the empty white side receives new events, and the gray side is read by the neuron as it updates from $t + h$ to $t + 2h$

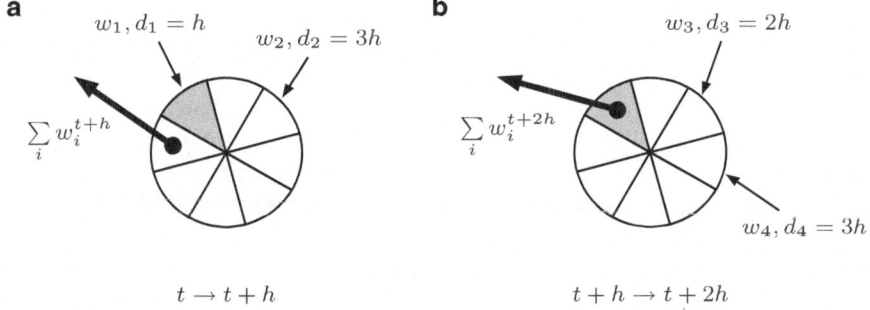

Fig. 10.4. A random access ring buffer, suitable for use in networks with delays that are integer multiples of h. The weights w_i of incoming events are written to the segments corresponding to their delays d_i, such that a delay $d = k \cdot h$ corresponds to the segment k along from the current read position: (**a**) Time step $(t, t + h]$: the read position of the buffer is the segment containing the summed weights of all the events which are to become visible at $t + h$; (**b**) Time step $(t + h, t + 2h]$: The read position has moved to the next segment (*gray*), containing the summed weights of all the events which are to become visible at $t + 2h$

Thus, the arc containing the data can be imagined as rotating around the ring as data is added and removed. In this way, a queue can be implemented without continually having to allocate fresh memory. For our purposes, we need something more like a random access ring buffer, as shown in Fig. 10.4 (see also [3]). Each segment of the ring corresponds to one time step. When the neuron is updating from t to $t + h$, it reads from the segment containing the summed weights of the events that become visible at $t+h$, and then clears this segment. Incoming events are sorted into the other segments depending on their delays: a spike with a delay of $k \cdot h$ would be sorted into the kth segment along from the current read position. After all the neurons have been updated, the read positions for all buffers are moved around one segment. The ring buffer needs to be appropriately sized if the correct order of events is to be maintained — it must be large enough to accommodate the largest propagation delay between neurons in the simulated system, $d_{\max} = k_{\max} \cdot h$, without 'wrapping'. Therefore the optimal size for the buffer is $k_{\max} + 1$. Depending on implementation, $d_{\max}$ may either be specified before the creation of the network, or, more elegantly, determined dynamically whilst the network is created.

10.3 Networks with Continuous Spike Times

In the systems discussed above, spike times were constrained to the discrete time grid. However, Hansel [4] showed that forcing spikes onto the grid can significantly distort the synchronization dynamics of certain networks. The integration error decreases only linearly with the computational step size, so a very small h is necessary to capture the dynamics accurately. An alternative solution is to interpolate the membrane potential between grid points and evaluate the effect of incoming spikes on the neuronal grid in continuous time [4, 5]. This concept was extended in [6] by combining it with exact integration of the subthreshold dynamics (see [7]). Here, we discuss how the scheme described in Sect. 10.2.2 can incorporate off-grid spike times.

Continuous spike times can easily be incorporated into discrete time simulations without having to implement a central queuing structure if the minimum propagation delay is greater than or equal to the computational step size h and an appropriate representation of time is used. In the networks discussed in Sect. 10.2.2, first the subthreshold dynamics is advanced by one time step from t to $t + h$, and then the spiking criteria are applied. If the neuron state passes the criteria (for example, by having a membrane potential above a threshold), the neuron emits a spike which is assigned to the time $t + h$. Now, let us assume that the actual spike time can be determined more precisely, either by interpolation of the membrane potential or by inverting the dynamics or by any other method, such that $t < t_{\text{spike}} \leq t + h$. If the propagation delay to the neuron's postsynaptic target is $k \cdot h$, the event should become visible at time $t_{\text{spike}} + k \cdot h$, which is in the update interval $(t + k \cdot h, t + (k + 1) \cdot h]$. An appropriate representation of the spike time therefore consists of an integer time stamp $t + h$ and a floating point offset $\delta = t_{\text{spike}} - t$. By definition, δ is in the interval $(0, h]$. This choice of representation allows the infrastructure described in Sect. 10.2.2 to be kept with only minimal changes. The propagation delay k can still be used to sort the event into the segment that will be read in the step $(t + k \cdot h, t + (k + 1) \cdot h]$, but the ring buffer is adapted to hold a vector of events in each segment instead of a single value. The weight w and offset δ of the event are appended to the vector. When the neuron performs this update step, the vector is first sorted in order of increasing δ. Note that the vector is just the simplest possible implementation and could be replaced by a more sophisticated data structure such as a calendar queue [8].

The subthreshold dynamics is then advanced from the beginning of the time step to the arrival time of the first event, at which point the neuron state is modified to take account of the first event. Then the dynamics is advanced between the arrival time of the first event and the arrival time of the second event, at which point the neuron state is modified to take account of the second event, and so on until all the events for that update step have been processed. Finally the dynamics is advanced from the arrival time of the final event to the end of the time step. Thus all incoming events have been processed in the correct temporal order.

The scheme described above is very general and can be applied to any kind of subthreshold dynamics, allowing the processing and generation of spikes in continuous time within a discrete time algorithm. No global queuing of events is required, as each neuron queues its events locally. In the case that the subthreshold dynamics is linear, this can be exploited such that not even local queuing is required. This involves a slightly more complicated mechanism for receiving spikes, see [6].

10.4 Distributed Networks

Neuronal network simulations can consume a huge amount of memory, especially if biologically realistic levels of connectivity are assumed. In the cortex, each neuron has of the order of 10^4 incoming synapses and a connection probability of about 0.1 in its local area [9]. Therefore, a network fulfilling both of these constraints must have at least 10^5 neurons. This is equivalent to about $1 \, \text{mm}^3$ of cortical tissue, and represents a threshold network size for simulations, as beyond this point, the number of synapses increases only linearly with the number of neurons, rather than quadratically as is the case for smaller systems. Such networks contain 10^9 synapses, which, even using an extremely simple synapse representation, require several gigabytes of RAM. This state of affairs has naturally prompted much interest in distributed computing, for example [3, 10, 11]. However, distribution raises new issues about maintaining causality in the simulated system. If neuron *pre* projects to neuron *post* with a delay of $k \cdot h$, a spike produced by neuron *pre* at time t should be visible to neuron *post* at time $t + k \cdot h$, no matter whether the two neurons are located on the same machine.

In [3], it was demonstrated that it is more efficient to distribute a neuronal simulation by placing the synapses on the machines of their postsynaptic neurons than of their presynaptic neuron. This is equivalent to distributing a neuron's axon but keeping its dendrite local. That way, when a neuron fires, only its index must be sent across the computer network, rather than a weight and delay for every one of its postsynaptic targets. For neuronal networks with biologically realistic levels of connectivity, this can represent a difference of several orders of magnitude in the amount of information being communicated. One way of ensuring that spikes are always delivered on time is to communicate in each time step, after all the neurons have been updated but before the read positions of their buffers has been incremented (see Sect. 10.2.2). However, this approach is sub-optimal with respect to communication efficiency. Communication between machines has an overhead, so it is more efficient to send one message of N bytes than N messages of one byte each. Fortunately, it is generally possible to communicate less often and still deliver the events correctly. For this, it is necessary to determine the minimum propagation delay between neurons in the simulated system, $d_{\min} = k_{\min} \cdot h$. As in the case of $d_{\max}$, described in Sect. 10.2.2, depending on implementation, $d_{\min}$ could either be specified before the creation of the neuronal network, or determined dynamically whilst the neurons are connected. By definition, a spike cannot have an effect on a postsynaptic neuron earlier than $k_{\min}$ time steps after generation. Therefore, as long as the temporal order of spikes is preserved, it is possible to communicate in intervals of $k_{\min}$ time steps. This can be a significant improvement, as the minimum delay can be considerably larger than the computational time step.

Preserving the temporal order of spikes has two parts: correct storage before communication, and correct delivery after communication. If spikes are constrained to the discrete time grid, for correct storage it is sufficient for each machine to store the indices of spiking neurons in a buffer, with tokens separating the indices of neurons which spiked in one time step from those which spiked in the previous or next step. Thus, at the end of $k_{\min}$ time steps, the buffer contains $k_{\min}$ blocks neuron indices separated by $k_{\min} - 1$ tokens. Then the machine sends this buffer of indices to all other machines, and receives buffers in turn. If the spikes are not constrained to the grid, as described in Sect. 10.3, in addition to the index of a spiking neuron, its spike offset δ must be buffered and communicated as well. For correct delivery, it is necessary to activate the synapses of the neurons registered in these buffers whilst taking the temporal order into consideration. The position of an index in the buffer represents the communication lag, k_{lag}, in delivering the information that a neuron has fired, i.e. $k_{\mathrm{lag}} = k_{\min}$ for an index in the first block of data, $k_{\mathrm{lag}} = k_{\min} - 1$ for an index in the second block of data and so on until $k_{\mathrm{lag}} = 1$ for indices in the last block of data. Note that this assumes that the read positions of all the ring buffers had been incremented before exchange of spike data (see Sect. 10.2.2). If the order is exchanged, then the communication lag ranges from $k_{\min} - 1$ for the first block to 1 for the last block. If the communication lag is subtracted from the propagation delay encoded in a synapse, then the event will become visible to the postsynaptic neuron at exactly the same time as it would in a serial simulation. For example, consider a synapse from neuron pre to neuron $post$ with a weight w and delay $k \cdot h$. In a serial simulation, if neuron pre emitted a spike at time t, w would be added to the ring buffer in the kth segment along from the current read position. In a distributed simulation, w would be added to neuron $post$'s ring buffer $k - k_{\mathrm{lag}}$ segments along from the current read position, with k_{lag} determined by the position of the index of neuron pre in the received index buffer. A slightly more sophisticated version of this approach is discussed in [3].

10.5 Conclusions and Perspectives

We have shown how relatively simple methods and data structures can be used to simulate networks of spiking neurons in discrete time whilst reliably maintaining the temporal order of events. If spike times are constrained to the discrete time grid, the framework is applicable to networks with no propagation delay and to networks with arbitrary delays that are integer multiples of the computation step size h. If spike times are not constrained to the grid, the framework is only applicable to networks with propagation delays greater than or equal to the computational step size h. In this case, the constraint that delays must be an integer multiple of h could be relaxed, because floating point offsets and delays can always be recombined on the fly to produce integer delays and floating point offsets. All these networks can be implemented in a distributed environment in a symmetrical fashion, i.e. the architecture is peer-to-peer rather than master-slave.

These networks can be simulated efficiently because the delivery time of an event in the simulated system has been decoupled from the arrival time at the postsynaptic neuron and the temporal resolution of the simulation. This is the concept that underlies both the ring buffers and the minimum delay intervals for communication. However, if delivery and arrival times are decoupled, this can be problematic for synaptic processes that depend on the state of the postsynaptic neuron, for example spike-timing-dependent plasticity [12, 13] (see also Chaps. 2 and 9). An algorithm has been developed that maintains the correct relationships if propagation delays are assumed to be predominantly dendritic [14]. However, if the propagation delays are predominantly axonal, the framework presented here is not sufficient and will have to be adapted.

Acknowledgements

We acknowledge constructive discussions with the members of the NEST initiative (http://www.nest-initiative.org). Partially funded by DIP F1.2, BMBF Grant 01GQ0420 to the Bernstein Center for Computational Neuroscience Freiburg and EU Grant 15879 (FACETS).

References

1. Brette, R. (2006). Exact simulation of integrate-and-fire models with synaptic conductances. *Neural Comput. 18*(8), 2004–2027.
2. Lytton, W. W., & Hines, M. L. (2005). Independent variable time-step integration of individual neurons for network simulations. *Neural Comput. 17*, 903–921.
3. Morrison, A., Mehring, C., Geisel, T., Aertsen, A., & Diesmann, M. (2005). Advancing the boundaries of high connectivity network simulation with distributed computing. *Neural Comput. 17*(8), 1776–1801.
4. Hansel, D., Mato, G., Meunier, C., & Neltner, L. (1998). On numerical simulations of integrate-and-fire neural networks. *Neural Comput. 10*(2), 467–483.
5. Shelley, M. J., & Tao, L. (2001). Efficient and accurate time-stepping schemes for integrate-and-fire neuronal networks. *J. Comput. Neurosci. 11*(2), 111–119.
6. Morrison, A., Straube, S., Plesser, H. E., & Diesmann, M. (2006). Exact subthreshold integration with continuous spike times in discrete time neural network simulations. *Neural Comput. 19*, 47–79.
7. Rotter, S., & Diesmann, M. (1999). Exact digital simulation of time-invariant linear systems with applications to neuronal modeling. *Biol. Cybern. 81*(5/6), 381–402.

8. Brown, R. (1988). Calendar queues: a fast O(1) priority queue implementation for the simulation event set problem. *Communications of the ACM 31*(10), 1220–1227.

9. Braitenberg, V., & Schüz, A. (1991). *Anatomy of the Cortex: Statistics and Geometry.* Berlin, Heidelberg, New York: Springer-Verlag.

10. Hammarlund, P., & Ekeberg, O. (1998). Large neural network simulations on multiple hardware platforms. *J. Comput. Neurosci. 5*(4), 443–459.

11. Harris, J., Baurick, J., Frye, J., King, J., Ballew, M., Goodman, P., & Drewes, R. (2003). A novel parallel hardware and software solution for a large-scale biologically realistic cortical simulation. Technical report, University of Nevada.

12. Bi, G.-q., & Poo, M.-m. (1998). Synaptic modifications in cultured hippocampal neurons: Dependence on spike timing, synaptic strength, and postsynaptic cell type. *J. Neurosci. 18*, 10464–10472.

13. Markram, H., Lübke, J., Frotscher, M., & Sakmann, B. (1997). Regulation of synaptic efficacy by coincidence of postsynaptic APs and EPSPs. *Science 275*, 213–215.

14. Morrison, A., Aertsen, A., & Diesmann, M. (2007). Spike-timing dependent plasticity in balanced random networks. *Neural Comp.* 19, 1437–1467.

11

Sequential and Parallel Implementation of Networks

Werner von Bloh

Potsdam Institute for Climate Impact Research, PO Box 601203, 14412 Potsdam, Germany
bloh@pik-potsdam.de

11.1 Implementation of Diffusive Coupled Networks

A diffusive coupled network of n neurons can be described by a state vector $N_i(t)$:

$$\frac{\mathrm{d}N_i(t)}{\mathrm{d}t} = f(N_i(t)) + \sum_{j=1}^{n} w_{ij} \times (N_j(t) - N_i(t)), \qquad (11.1)$$

where the function f describes the evolution of the neuron and the weight matrix w_{ij} the coupling strength between the neurons. For simplicity, we assume that the state of the neuron can be described by a single scalar. The matrix w_{ij} has the following properties:

$$w_{ij} \begin{cases} > 0 \text{ excitatory coupling} \\ = 0 \text{ no coupling} \\ < 0 \text{ inhibitory coupling} \end{cases} \qquad (11.2)$$

The matrix can be implemented by a two-dimensional array. The size of this array scales with $O(n^2)$ independent of the number of couplings (equivalent to non-zero elements in w_{ij}). This is in particular inefficient for sparsely coupled neurons. In this case, it is better to use the list structure (see Fig. 11.1) [1].

From now, on we will apply the Morris-Lecar (ML) neuron model [2] (see Chaps. 1, 9, 13, 14). The dynamics can be described by two coupled ordinary differential equations

$$c\frac{\mathrm{d}v}{\mathrm{d}t} = f_v(v, w) = I - g_L(v - v_L) - g_K w(v - v_K)$$
$$-g_{Ca}m_v(v)(v - v_{Ca}) \qquad (11.3)$$

$$\frac{\mathrm{d}w}{\mathrm{d}t} = f_w(v, w) = \lambda(v)(w_v(v) - w), \qquad (11.4)$$

$$m_v(v) = \frac{1}{2}(1 + \tanh((v - v_1)/v_2)),$$

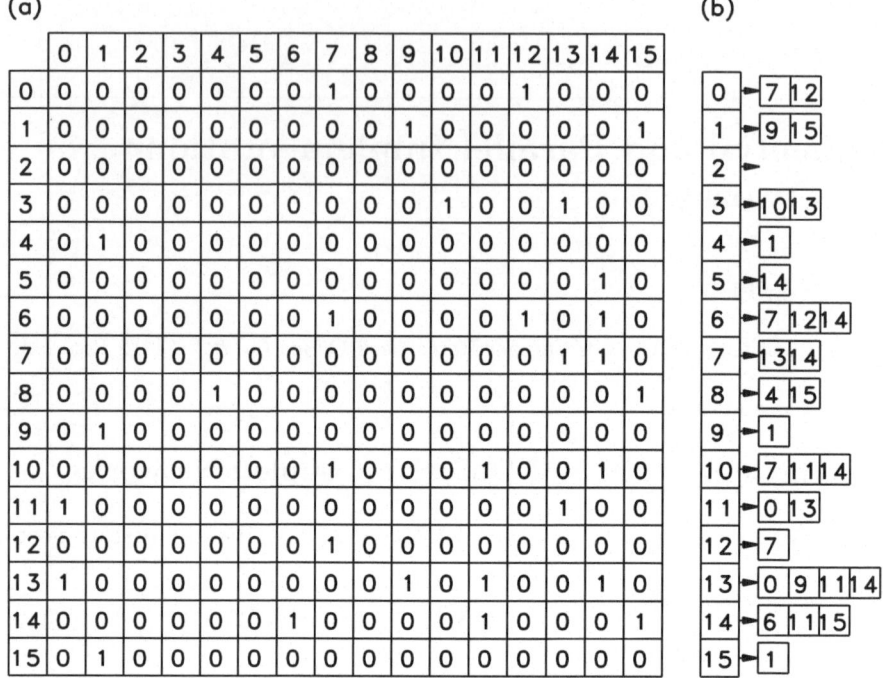

Fig. 11.1. (a) Adjacency matrix and; **(b)** list representation of a random network of 16 neurons with connection probability of $p = 0.15$

$$m_w(v) = \frac{1}{2}(1 + \tanh((v - v_3)/v_4)),$$
$$\lambda(v) = \theta \cosh((v - v_3)/(2v_4)).$$

The diffusive coupling (corresponding to electrical synapses; cf. Chaps. 2 and 9) is achieved with the component v, i.e

$$c\frac{\mathrm{d}v_i}{\mathrm{d}t} = f_v(v_i, w_i) + \sum_{j=1}^{n} w_{ij} \times (v_j - v_i). \tag{11.5}$$

11.1.1 Sequential Code

The implementation of the list and the dynamics of the network is performed with the help of the programming language C. It allows the definition of abstract datatypes, where definition and implementation are separate [3]. An introduction to the programming language C is given in [4]. A neuron can be declared as a datatype Neuron in the following way (in the declaration header file neuron.h):

```
typedef struct
{
  float v[2];    /* state of ML neuron */
  int inhib;     /* inhibitory (TRUE) or not (FALSE) */
  List connect;  /* list of connections */
} Neuron;
```

The datatype of lists List used by Neuron is declared in the header file list.h:

```
typedef struct
{
  int *index; /* pointer to list of indices itself */
  int len;    /* length of list */
} List;

/* Declaration of functions */

/* initialize empty list to  */
extern void initlist(List *);
/* add connection to list */
extern int  addlistitem(List *,int);
```

The header file contains the declaration of the structure List and the proto-types of the functions for initialization and adding a connection to the net-work. Lists can be implemented in two ways: (1) as a variable sized array, and (2) with linked pointers. The advantage of a implementation using arrays is the better storage efficiency. A pointer implementation uses for every element an additional pointer causing some memory overhead. Adding an element to an array, however, is performed by a copying process of the full list not neces-sary in the pointer implementation. If the network is not changed often during runtime, an array implementation is usually more efficient. The initialization function sets the length of the list to zero and initializes the index array to the NULL pointer.

```
void initlist(List *list)
{
  list->len=0;
  list->index=NULL;
} /* of 'newlist' */
```

An item can be added to the list by a call to the addlistitem function. The updated length of the list is returned by this function. The memory allocation can be done by the realloc function of the standard C library. It increases the allocated memory by a certain number of bytes.

```
int addlistitem(List *list,int item)
{
  /* add item to index vector */
```

```
        list->index=(int *)realloc(list->index,
                               sizeof(int)*(list->len+1));
        list->index[list->len]=item;
        list->len++;
        return list->len;
      } /* of 'addlistitem' */
```

A random network can be built up with the function randomnet. For each element of the neuron array, the list is first initialized and then random connections are added to the list with a probability p_{conn}. The neuron is marked as an inhibitory neuron with a probability p_{inhib}.

```
      void randomnet(Neuron net[], /* array of neurons */
                     int n,        /* number of neurons */
                     float p_conn,  /* probability of establishing
                                       connection */
                     float p_inhib /* probability of a inhibitory
                                       neuron */
                    )
      {
        int i,j;
        for(i=0;i<n;i++) /* iterate over all neurons */
        {
          initlist(&net[i].connect);
          net[i].inhib=(drand48()<p_inhib);
          for(j=0;j<n;j++)
            if(i!=j && /* avoid self connections */
               drand48()<p_conn)
              addlistitem(&net[i].connect,j);
        }
      } /* of 'randomnet' */
```

The drand48 function is a generator of uniformly distributed pseudo-random numbers defined in stdlib.h. The update of the ML model can be implemented by the following updateml function:

```
      /* constants for ML model */
      #define   c 1.0
      #define   gL 0.5
      #define   gK 2.0
      #define   gCa 1.0
      #define   vL (-0.5)
      #define   vK (-0.7)
      #define   vCa 1.0
      #define   v1 (-0.01)
      #define   v2 0.15
      #define   v3 0.1
```

```
#define    v4 0.145
#define    theta (1.0/3.0)
void updateml(float dv[2], /* derivatives dv/dt  */
              float v[2],  /* state vector v, v[0]=v,
                              v[1}=w */
              float I      /* applied current */
              )
{
   float mv,wv,lambda;
   mv=0.5*(1+tanh((v[0]-v1)/v2));
   wv=0.5*(1+tanh((v[0]-v3)/v4));
   lambda=theta*cosh((v[0]-v3)/(2*v4));
   dv[0]=I-gL*(v[0]-vL)-gK*v[1]*(v[0]-vK)-gCa*mv*(v[0]-vCa);
   dv[1]=lambda*(wv-v[1]);
} /* of 'updateml' */
```

Then the update of the state of all coupled neurons can be performed by the update function. A simple explicit Euler scheme is used to solve the ordinary differential equations:

$$v_i(t + \Delta t) = v_i(t) + \frac{\Delta t}{c} \times f_v(v_i(t), w_i(t)),$$
$$w_i(t + \Delta t) = w_i(t) + \Delta t \times f_w(v_i(t), w_i(t)). \tag{11.6}$$

```
void update(Neuron net_new[], /* updated array of neurons */
            Neuron net[],      /* array of neurons */
            int n,             /* number of neurons */
            float I,           /* applied current */
            float w_in,        /* inhibitory coupling
                                  strength */
            float w_ex,        /* excitatory coupling
                                  strength */
            float h            /* time step */
            )
{
   int i,j,index;
   float sum,dv[2];
   for(i=0;i<n;i++) /* calculate coupling */
   {
      sum=0;
      for(j=0;j<net[i].connect.len;j++)
      {
         index=net[i].connect.index[j];
         if(net[index].inhib)
            /* inhibitory neuron */
            sum-=w_in*(net[index].v[0]-net[i].v[0]);
```

```
          else
            /* excitatory neuron */
            sum+=w_ex*(net[index].v[0]-net[i].v[0]);
        }
        updateml(dv,net[i].v,I);
        /* apply simple Euler scheme */
        net_new[i].v[0]=net[i].v[0]+(h/c)*(dv[0]+sum);
        net_new[i].v[1]=net[i].v[1]+h*dv[1];
    } /* of 'update' */
```

11.1.2 Parallel Code

The sequential code is limited by the speed and storage capacity of a single workstation. In particular, a large network of neurons uses $O(n^2)$ memory for the connections, quite easily exceeding the storage of a typical workstation. Instead of running the code on a faster computer with more memory, it is usually more cost efficient to run the code in parallel on a network of computers. There are two different parallel computational models: shared memory and distributed memory. In the shared memory model, the processors share the same memory and address space. The memory bandwidth, however, limits the achievable maximum number of processors. In the distributed memory model, each processor has its own local memory. Data exchange is performed via a communication network. This approach allows parallel computers consisting of several thousands of processors.

Message-passing Paradigm

The parallelization of the code uses the message-passing paradigm based on the distributed memory model. The message-passing model consists of a set of processes or tasks that only have local memory but are able to communicate with other tasks by sending and receiving messages. The data transfer from the local memory of one task to the local memory of another task requires operations to be performed on both processes. A portable implementation running on different platforms and architectures is provided by the message-passing interface (MPI). An introduction to MPI is given in [5]. The basic functions of MPI are listed in Table 11.1. The datatypes and message-passing functions are declared in the header mpi.h: A simple *hello world* program in MPI looks like:

```
#include <stdio.h>
#include <mpi.h>
int main(int argc,char **argv)
{
    int mytask,ntask;
```

Table 11.1. The basic six-function version of MPI

Routine	Description
MPI_Init	Initialize MPI
MPI_Comm_size	Find out how many tasks there are
MPI_Comm_rank	Find out which task I am
MPI_Finalize	Finish MPI
MPI_Send	Send a message
MPI_Recv	Receive a message

```
MPI_Init(&argc,&argv);
MPI_Comm_size(MPI_COMM_WORLD,&ntask);
MPI_Comm_rank(MPI_COMM_WORLD,&mytask);
printf("Hello! I am task %d out of %d tasks\n",
       mytask,ntask);
MPI_Finalize();
return 0;
}  /* of 'main' */
```

This code can be run in parallel on a arbitrary number of processors. On four processors, it will produce the following output:

```
% mpirun -np hello
Hello! I am task 0 out of 4 tasks
Hello! I am task 2 out of 4 tasks
Hello! I am task 1 out of 4 tasks
Hello! I am task 3 out of 4 tasks
```

The task belonging to a group of n tasks is identified by a number ranging from 0 to $n-1$. The default group containing all tasks is named MPI_COMM_WORLD. There are two types of communication routines. The first class are point to point routines, like MPI_Recv and MPI_Send in order to send to/receive from a specified task (Fig 11.2). The basic send operation in MPI is declared as

```
MPI_Send(address,count,datatype,destination,tag,comm),
```

where

- address, count, datatype describe count occurrences of items of the form datatype starting at address.
- destination is the task identifier of the destination in the group associated with the communicator comm.
- tag is an integer used for message matching.
- comm identifies a group of tasks and a communication context.

The corresponding receive is

```
MPI_Recv(address,maxcount,datatype,source,tag,comm,status)
```

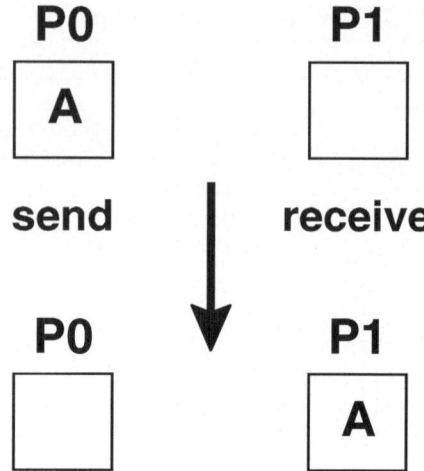

Fig. 11.2. Basic message-passing function of sending information from task P0 to task P1

where

- `address`, `maxcount`, `datatype` describe the receive buffer as they do in the case of `MPI_Send`.
- `source` is the task identifier of the source of the message in the group associated with the communicator `comm`.
- `status` holds information about the actual message size, source and tag.

MPI has predefined datatypes of the objects sent to or received from remote tasks. They are listed in Table 11.2. The size of the MPI datatypes can be calculated by a call of the `MPI_Type_extent` routine.

The second class of MPI functions are collective operations that are called by all processors simultaneously. The most important collective routines are summarized in Table 11.3. Their communication patterns are graphically represented in Fig. 11.3. Finally, the C bindings of all MPI routines used are given in Table 11.4.

Table 11.2. Subset of basic (predefined) MPI datatypes in C

MPI Datatype	C Datatype
MPI_BYTE	signed char
MPI_DOUBLE	double
MPI_FLOAT	float
MPI_INT	int

Table 11.3. Collective operations of the MPI

Routine	Description
MPI_Bcast	Broadcast data to all tasks
MPI_Gather	Gather all data to a single task
MPI_Reduce	Reduce data to one task
MPI_Alltoall	All to all communication
MPI_Alltoallv	All to all communication with variable size
MPI_Scatter	Scatter data to all tasks

Distributing the Network

In a parallel application, the neurons have to be distributed evenly on all tasks. The parallel algorithm is described in [6] in detail (cf. Chap. 10). They use basic send/receive routines, while our implementation is based on collective MPI operations. An algorithm based on send/receive must use a complete pairwise exchange algorithm [7] in order to prevent deadlocks.

In a distributed network, the connection list contains entries to remote neurons. Then, the state of the neuron has to be transferred to the remote

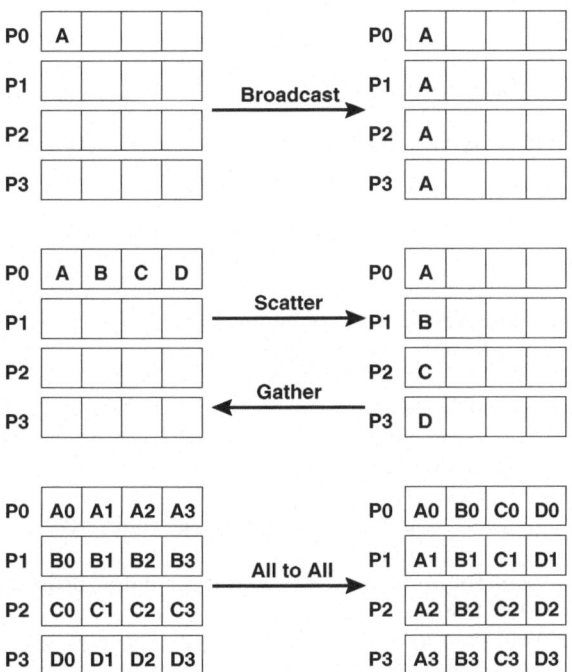

Fig. 11.3. Communication patterns for the MPI collective operations MPI_Bcast, MPI_Gather/MPI_Scatter, and MPI_Alltoall on four tasks P0–P3

neuron stored in a different task. The algorithm for setting up the communication structure works in the following way:

(i) For each task, a sorted list of neuron indices to which a connection exists has to be created. Sorting is necessary in order to delete duplicate entries.
(ii) It has to be determined how many neuron states have to be sent to remote neurons of all tasks. This defines the length of the output buffers.
(iii) This information is distributed to all tasks via a `MPI_Alltoall` collective operation call.
(iv) The indices of the neurons that have to be sent to a specific task must be distributed from all tasks to all tasks via a `MPI_Alltoallv` call. The length of the packets has been determined by step 2 and 3.
(v) Two index vectors are built up containing the mapping from the input buffer to the connection lists and the neuron indices to the output buffer.

Then, the exchange of the neuron states can be performed in three steps:

(i) Copy the state of the neurons to be sent to other tasks to the output buffer.
(ii) Distribute the information with a call of `MPI_Alltoallv`. The `MPI_Alltoallv` collective operation transports the output buffers to the corresponding input buffers.
(iii) The input buffer must be mapped to the connection list.

Table 11.4. C bindings for the MPI functions used in the parallelization of networks

---

int **MPI_Alltoall**(void *sendbuf,int sendcount,MPI_Datatype sendtype,
             void *recvbuf,int recvcount,MPI_Datatype recvtype,
             MPI_Comm comm)
int **MPI_Alltoallv**(void *sendbuf,int *sendcounts,int *sdispls,
             MPI_Datatype sendtype,void *recvbuf,int *recvcounts,
             int *rdispls,MPI_Datatype recvtype,MPI_Comm comm)
int **MPI_Comm_size**(MPI_Comm comm,int *size)
int **MPI_Comm_rank**(MPI_Comm comm,int *rank)
int **MPI_Finalize**()
int **MPI_Gather**(void *sendbuf,int sendcount,MPI_Datatype sendtype,
             void *recvbuf,int recvcount,MPI_Datatype recvtype,int root,
             MPI_Comm comm)
int **MPI_Init**(int *argc,char ***argv)
int **MPI_Reduce**(void *sendbuf,void *recvbuf,int count,MPI_Op op,int root,
             MPI_Comm comm)
int **MPI_Recv**(void *buf,int count,MPI_Datatype datatype,int source,int tag,
             MPI_Comm comm)
int **MPI_Send**(void *buf,int count,MPI_Datatype datatype,int source,int tag,
             MPI_Comm comm,MPI_Status *status)
int **MPI_Type_extent**(MPI_datatype datatype,MPI_Aint *extent)

The datatype of a distributed network Pnet can be defined by the following structure:

```
typedef struct {
  int n;  /* total number of neurons */
  int lo; /* lower bound of subarray */
  int hi; /* upper bound of subarray */
  int ntask; /* number of tasks */
  int taskid;  /* my task identifier */
  int outsize; /* size of output buffer */
  int insize; /* size of input buffer */
  int *outdisp; /* displacement vector  for output */
  int *indisp; /* displacement vector for input */
  int *inlen,*outlen; /* vector length for input/output */
  MPI_Datatype datatype; /* datatype  of input/output
                             buffer */
  void *outbuffer,*inbuffer; /* input/output buffer of
                                /* generic type void */
  int *outindex; /* index vector of output */
  List *connect;  /* list of connections */
} Pnet;
```

The topology of the network is now part of this datatype and not part of neuron. The basic functions for the datatype Pnet are:

```
/* Initialization of datatype */
extern Pnet *pnet_init(MPI_Datatype,int);
/* Random network setup */
extern void pnet_random(Pnet *,float);
/* Creating communication structure */
extern void pnet_setup(Pnet *);
/* Exchange information */
extern void pnet_exchg(Pnet *);
/* Macros for convenience */
/* iterator of subarray */
#define pnet_foreach(pnet,i) for(i=(pnet)->lo;
        i<=(pnet)->hi;i++)
/* allocating an array ar[lo:hi] of datatype type */
#define newvec(type,lo,hi) \
   (type *)malloc(sizeof(type)*(hi-(lo)+1)-(lo))
#define freevec(ptr,lo) free(ptr+(lo))
```

The datatype is initialized by a call to pnet_init. The function has two arguments. The first argument defines the datatype of the data to be distributed between the different tasks. The second argument defines the total number of neurons. In the first part of the function, the necessary arrays are allocated, the number of tasks and the task identifier are determined. In the next part, the lower and upper bounds of the neuron array are calculated. In particular, it must be considered that the total number of neurons cannot be divided by the total number of tasks. Finally, an array for the neuron connections is allocated and initialized:

```
Pnet *pnet_init(MPI_Datatype datatype, /* MPI datatype */
                int n /* total number of neurons */
                )          /* returns allocated struct */
{
  int slice,rem,i;
  Pnet *pnet;
  pnet=(Pnet *)malloc(sizeof(Pnet));
  pnet->n=n;
  pnet->datatype=datatype;
  MPI_Comm_size(MPI_COMM_WORLD,&pnet->ntask);
  MPI_Comm_rank(MPI_COMM_WORLD,&pnet->taskid);
  /* calculate lower and upper bound of subarray */
  slice=pnet->n/pnet->ntask;
  pnet->lo=pnet->taskid*slice;
  pnet->hi=(pnet->taskid+1)*slice-1;
  rem=pnet->n % pnet->ntask;
  /* distribute the remainder evenly on all tasks */
  if(pnet->taskid<rem)
  {
    pnet->lo+=pnet->taskid;
    pnet->hi+=pnet->taskid+1;
  }
  else
  {
    pnet->lo+=rem;
    pnet->hi+=rem;
  }
  /* allocate arrays */
  pnet->outdisp=(int *)malloc(sizeof(int)*pnet->ntask);
  pnet->indisp=(int *)malloc(sizeof(int)*pnet->ntask);
  pnet->outlen=(int *)malloc(sizeof(int)*pnet->ntask);
  pnet->inlen=(int *)malloc(sizeof(int)*pnet->ntask);
  pnet->connect=newvec(List,pnet->lo,pnet->hi);
  pnet_foreach(pnet,i)
    initlist(pnet->connect+i);
  return pnet;
} /* of 'pnet_init' */
```

A random network is set up by calling pnet_random. The function uses the macro pnet_foreach, ensuring that only the local subarray of each task is accessed.

```
void pnet_random(Pnet *pnet,
                 float p_conn /* connection probability */
                 )
{
  int i,j;
  pnet_foreach(pnet,i)
    for(j=0;j<pnet->n;j++)
      if(i!=j && drand48()<p_conn)
        addlistitem(pnet->connect+i,j);
} /* of 'pnet_random' */
```

The setup of the necessary communication patterns between the different tasks is performed by the pnet_setup function. Each task has to know the upper and lower bounds of the subarrays of all other tasks. This information is stored in the arrays lo and hi:

```
void pnet_setup(Pnet *pnet)
{
  int *lo,*hi;
  int i,j,k,*in,size,slice,rem,task;
  slice=pnet->n/pnet->ntask;
  rem=pnet->n % pnet->ntask;
  lo=newvec(int,0,pnet->ntask-1);
  hi=newvec(int,0,pnet->ntask-1);
  for(i=0;i<pnet->ntask;i++)
  /* calculate boundaries of all tasks for n mod ntask<>0 */
  {
    lo[i]=i*slice;
    hi[i]=(i+1)*slice-1;
    if(i<rem)
    {
      lo[i]+=i;
      hi[i]+=i+1;
    }
    else
    {
      lo[i]+=rem;
      hi[i]+=rem;
    }
  }
```

Then, the total number of connections and their indices are calculated. The array in stores the neuron indices of all connections (Fig. 11.4).

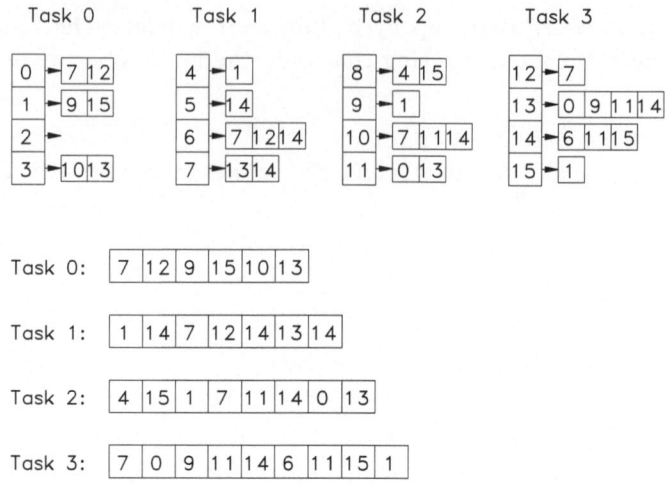

Fig. 11.4. Network of 16 neurons distributed to 4 tasks (*upper figure*). These are the contents of `pnet->connect` before the call of `pnet_setup`. The lower figure shows the combined lists of network connections for each task 0–3

```
for(i=0;i<pnet->ntask;i++)
  pnet->outlen[i]=pnet->inlen[i]=0;
/* calculating total length of connection list */
size=0;
pnet_foreach(pnet,i)
  size+=pnet->connect[i].len;
in=(int *)malloc(sizeof(int)*size);
k=0; /* concatenating connection lists */
pnet_foreach(pnet,i)
  for(j=0;j<pnet->connect[i].len;j++)
    in[k++]=pnet->connect[i].index[j];
```

This array is sorted and duplicated entries are deleted (see Fig. 11.5).

```
/* sort connection list */
qsort(in,size,sizeof(int),
      (int (*)(const void *,const void *))compare);
pnet->insize=1; /* delete duplicate entries */
for(i=1;i<size;i++)
  if(in[i]!=in[i-1]) /* same indices? */
  {
    /* no, increase insize by one */
    in[pnet->insize]=in[i];
    pnet->insize++;
  }
```

1 2 3
Task 0: 7 9 10 12 13 15
0 1 3
Task 1: 1 7 12 13 14
0 1 2 3
Task 2: 0 1 4 7 11 13 14 15
0 1 2 3
Task 3: 0 1 6 7 9 11 14 15

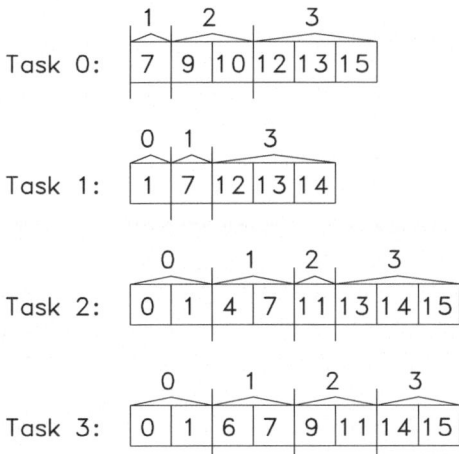

Fig. 11.5. List of network connections for each task after sorting and deleting duplicate entries. The numbers above the lists are the corresponding task identifiers to which the information must be sent

After this step, the length of the vector sent to all tasks has to be calculated. This can be performed by counting all neuron indices inside the lower and upper bound for each task (Fig. 11.6).

```
task=0;
pnet->inlen[task]=0;
/* calculating inlen vector */
for(i=0;i<pnet->insize;)
  if(in[i]<=hi[task]) /* inside the boundaries of task? */
  {
    /* yes, increment inlen by one */
    pnet->inlen[task]++;
    i++;
  }
```

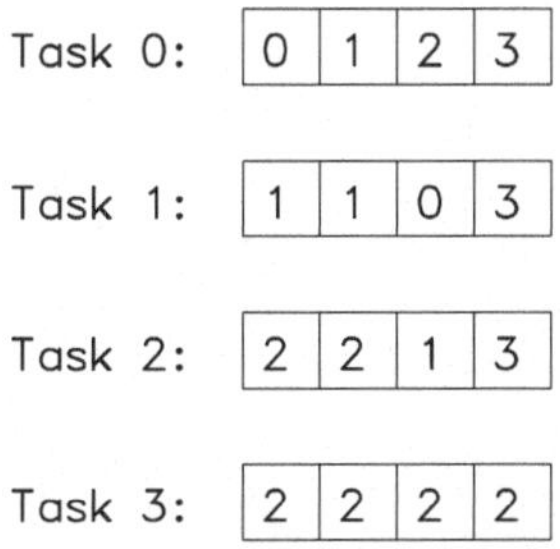

Task 0: 0 1 2 3
Task 1: 1 1 0 3
Task 2: 2 2 1 3
Task 3: 2 2 2 2

Fig. 11.6. The length of the communication packets needed by `MPI_Alltoallv`. The information has to be distributed by `MPI_Alltoall` to all tasks

```
else
{
    /* no, goto next task, set new inlen to zero */
    task++;
    pnet->inlen[task]=0;
}
```

It is important to map the connection from the input buffer to the connection list of the local neurons. This information is stored again in the array of lists pnet->connect, because the original connection lists are not needed any more (Fig. 11.7). The current implementation uses a linear search algorithm to find the indices in the connection list leading to a runtime characteristic of $O(n^2)$. By using a better algorithm (e.g. binary search), the runtime can be significantly reduced. The setup function, however, is called only once during initialization of the network.

```
/* calculating mapping from input buffer to connection */
pnet_foreach(pnet,i)
{
    for(j=0;j<pnet->connect[i].len;j++)
        /* search for index in array in */
        for(k=0;k<pnet->insize;k++)
            if(in[k]==pnet->connect[i].index[j])
            {
                /* index found and stop searching */
                pnet->connect[i].index[i]=k;
                break;
            }
} /* of pnet_foreach */
```

The information of the lengths of the communication packets to be received from other tasks has to be distributed by a call to MPI_Alltoall. Then, the pnet->outlen array contains the length of the outgoing packets. This information is needed by the MPI_Alltoallv function. MPI_Alltoallv sends a distinct message from each task to every task, where the messages can

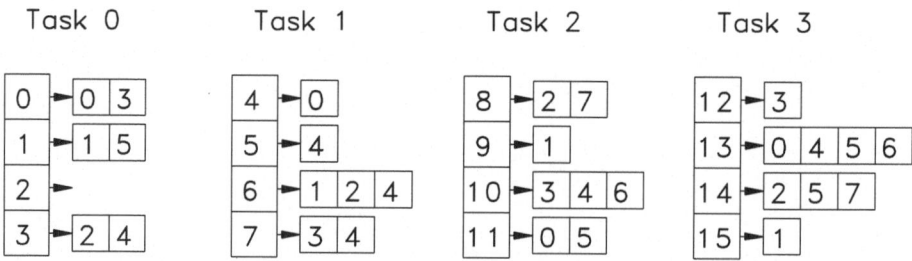

Fig. 11.7. Mapping of the input buffer to the connection list. These are the contents of pnet->connect after the call of **pnet_setup**

have different sizes and displacements. The displacement is the offset from the first element of the array to the first element of the message and can be simply calculated by summing up the `pnet->inlen` array. After calling `MPI_Alltoallv`, the array `pnet->outindex` contains the indices of neurons that must be sent to other tasks (see Fig. 11.8).

```
MPI_Alltoall(pnet->inlen,1,MPI_INT,pnet->outlen,1,
             MPI_INT,MPI_COMM_WORLD);
/* calculating displacements needed by MPI_Alltoallv */
pnet->indisp[0]=pnet->outdisp[0]=0;
for(k=1;k<pnet->ntask;k++)
{
  pnet->indisp[k]=pnet->inlen[k-1]+pnet->indisp[k-1];
  pnet->outdisp[k]=pnet->outlen[k-1]+pnet->outdisp[k-1];
}
pnet->outsize=0;
for(i=0;i<pnet->ntask;i++)
  pnet->outsize+=pnet->outlen[i];
/* allocating outindex */
pnet->outindex=(int *)malloc(sizeof(int)*pnet->outsize);
/* information is moved from in to pnet->outindex calling
   the collective operation MPI_Alltoallv */
MPI_Alltoallv(in,pnet->inlen,pnet->indisp,MPI_INT,
              pnet->outindex,pnet->outlen,
              pnet->outdisp,MPI_INT,MPI_COMM_WORLD);
```

Finally, we allocate the input and output buffers of type stored in `pnet->datatype`. The function `MPI_Type_extent` returns the number of bytes for this datatype. These buffers of generic type `void` have to be cast by the user to the appropriate type.

```
/* allocating input and output buffer */
MPI_Type_extent(pnet->datatype,&size);
pnet->outbuffer=malloc(pnet->outsize*size);
```

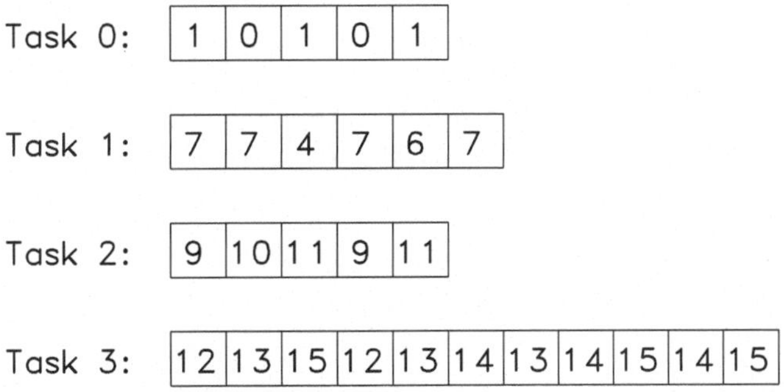

Task 0: | 1 | 0 | 1 | 0 | 1 |

Task 1: | 7 | 7 | 4 | 7 | 6 | 7 |

Task 2: | 9 | 10 | 11 | 9 | 11 |

Task 3: | 12 | 13 | 15 | 12 | 13 | 14 | 13 | 14 | 15 | 14 | 15 |

Fig. 11.8. `pnet->outindex` describes the mapping of the output buffer to the neuron indices

```
    pnet->inbuffer=malloc(pnet->insize*size);
    /* free auxiliary storage */
    free(in);
    free(lo);
    free(hi);
} /* of 'pnet_setup' */
```

The compare function is needed for the quicksort C library function qsort. It sorts integer values in ascending order:

```
static int compare(const int *a,const int *b)
{
    return *a-*b;
} /* of 'compare' */
```

Then, the exchange is done by the pnet_exchg function calling only the MPI_Alltoallv function. Information is copied from the output buffers to the input buffers.

```
void pnet_exchg(Pnet *pnet)
{
    MPI_Alltoallv(pnet->outbuffer,
                  pnet->outlen,pnet->outdisp,
                  pnet->datatype,pnet->inbuffer,
                  pnet->inlen,pnet->indisp,
                  pnet->datatype,MPI_COMM_WORLD);
} /* of 'pnet_exchg' */
```

The parallel initialization of the network is performed by init:

```
Neuron *init(Pnet **pnet,
             int n, /* number of neurons */
             float p_conn /* connection probability */
             ) /* returns allocated array of neurons */
{
    Neuron *net;
    int i;
    *pnet=pnet_init(MPI_FLOAT,n);
    pnet_random(*pnet,p_conn);
    pnet_setup(*pnet);
    net=newvec(Neuron,(*pnet)->lo,(*pnet)->hi);
    /* random initial state of ML neurons */
    pnet_foreach(*pnet,i)
    {
        net[i].v[0]=-0.02*0.01*drand48();
        net[i].v[1]=0.05+0.20*drand48();
    }
    return net;
} /* of 'init' */
```

Then, the parallel update for diffusive coupling can be done in the following way. For simplicity, only excitatory coupling is used. First, the state of the neuron has to be copied to the output buffers. After calling pnet_exchg, the information is moved to the input buffer. The list pnet->connect provides the information about the mapping of the input buffer to the connections.

```
void update(Pnet *pnet,
            Neuron net[], /* subarray of neurons */
            float I,      /* applied current */
            float w,      /* coupling strength */
            float h       /* time step */
            )
{
  int i;
  float sum,*buffer,dv[2];
  /* cast outbuffer to float pointer */
  buffer=(float *)pnet->outbuffer;
  /* copy state to output buffer */
  for(i=0;i<pnet->outsize;i++)
    buffer[i]=net[pnet->outindex[i]].v[0];
  /* Exchange of necessary information to all tasks */
  pnet_exchg(pnet);
  /* cast inbuffer to float pointer */
  buffer=(float *)pnet->inbuffer;
  pnet_foreach(pnet,i)
  {
    sum=0;
    for(j=0;j<pnet->connect[i].len;j++)
      sum+=buffer[pnet->connect[i].index[j]]-net[i].v[0];
    updateml(dv,net[i].v,I);
    /* performing Euler step */
    net[i].v[0]+=(h/c)*(dv[0]+w*sum);
    net[i].v[1]+=h*dv[1];
  }
} /* of 'update' */
```

Efficiency of Parallelization

Communication is necessary after every time step in the case of diffusive coupling. This limits the efficiency of the parallel code, in particular for large networks with dense couplings. Efficiency E is defined by

$$E(p) := \frac{T(1)}{p \times T(p)}, \tag{11.7}$$

where $T(p)$ denotes the computation time running on p parallel tasks. The computation time can be divided into two parts:

$$T = T_{\text{calc}} + T_{\text{comm}}, \tag{11.8}$$

where T_{calc} denotes the computation part and T_{comm} the communication part. In the case of a fully coupled network of n neurons, T_{comm} scales as $O(n^2)$, while the computation part scales as $O(n^2/p)$:

$$T(p) \sim \frac{n^2}{p} T'_{\text{calc}} + n^2 T'_{\text{comm}}, \tag{11.9}$$

Thus,

$$E(p) = \frac{T'_{\text{calc}}}{T'_{\text{calc}} + p T'_{\text{comm}}} \tag{11.10}$$

resembling Amdahl's law [8]. For large p, the total runtime is dominated by the communication time and the efficiency tends to zero.

11.2 Non-diffusive Coupling

The coupling of neurons can also be done in a different way: If the integrated postsynaptic potentials (PSP) reach a certain threshold, a spike train is sent to remote neurons with a delay time t_{delay}. The PSP evoked by one single spike can be parameterized by a gain function $g(t)$ with a rise time τ_1 and decay time τ_2 (cf. Chap. 1):

$$g(t) = \frac{\exp(-t/\tau_1) - \exp(-t/\tau_2)}{\tau_1 - \tau_2} \tag{11.11}$$

For $t > t_{\max} \approx 15\text{ms}$ and the chosen parameter $\tau_1 = 1$ ms and $\tau_2 = 2$ ms, the PSP gain function g is nearly zero (Fig. 11.9).

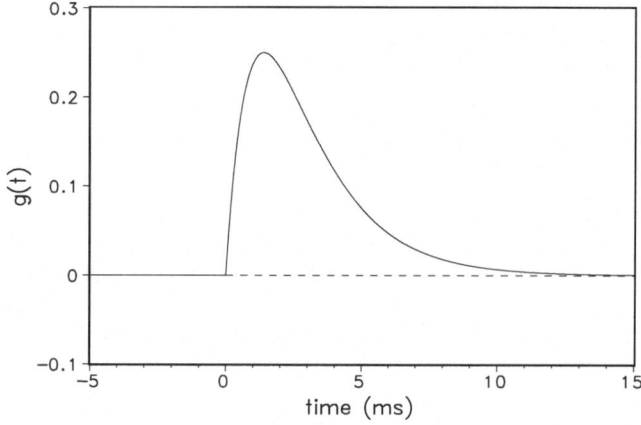

Fig. 11.9. PSP gain function $g(t)$ with rise time $\tau_1 = 1$ ms and decay time $\tau_2 = 2$ ms

A spike is generated if v exceeds the critical threshold of zero at time t_{spike}. The coupled model of n neurons is defined by

$$c\frac{dv_i}{dt} = f_v(v_i, w_i) + \sum_{j=1}^{n}\sum_{k=1}^{m_j} w_{\text{in,ex}}g(t - t_{\text{spike},j,k} - t_{\text{delay},i,j})(v_i - v_{\text{in,ex}}),$$

$$\frac{dw_i}{dt} = f_w(v_i, w_i) + \sigma\xi, \tag{11.12}$$

$$i = 1, \ldots, n,$$

where $t_{\text{spike},j,k}, k = 1, \ldots, m_j$ are the m_j spike times of neuron j, $w_{\text{in,ex}}$, $v_{\text{in,ex}}$ are weights for inhibitory/excitatory coupling, and ξ is additional Gaussian white noise with amplitude σ. The numerical discretization using an Euler scheme for a stochastic differential equation is:

$$v_i(t + \Delta t) = v_i(t) + \frac{\Delta t}{c} \times f_v(v_i(t), w_i(t)),$$

$$w_i(t + \Delta t) = w_i(t) + \Delta t \times f_w(v_i(t), w_i(t)) + \sqrt{\Delta t} \times \xi. \tag{11.13}$$

11.2.1 Sequential Version

In order to speed up the evaluation of the gain function $g(t)$, a lookup table is used created by the function getg:

```
#define tau_1 1.0 /* rise time (ms) */
#define tau_2 2.0 /* decay time (ms) */

float *getg(int tmax, /* size of lookup table */
            float h   /* time step (ms) */
            )         /* returns calculated lookup table */
{
  float *g,gmax;
  int t;
  g=(float *)malloc(sizeof(float)*tmax);
  gmax=0;
  for(t=0;t<tmax;t++)
  {
    g[t]=1/(tau_1-tau_2)*(exp(-t*h/tau_1)-exp(-t*h/tau_2));
    if(g[t]>gmax)
      gmax=g[t];
  }
  /* normalize g */
  for(t=0;t<tmax;t++)
    g[t]/=gmax;
  return g;
} /* of 'getg' */
```

An efficient implementation of such coupling uses a finite event buffer where the time of spike events are stored (Chap. 10). It is assumed that the maximum number of overlapping spikes is limited by the size of the buffer. The datatype Rbuffer can be defined by using an array that stores a limited number of events. After the maximum number of spikes is reached, the oldest event is overwritten. This can be implemented by a ring topology using the modulo operator. A graphical representation of datatype Rbuffer is shown in Fig. 11.10.

```
typedef struct
{
  int len;  /* length of ring buffer */
  int top;  /* index of first element in ring buffer */
  int size; /* maximum length of ring buffer */
  int *events; /* stores up to size latest events */
} Rbuffer;
void initrbuffer(Rbuffer *,int);
void addrbuffer(Rbuffer *,int);
```

A macro is defined in order to access an event at position pos from the top of the ring buffer:

```
#define getrbuffer(rbuf,pos) \
  (rbuf)->events[((rbuf)->top-1-(pos)+(rbuf)->size) % (rbuf)
  ->size]
```

The datatype is initialized by a call to initrbuffer. The parameter size sets the maximum number of events that can be stored.

```
void initrbuffer(Rbuffer *rbuf,int size)
{
  rbuf->len=rbuf->top=0;
  rbuf->events=(int *)malloc(sizeof(int)*size);
  rbuf->size=size;
} /* of 'initrbuffer' */
```

The implementation for adding an event to the ring buffer is straightforward. The modulo operator is used after incrementing the index to the first element in the buffer top defining a ring topology.

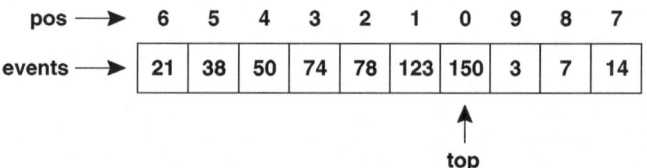

Fig. 11.10. Graphical representation of datatype Rbuffer with size equal to 10

```
void addrbuffer(Rbuffer *rbuf,int event)
{
  /* have we reached the maximum number of events? */
  if(rbuf->len<rbuf->size)
    rbuf->len++; /* no, increase len by one */
  rbuf->events[rbuf->top]=event;
  /* increase top by one, then perform  modulo operator */
  rbuf->top=(rbuf->top+1) % rbuf->size;
} /* of 'addrbuffer' */
```

For the ML model, the datatype Neuron has to be modified:

```
typedef struct
{
  float v[2];    /* state of neuron (ML) */
  Rbuffer spikes;  /* times of last spikes */
  List connect;  /* list of connections */
  int inhib; /* inhibitory (TRUE) or excitatory (FALSE) */
} Neuron;
```

The neurons are initialized and the network is established by the function init:

```
#define RBUFF_LEN 10 /* maximum length of ring buffer */

void init(Neuron net[], /* array of neurons */
          int n,         /* total number of neurons */
          float p_conn, /* probability of establishing
                           connection */
          float p_inhib /* probability of a inhibitory
                           coupling */
          )
{
  int i,j;
  /* Setup of random network */
  randomnet(net,n,p_conn,p_inhib);
  for(i=0;i<n;i++)
  {
    /* Set initial conditions of neurons */
    net[i].v[0]=-0.02*0.01*drand48();
    net[i].v[1]=0.05+0.20*drand48();
    /* Initializing event ring buffer */
    initrbuffer(&net[i].spikes,RBUFF_LEN);
  }
} /* of 'init' */
```

The update function uses the datatype Rbuffer and List. The Gaussian white noise is generated by a call of the gasdev function as part of the numerical recipes library [9]. The function has been slightly modified and uses the drand48 random number generator.

```c
/* constants for ML model */
#define   c 1.0
/* coupling constants */
#define V_in (-0.55)
#define V_ex 0.05

void update(FILE *file, /* output file for spikes */
            Neuron net[], /* array of neurons */
            int n, /* number of neurons */
            int time, /* integer time */
            float h, /* time step */
            float *g, /* lookup table */
            int t_max, /* size of lookup table */
            float w_in,float w_ex,int delay,
            float I, /* applied current */
            float sigma /* amplitude of Gaussian white
                            noise */
            )
{
  int i,j,k,index;
  float sum,v_new[2],dv[2];
  int spike;
  for(i=0;i<n;i++)
  {
    sum=0; /* calculate interactions */
    for(j=0;j<net[i].connect.size;j++)
    {
      index=net[i].connect.index[j];
      /* iterating over the ring buffer of neuron index */
      for(k=0;k<net[index].spikes.len;k++)
      {
        spike=getrbuffer(&net[index].spikes,k)+delay;
        /* testing whether spike is active */
        if(time>=spike && time<spike+t_max)
        {
          /* yes */
          if(net[index].inhib)
            /* inhibitory coupling */
            sum-=w_in*g[time-spike]*(net[i].v[0]-V_in);
          else
            /* excitatory coupling */
            sum-=w_ex*g[time-spike]*(net[i].v[0]-V_ex);
```

```
      }
      else
        /* no, break if all spikes are not active
           anymore */
        if(spike+t_max<time)
          break;
    } /* of for(k=...) */
  } /* of for(j=...) */
  /* update of ML */
  updateml(dv,net[i].v,I);
  v_new[0]=net[i].v[0]+(h/c)*(dv[0]+sum);
  v_new[1]=net[i].v[1]+h*dv[1]+sqrt(h)*sigma*gasdev();
  /* critical threshold of zero reached for v? */
  if(net[i].v[0]<0 && v_new[0]>0)
  {
    /* yes, we have a spike event, add to my ring buffer */
    addrbuffer(&net[i].spikes,time);
    fprintf(file,"%g %d\n",time*h,i);
  }
  net[i].v[0]=v_new[0];
  net[i].v[1]=v_new[1];
  } /* of for(i=...) */
} /* of 'update' */
```

The mean value of v_i over the n neurons as a diagnostic variable, $\bar{v} = \frac{1}{n} \sum_{i=1}^{n} v_i$, can be calculated using the mean function:

```
float vmean(const Neuron net[],int n)
{
  int i;
  float sum;
  sum=0;
  for(i=0;i<n;i++)
    sum+=net[i].v[0];
  return sum/n;
} /* of 'vmean' */
```

The main program calculating the dynamics of the network is as follows:

```
#include "list.h"
#include "neuron.h"
#define n 100 /* number of neurons */
#define I 0.1 /* input current */
#define sigma 0.0 /* amplitude of Gaussian white noise */
#define t_max 20.0 /* duration of spike (ms) */
#define h 0.01  /* time step  (ms) */
#define p_conn 1.0 /* fully coupled network */
```

```
#define p_inhib 0.0 /* no inhibitory coupling */
#define t_end 100.0 /* simulation time (ms) */
#define delay 0 /* no delay */

int main(int argc,char **argv)
{
  Neuron net[n];
  float *g,g1,w_in,w_ex;
  int t,nstep,ostep,tmax;
  FILE *file,*log;
  init(net,n,p_conn,p_inhib);
  w_in=0.1/((n-1)*p);
  w_ex=0.1/((n-1)*p);
  nstep=t_end/h;
  ostep=nstep/100;
  tmax=t_max/h;
  g=getg(tmax,h); /* creating lookup table for g */
  file=fopen("neuron.spike","wb");
  log=fopen("neuron.mean","w");
  for(t=0;t<nstep;t++) /* time loop */
  {
    if(t % ostep==0)
      /* write to output file every ostep time steps */
      fprintf(log,"%g %g\n",t*h,vmean(net,n));
    update(file,net,n,t,h,g,t_max,w_in,w_ex,delay,I,sigma);
  } /* of  time loop */
  fclose(file);
  fclose(log);
  return 0;
} /* of 'main' */
```

11.2.2 Parallel Version

The parallelization is based on Pnet. The data to be exchanged now have the
type MPI_INT. In order to incorporate the ring buffer storing the spike events,
a new datatype Rnet has to defined. inhib is not part of datatype neuron,
because information about remote neurons is also needed.

```
typedef struct
{
  Pnet *pnet;     /* parallel network datatype */
  Rbuffer *rbuffer; /* event ring buffer */
  int *inhib;     /* inhibitory list */
} Rnet;
```

The datatype Neuron now contains only the last spike event of the neuron itself:

```
typedef struct
{
  float v[2];    /* state of neuron (ML) */
  int spike;     /* time of last spike */
} Neuron;
```

Both the initialization of the neuron and the parallel communication pattern are organized by the function init. The information about whether a remote neuron is inhibitory or not is distributed by a call to pnet_exchg. The output buffer contains boolean values.

```
#define NOFIRE -1
Neuron *init(Rnet *rn,
             int n, /* total number of neurons */
             float p_conn, /* probability of establishing
                              connection */
             float p_inhib /* probability of a inhibitory
                              neuron */
             ) /* returns allocated subarray of neurons */
{
  Neuron *net;
  int i,*buffer;
  int *inhib;
  rn->pnet=pnet_init(MPI_INT,n);
  /* setup of random network */
  pnet_random(rn->pnet,p_conn);
  /* allocate subarray of neurons and temp. inhibitory
     array*/
  net=newvec(Neuron,rn->pnet->lo,rn->pnet->hi);
  inhib=newvec(int,rn->pnet->lo,rn->pnet->hi);
  pnet_foreach(rn->pnet,i)
  {
    net[i].spike=NOFIRE;
    /* Set initial condition of neuron */
    net[i].v[0]=-0.02*0.01*drand48();
    net[i].v[1]=0.05+0.20*drand48();
    inhib[i]=(drand48()<p_inhib);
  }
  pnet_setup(rn->pnet);
  /* initialization of ring buffer  */
  rn->rbuffer=(Rbuffer *)malloc(sizeof(Rbuffer)*rn->pnet->
      insize);
  for(i=0;i<rn->pnet->insize;i++)
    initrbuffer(rn->rbuffer+i,RBUFF_LEN);
```

```
      /* allocating information about inhibitory neurons */
      rn->inhib=(int *)malloc(sizeof(int)*rn->pnet->insize);
      /* mapping inhibitory array to output buffer;
      buffer=(int *)rn->pnet->outbuffer;
      for(i=0;i<rn->pnet->outsize;i++)
        buffer[i]=inhib[rn->pnet->outindex[i]];
   /* distributing inhibitory vector */
      pnet_exchg(rn->pnet);
      buffer=(int *)rn->pnet->inbuffer;
      for(i=0;i<rn->pnet->insize;i++)
        rn->inhib[i]=buffer[i];
      /* free temporary storage */
      freevec(inhib,rn->pnet->lo);
      return net;
   } /* of 'init' */
```

In order to write out the timing and the corresponding index of the firing neuron in a sequential way, a datatype Slist has to be defined:

```
   typedef struct
   {
     int time,neuron;
   } Spike;
   typedef struct
   {
     Spike *index;
     int len;  /* length of list */
   } Slist;

   /* Declaration of functions */

   /* Initialize empty list */
   extern void initspikelist(Slist *);
   /* Add spike to list */
   extern int  addspikelistitem(Slist *,Spike);
   /* Empty list */
   extern void emptyspikelist(Slist *);
```

Implementation of Slist is identical to List. If neuron fires, then the index together with the timing of the event is stored in the Spike structure and added to the spike list. The parallel update function can be written as:

```
   void update(Slist *spikelist, /* Spike list */
               Rnet *rn,
               Neuron net[], /* array of neurons */
               int time, /* integer time */
               float h,  /* time step (ms) */
```

```
              float *g, /* lookup table */
              int t_max, /* size of lookup table */
              float w_in,float w_ex,int delay,
              float I, /* applied current */
              float sigma /* amplitude of Gaussian white
                             noise */
              )
{
  int i,j,k,index;
  float sum,v_new[2],dv[2];
  int spike;
  Spike event;
  pnet_foreach(rn->pnet,i)
  {
    sum=0; /* calculate interactions */
    for(j=0;j<rn->pnet->connect[i].len;j++)
    {
      index=rn->pnet->connect[i].index[i];
      for(k=0;k<rn->rbuffer[index].len;k++)
      {
        spike=getrbuffer(rn->rbuffer+index,k)+delay;
        if(time>=spike && time<spike+t_max)
        {
          if(rn->inhib[index])
            sum-=w_in*g[time-spike]*(net[i].v[0]-V_in);
          else
            sum-=w_ex*g[time-spike]*(net[i].v[0]-V_ex);
        }
        else if(spike+t_max<time)
          break;

      }
    } /* of for(j=..) */
    updateml(dv,net[i].v,I);
    v_new[0]=net[i].v[0]+(h/c)*(dv[0]+sum);
    v_new[1]=net[i].v[1]+h*dv[1]+sqrt(h)*sigma*gasdev();
    /* critical threshold of zero reached for v? */
    if(net[i].v[0]<0 && v_new[0]>0)
    {
      /* yes, store time */
      net[i].spike=time;
      event.time=time;
      event.neuron=i;
      addspikelist(spikelist,event); /* add spike to list */
    }
```

```
    net[i].v[0]=v_new[0];
    net[i].v[1]=v_new[1];
  } /* of for(i=..) */
} /* of 'update' */
```

The exchange of spike timings is performed by a call of the exchg function:

```
void exch(Rnet *rn,Neuron net[])
{
  int i,*buffer;
  /* write time of last spike to output buffer */
  buffer=(int *)rn->pnet->outbuffer;
  for(i=0;i<rn->pnet->outsize;i++)
    buffer[i]=net[rn->pnet->outindex[i]].spike;
  pnet_exchg(rn->pnet); /* Communication */
  /* add times of last spike to the corresponding ring
      buffer */
  buffer=(int *)rn->pnet->inbuffer;
  for(i=0;i<rn->pnet->insize;i++)
    if(buffer[i]!=NOFIRE) /* spike occured */
      if(rn->rbuffer[i].len==0 ||
          rn->pnet->inbuffer[i]!=getrbuffer(rn->rbuffer+i,0))
          /* we have a new spike */
          addrbuffer(rn->rbuffer+i,buffer[i]);
} /* of 'exch' */
```

The exchange is only necessary after time t_{delay}. Therefore, communication is significantly reduced in comparison to diffusive coupled neurons.

Each task contains a list of all spike events occured stored locally in the Slist datatype. The serialized output of the spike events collected from all tasks is achieved by the fwritespikes function. All tasks initially send the number of events recorded via the collective MPI_Gather operation to task zero. Then the content of the spike list is sent via MPI_Send to task zero.

```
#define MSG_TIME 99 /* message tag used by send/recv */

void fwritespikes(FILE *file,Pnet *pnet,Slist *list)
{
  int len,i,*list_len;
  Spike *vec;
  MPI_Status status; /* needed by MPI_Recv */
  len=list->len;
  list_len=newvec(int,0,pnet->ntask-1);
  /* Gather number of spikes from all tasks */
  MPI_Gather(&len,1,MPI_INT,list_len,1,MPI_INT,
              0,MPI_COMM_WORLD);
  if(pnet->taskid==0)
```

```
{
  /* write spike events of task 0 */
  fwrite(list->index,sizeof(Spike),len,file);
  /* collect spike events from all other tasks */
  for(i=1;i<pnet->ntask;i++)
    if(list_len[i]>0) /* spike occured in task i? */
    {
      /* yes, allocate temporal storage */
      vec=newvec(Spike,0,list_len[i]-1);
      /* receive spike list from task i */
      MPI_Recv(vec,sizeof(Spike)*list_len[i],
               MPI_BYTE,i,MSG_TIME,
               MPI_COMM_WORLD,&status);
      fwrite(vec,sizeof(Spike),list_len[i],file);
      free(vec);
    }
}
else if (len>0) /* spike occured in my task */
  /* send to task zero */
  MPI_Send(list->index,sizeof(Spike)*len,MPI_BYTE,
           0,MSG_TIME,MPI_COMM_WORLD);
free(list_len);
} /* of 'fwritespikes' */
```

The function for calculating the mean value $\bar{v}$ in parallel uses the global reduction function MPI_Reduce. The reduction function MPI_SUM adds the values of all tasks. The function returns in task zero the global sum:

```
float vmean(Rnet *rn,
            const Neuron net[] /* subarray of neurons */
            ) /* returns mean value of v on task zero */
{
  int i;
  float sum,globalsum;
  sum=0;
  pnet_foreach(rn->pnet,i)
    sum+=net[i].v[0];
  /* global reduction of sum o globlasum on task zero
     using add operator */
  MPI_Reduce(&sum,&globalsum,1,MPI_FLOAT,MPI_SUM,
             0,MPI_COMM_WORLD);
  return globalsum/rn->pnet->n;
} /* of 'vmean' */
```

The main program of the parallel version is:

```
#include <stdlib.h>
```

```
#include <stdio.h>
#include <mpi.h> /* MPI prototypes */
#include "list,h" /* list datatype */
#include "pnet.h"
#include "rnet.h"
#include "neuron.h"
int main(int argc,char **argv)
{
  Neuron *net;
  float *g,g1,w_in,w_ex,v;
  int t,nstep,ostep,tmax;
  FILE *file,*log;
  Rnet rnet;
  Slist spikelist;
  MPI_Init(&argc,&argv); /* initialize MPI */
  net=init(&rnet,n,p_conn,p_inhib);
  /* set random seeds differently for each task */
  srand48(22892+38*rnet.pnet->taskid);
  /* setting inhibitory and excitatory coupling strength */
  w_in=0.1/((n-1)*p_conn);
  w_ex=0.1/((n-1)*p_conn);
  nstep=t_end/h;
  ostep=nstep/100;
  tmax=t_max/h;
  g=getg(tmax,h);
  if(rnet.pnet->taskid==0)
  {
    /* opening output files on task 0 */
    file=fopen("neuron.spike","wb");
    log=fopen("neuron.mean","w");
  }
  initspikelist(&spikelist);
  for(t=0;t<nstep;t++) /* time loop */
  {
    if(t % ostep==0)
    {
      /* write mean value of v to output file every
         ostep time steps */
      v=vmean(&rnet,net);
      if(rnet.pnet->taskid==0)
        fprintf(log,"%g %g\n",t*h,v);
    }
    update(&spikelist,&rnet,net,t,h,g,t_max,
           w_in,w_ex,delay,I,sigma);
    if(delay==0 || t % (delay-1)==0)
```

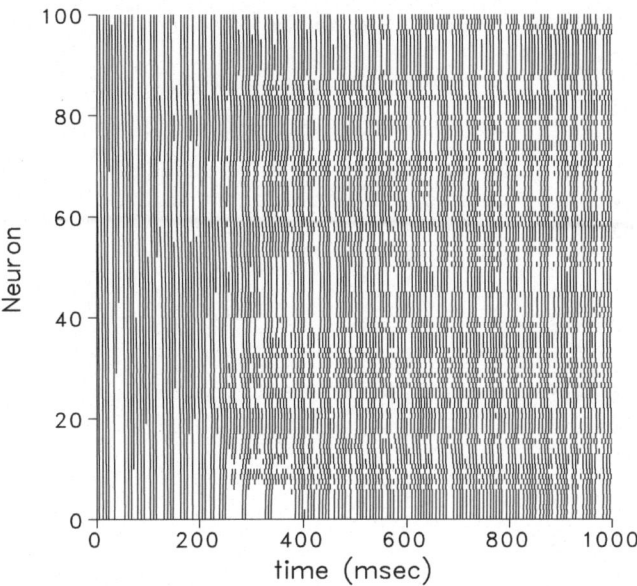

Fig. 11.11. Spike pattern for a network of 100 neurons derived from file `neuron.spike`

```
    { /* exchange necessary every delay time steps */
      exch(&rnet,net);
      /* write out spike timings */
      fwritespikes(file,rnet.pnet,&spikelist);
      emptyspikelist(&spikelist);
    }
  } /* of time loop */
  if(rnet.pnet->taskid==0)
  {
    fclose(file);
    fclose(log);
  }
  MPI_Finalize(); /* end MPI */
  return 0;
} /* of 'main' */
```

Sample output of the spiking times of a ML network is shown in Fig. 11.11.

11.3 Connection Dependent Coupling Strengths and Delays

For the sake of simplicity, we have up to now only considered globally uniform values for the coupling strength $w_{\text{in,ex}}$ and delays t_{delay}. In general, these are connection dependent values. This can be implemented by defining a new datatype for connections:

```
typedef struct
{
  int index; /* index of neuron */
  float w;   /* connection dependent weight */
  int delay; /* delay */
} Conn;

typedef struct
{
  Conn *conns; /* array of connections */
  int size;    /* number of connections */
} Connlist;
```

Then, the datatype neuron is defined as:

```
typedef struct
{
  float v[2];         /* state of neuron (ML) */
  Rbuffer spikes;     /* times of last spikes */
  int inhib;          /* inhibitory or not */
  Connlist connect;   /* connection list */
} Neuron;
```

Using this data structure, it is possible to model spike-timing-dependent plasticity (STDP), i.e. the weights are modified differently, dependent on the times of the pre- and postsynaptic spike arrival times t_i and t_j (Chaps. 2 and 9). The weight w_{ij} of a connection is increased or decreased by Δw_{ij} according to:

$$\Delta w_{ij} = \begin{cases} A_+ \exp(\Delta t/t_+) \text{ for } \Delta t > 0 \\ A_- \exp(\Delta t/t_-) \text{ for } \Delta t < 0 \end{cases}, \tag{11.14}$$

where $A_+ > 0$, $A_- < 0$ and $\Delta t = t_i - t_j$. The function Δw_{ij} (Fig. 11.12) is implemented in the following way:

```
#define A_plus 0.01
#define A_minus (-0.012)
#define t_plus 20.0
#define t_minus 20.0

float deltaw(float deltat)
{
  return (deltat>0) ? A_plus*exp(-deltat/t_plus)
                    : A_minus*exp(deltat/t_minus);
} /* of 'deltaw' */
```

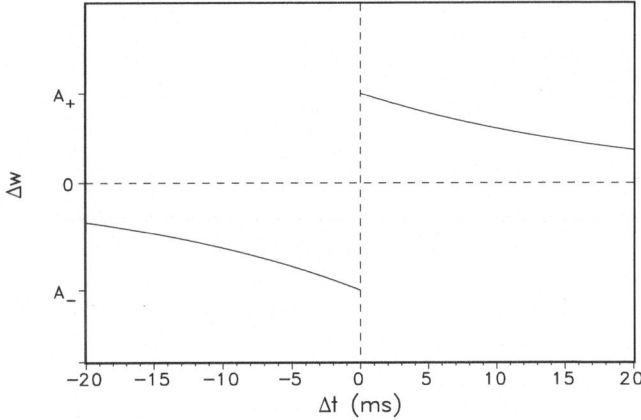

Fig. 11.12. Plasticity Δw as a function of difference between pre- and postsynaptic spike arrival Δt

Then the sequential update function with plasticity is:

```
void update(FILE *file, /* output file for spikes */
            Neuron net[], /* array of neurons */
            int n, /* number of neurons */
            int time, /* integer time */
            float h, /* time step */
            float *g, /* lookup table */
            int t_max, /* size of lookup table */
            float I, /* applied current */
            float sigma /* amplitude of Gaussian white
                            noise */
            )
{
  int i,j,k,index,last;
  float sum,v_new[2],dv[2];
  int spike;
  for(i=0;i<n;i++)
  {
    sum=0; /* calculate interactions */
    for(j=0;j<net[i].connect.size;j++)
    {
      index=net[i].connect.conns[j].index;
      /* iterating over the ring buffer of neuron index */
      for(k=0;k<net[index].spikes.len;k++)
      {
        spike=getrbuffer(&net[index].spikes,k)+
              net[i].connect.conns[j].delay;
        /* testing whether spike is active */
        if(time>=spike && time<spike+t_max)
```

```
      {
        /* yes */
        if(net[index].inhib)
          /* inhibitory coupling */
          sum-=net[i].connect.conns[j].w*
              g[time-spike]*(net[i].v[0]-V_in);
        else
          /* excitatory coupling */
          sum-=net[i].connect.conns[j].w*
              g[time-spike]*(net[i].v[0]-V_ex);
        /* plasticity */
        if(time==spike)
        {
          /* previous spike occured? */
          if(net[i].spikes.len>0)
          {
            /* yes, change weight of connection, delta
               t<0 */
            last=getqueue(net[i].spikes,0);
            net[i].connect.conns[j].w+=
                deltaw(-(spike-last)*h);
          }
        }
        else
          /* no, break if all spikes are not active
             anymore */
          if(spike+t_max<time)
            break;
    } /* of for(k=...) */
  } /* of for(j=...) */
  /* update of ML */
  updateml(dv,net[i].v,I);
  v_new[0]=net[i].v[0]+(h/c)*(dv[0]+sum);
  v_new[1]=net[i].v[1]+h*dv[1]+sqrt(h)*sigma*gasdev();
  /* critical threshold of zero reached for v? */
  if(net[i].v[0]<0 && v_new[0]>0)
  {
    /* yes, we have a spike event, add to my ring buffer */
    addrbuffer(&net[i].spikes,time);
    fprintf(file,"%g %d\n",time*h,i);
    /* plasticity */
    for(j=0;j<net[i].connect.len;j++)
    {
      index=net[i].connect.conns[j].index;
      for(k=0;k<net[index].spikes.len;k++)
```

```
      {
        spike=getqueue(net[index].spikes,k)+
              net[i].connect.conns[j].delay;
        if(spike<time)
        {
          /*change weight of connection, delta t>0 */
          net[i].connect.conns[j].w+=deltaw((time-spike)
          *h);
          break;
        }
      } /* of for(k=...) */
    } /* of for(j=...) */
  }
  net[i].v[0]=v_new[0];
  net[i].v[1]=v_new[1];
  } /* of for(i=...) */
} /* of 'update' */
```

The initialization of the network has to include setting up the connection-dependent weights and delays. The parallel version of the code can be implemented analogously. The parallel exchange performed by pnet_exch is only necessary every $t_{\min} \times \Delta t$ time steps, where $t_{\min}$ is defined as

$$t_{\min} = \min_{i,j=1...n} t_{\text{delay},i,j}. \tag{11.15}$$

References

1. R. Sedgewick: *Algorithms in C, Part 5: Graph algorithms*, 3rd edn (Addison Wesley Reading MA 2001)
2. C. Morris, H. Lecar: Biophys. J., **35**, 193 (1981)
3. T. H. Cormen, C. E. Leiserson, R. L. Rivest, C. Stein: *Introduction to algorithms*, 2nd edn (MIT Press Cambridge MA 2001)
4. B W. Kernighan, D. Ritchie: *The C programming language* (Prentice Hall 1988)
5. W. Gropp, E. Lusk, A. Skjellum: *Using MPI: Portable parallel programming with the message-passing interface*, 2nd edn (MIT Press Cambridge MA 1999)
6. A. Morrison, C. Mehring, T. Geisel, A. Aertsen, M. Diesmann: Neural Comput., **17**, 1776 (2005)
7. A. Tam, C. Wang: Efficient scheduling of complete exchange on clusters. In: *13th international conference in parallel and distributed computing systems* (PCDS Las Vegas 2000)
8. G. M. Amdahl: Validity of the single-processor approach to achieving large scale computing capabilities. In: *AFIPS Conference Proceedings*, Vol. 30, (AFIPS Press Reston VA 1967) pp. 483–485.
9. W. H. Press, S. A. Teukolsky, W. T. Vetterling, B. P. Flannery: *Numerical recipes in C: The art of scientific computing*, 2nd edn (Cambridge University Press Cambridge MA 1992)

Part V

Applications

Parametric Studies on Networks of Morris-Lecar Neurons

Steffen Tietsche[1], Francesca Sapuppo[2] and Petra Sinn[1]

[1] Institute of Physics, Potsdam University, Am Neuen Palais 10, 14469 Potsdam, Germany
[2] Dipartimento di Ingegneria Elettrica, Elettronica e dei Sistemi, Universita' degli Studi di Catania, V.le A.Doria 6, 95125, Catania, Italy
fsapuppo@diees.unict.it

12.1 Introduction

The properties of a network are determined by the network topology, including the connectivity and the coupling strength. Our network consists of neurons modeled by the Morris-Lecar equations. We study the influence of some important parameters on the network dynamics. The parameters we vary are the network topology, the global coupling strengths for excitatory and inhibitory neurons g_{ex} and g_{in} and the variance of an internal noise term.

In Sect. 12.2, the mean spike rate is taken as a measure for network activity, and scans through the g_{ex}–g_{in} parameter plane for different noise strengths reveal how the network activity typically depends on the coupling strength. Most of the study is concerned with ER random networks, but in Sect. 12.2.3, it is shown that the typical dependencies also hold for small-world networks.

In Sect.12.3, the network behavior near the lower critical coupling strength, where network activity sets in, is inspected more closely. Raster plots of random networks as well as small-world networks show the time series of spiking activity.

12.1.1 The Morris-Lecar Neuron Model

To simulate our neural network, we use the Morris-Lecar model (see Chaps. 1, 9, 11, 14). This model represents an electrical circuit similar to a cellular membrane. It consists of three general synaptic currents and an additional external current I_{ext}. Equation (12.1) describes the development of the membrane potential V, with instantaneous activation of the inward Ca^{2+} current, slower activation of the outward K^+ current and a leakage current.

$$\dot{V} = -g_{Ca}n_\infty(V - V_{Ca}) - g_K w(V - V_K) - g_{leak}(V - V_{leak}) + I_{ext} \quad (12.1)$$

$$\text{with } n_\infty = 0.5\left[1 + \tanh\left(\frac{V - V_1}{V_2}\right)\right]$$

$$\dot{w} = \phi\cosh\left(\frac{V - V_3}{2V_4}\right)(w_\infty - w) + \sigma\eta(t)\sqrt{h} \quad (12.2)$$

$$\text{with } w_\infty = 0.5\left[1 + \tanh\left(\frac{V - V_3}{V_4}\right)\right]$$

This is the normalized form of the Morris-Lecar equations used by Rinzel and Ermentrout [1]. The parameters g_{Ca}, g_K and g_{leak} are the maximal conductances for each synaptic current and V_{Ca}, V_K and V_{leak} are the corresponding reversal potentials. The number of open Ca^{2+} channels is n_∞. The number of activated K^+ channels is w, its time dependence being described in (12.2). Equation (12.2) contains an internal noise term $\sigma\eta(t)\sqrt{h}$, which is additive Gaussian white noise with variance σ^2. The external current I_{ext} is given by

$$I_{ext} = I + I_{syn},$$

where I is a constant external current and I_{syn} the synaptic current coming from connected neurons (see below). For a discussion of the equations and their parameters, refer to [2,4], or Chap. 1.[3]

12.1.2 Network Setup

The statistical properties of the networks (e.g. connection probability) are fixed, but the network behavior may vary for different realizations of the setup. We set up a new network for every simulation run, thereby ensuring that most of our data points are from typical realizations of the network setup.

The coupling of the neurons enters the Morris-Lecar equations through the synaptic current I_{syn}, which is defined as

$$I_{syn} = -\sum_{j=1}^{N}\sum_{k=1}^{\infty} g(t - t_{spike}(j,k))(V - V_j) \quad (12.3)$$

with

$$g(t) = \frac{g_j}{\tau_1 - \tau_2}(e^{-t/\tau_1} - e^{-t/\tau_2}) \quad (12.4)$$

and the replacement rule

---

[3] The parameter values were fixed to: $V_{Ca} = 1.0$, $V_K = -0.7$, $V_{leak} = -0.5$, $V_1 = -0.01$, $V_2 = 0.15$, $V_3 = 0.1$, $V_4 = 0.145$, $V_{inh} = -0.55$, $V_{ex} = 0.05$, $I = 0.08$, $\phi = 0.33$, $g_{Ca} = 1.0$, $g_K = 2.0$, $g_{leak} = 0.5$, $h = 0.01$ ms, $\tau_1 = 1.0$ ms, $\tau_2 = 2.0$ ms, $N = 400$.

$$\begin{aligned}
\text{neuron j is excitatory} &\iff V_j = V_{\text{ex}}, \quad g_j = g_{\text{ex}}, \\
\text{neuron j is inhibitory} &\iff V_j = V_{\text{in}}, \quad g_j = g_{\text{in}}.
\end{aligned} \tag{12.5}$$

In (12.3), N is the number of neurons in the network. The function $g(t)$ modulates the weight of a spike depending on the time that has passed. $V_{\text{ex}} > 0$ is the resting potential for excitatory neurons, $V_{\text{in}} < 0$ the resting potential for inhibitory neurons.

In addition to the spikes produced by the Morris-Lecar equations, we feed spikes to each neuron independently at random times with a mean frequency of 3 Hz. This Poisson process serves as an external forcing of the network and stimulates activity [3].

In our study, the network consists of 400 neurons. 90% of the neurons are excitatory and 10% are inhibitory. The connections between neurons are bidirectional and the connection strengths are uniform. Excitatory neurons are connected with a connection strength g_{ex}, and inhibitory neurons have connection strength g_{in}. The following two network topologies are considered:

(i) An ER random network, in which each neuron is connected to every other neuron with a probability of $p = 0.2$, leading to a connectivity of 20%.
(ii) A small-world network, which is set up by first forming a ring where every neuron is connected to the next 40 neighbors to each side and then replacing 5% of those connections by connections to random neurons. The resulting connectivity is also 20%.

Both topologies have properties that are important for cortical networks: the random networks has a short pathlength but low clustering, and the small-world network has short pathlengths and high clustering.

12.2 Influence of Coupling Strengths and Noise on the Network Activity

The dynamics of networks of neurons is affected by excitatory and inhibitory coupling strengths (g_{ex} and g_{in}, respectively). We take the spike rate as a measure for the global behavior of the network.

For any pair of g_{ex} and g_{in}, the spike rate per neuron is computed by running the simulation, counting the total number of spikes, and dividing by simulation time (typically 3 s) and the number of neurons.

We take the code from Chap. 11, make it parallel using MPI and run it on a cluster of 16 nodes. The parameter space ($g_{\text{ex}}, g_{\text{in}}$) is divided into subareas, with each area being assigned to one processor. The parameter g_{ex} is varied between 0 and 1 with a step size of 0.008, while g_{in} is varied between 0 and 0.4 with a step size of 0.01. We use a smaller incremental step for g_{ex} than for g_{in}, since we expect that the network dynamics depends more strongly on the excitatory coupling strength than on the inhibitory one, because only 10% of the neurons are inhibitory.

12.2.1 Characteristic Features of the Activity Function

We study a network of $N = 400$ neurons, which are randomly connected with a probability $p = 0.2$. In this first study, the noise level in (12.2) is set to $\sigma = 0$. This initial choice allows us to study the network behavior in the case of deterministic neuron equations.

Figure 12.1 shows the 3-D plot of the spike rate as a function of the coupling strengths g_{in} and g_{ex}. For small g_{ex}, the spike rate is rather low (3 Hz). This is the rate of spikes that are externally induced by the Poisson process. As expected, there is no self-sustained spiking activity above the input level if the coupling between the neurons is too weak.

At $g_{ex} \approx 0.05$, there is an abrupt increase in network activity. The value of the critical threshold only weakly depends on the inhibitory coupling strength, as Fig. 12.2 shows. As the excitatory coupling strength increases further, the spike rate reaches a maximum of about 130 Hz at $g_{ex} \approx 0.15$ and then decreases again. The most obvious difference when changing g_{in} is the behavior for large g_{ex}. For small g_{in}, the spike rate quickly drops to a constant value of 50 Hz, whereas for larger g_{in}, the decrease is slower, and saturation is not yet reached for the highest g_{ex} in our data.

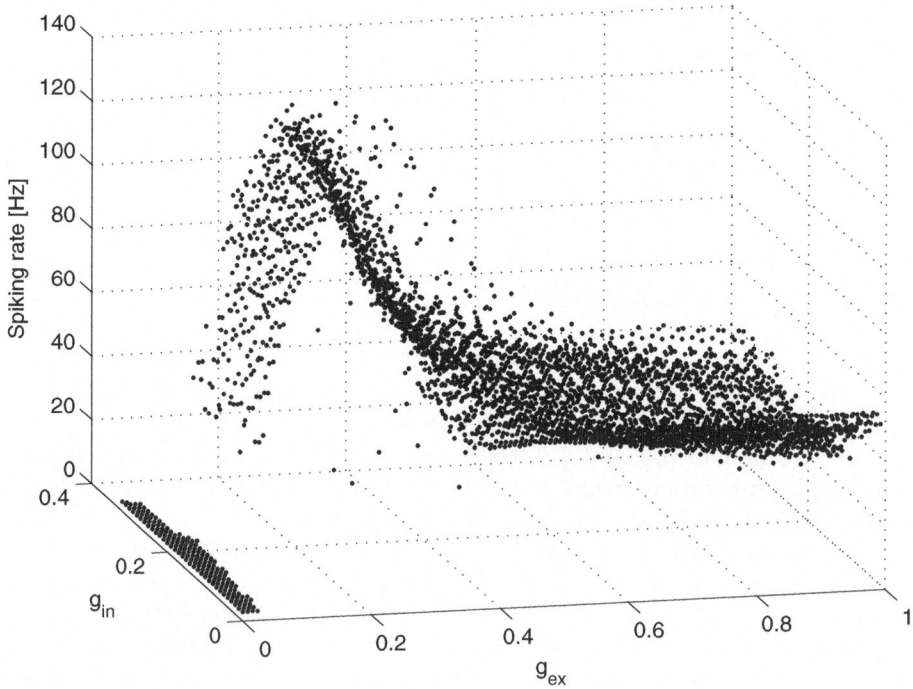

Fig. 12.1. Random network: spike rate per neuron as a function of coupling strengths g_{in}, g_{ex}

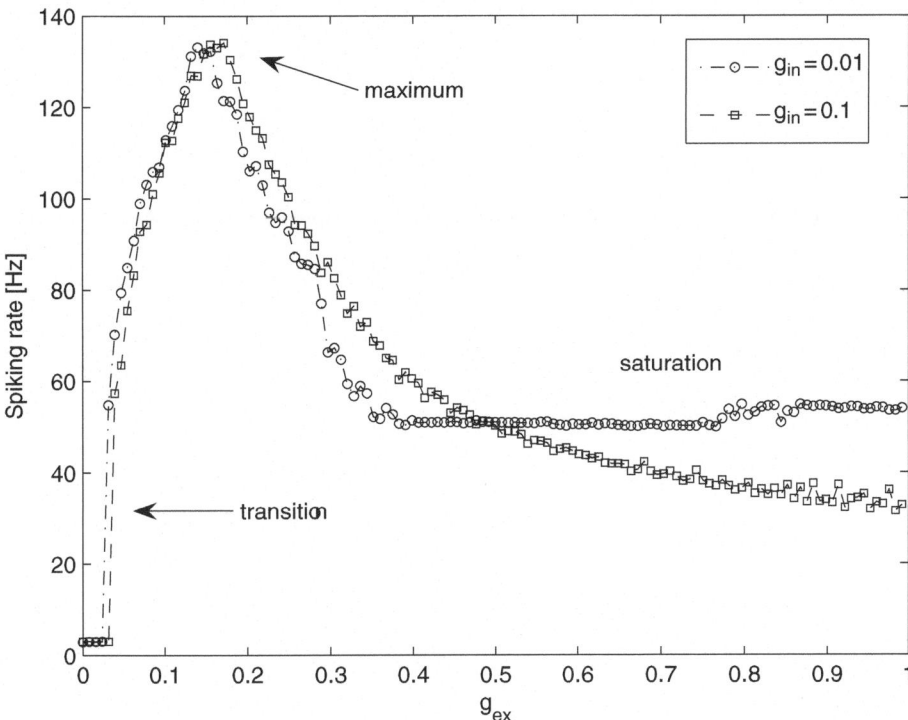

Fig. 12.2. Random network: spike rate per neuron as a function of coupling strength g_{ex}

12.2.2 Effects of Additional Gaussian Noise

Next, we study the effects of additional noise on the global behavior of a random network. The crucial parameter is the noise strength σ in the Morris-Lecar equation (12.2). In Fig. 12.3, the activity for $\sigma = 0.05$ is compared to the case where $\sigma = 0$.

The $\sigma = 0.05$ curve does not present the sharp increase in spike rate at a certain excitatory coupling strength like the one we observe for $\sigma = 0$. We speculate that the reason for this is the disturbance of collective behavior by the independent random signals fed to each neuron.

Furthermore, the peak of maximal activity moves towards higher coupling strengths. We compare the spike rates at a fixed value of $g_{\mathrm{in}} = 0.1$: for $\sigma = 0.05$ the maximum is at $g_{\mathrm{ex}} = 0.25$, and for $\sigma = 0$ it is at $g_{\mathrm{ex}} = 0.16$. Similarly to $\sigma = 0$, the range of g_{ex} considered is too small to observe saturation.

12.2.3 Activity Function for a Small-world Network

The study of the dynamics of a small-world network with $N = 400$ neurons, probability of long range connection $p = 0.05$ is performed with the same

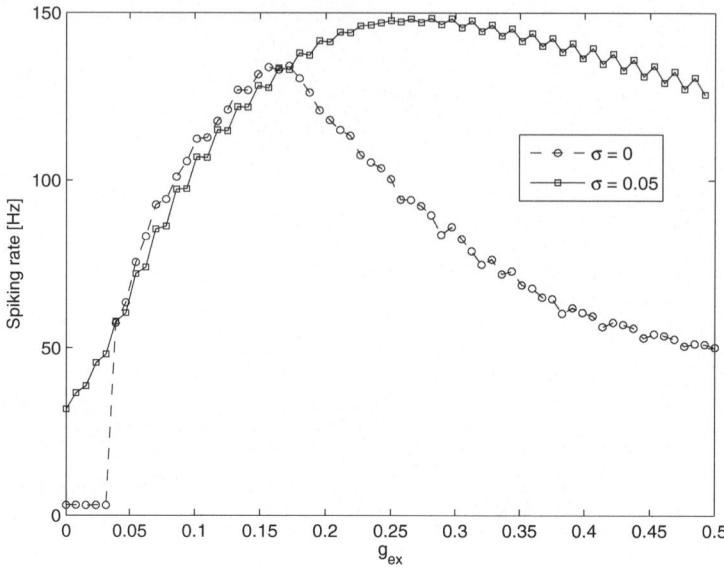

Fig. 12.3. Random network: spike rates per neuron for noisy Morris-Lecar neurons compared to the curve $\sigma = 0$. For both curves $g_{\text{in}} = 0.1$

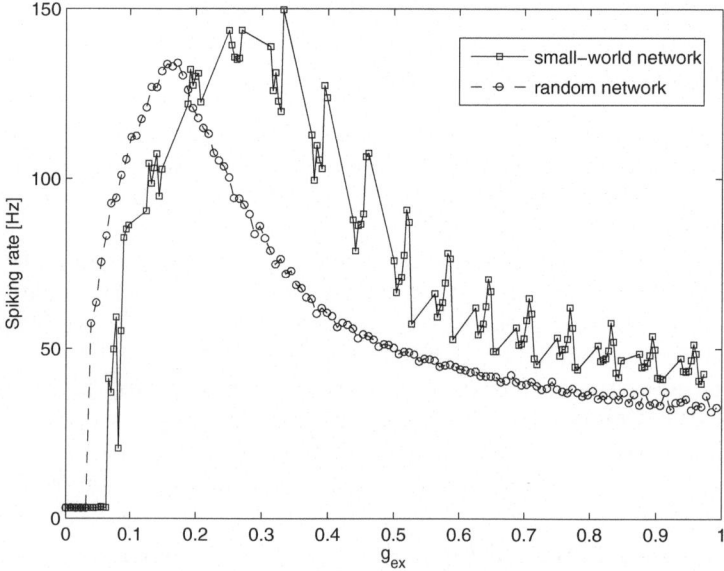

Fig. 12.4. Comparison between spike rates for small-world and random networks. For the random network, $g_{\text{in}} = 0.15$, for the small-world network, $g_{\text{in}} = 0.4$

global connectivity of 20% and the same Poissonian input process. Fig. 12.4 shows the spike rate curves for a random as well as small-world network. For both network types, there is a transition point at some value of the excitatory coupling strength g_{ex}, where a sharp increase in network activity occurs.

12.3 Network Dynamics Near the Critical Coupling Strength

As shown in the previous section, for both network types, there is a sharp increase in network activity at a certain coupling strength g_{ex}. This section will show that close to the critical coupling strength, the network dynamics develops complex features on the time scale of a few seconds (cf. Fig. 12.5).

To analyze the network behavior, the spike events are presented in *raster plots*. This type of plot has time on the horizontal axis, neuron index on the vertical axis, and contains a dot for every spike event. From this, temporal

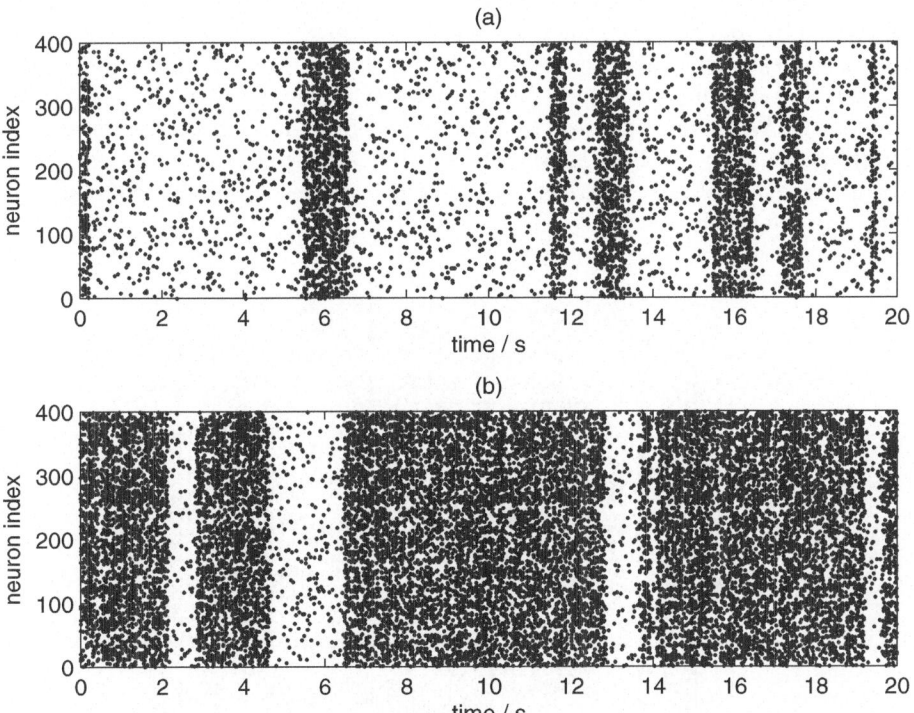

Fig. 12.5. Raster plots of a random network of Morris-Lecar neurons for two different coupling strengths that are in the critical range: (a) $g_{ex} = 0.062$; (b) $g_{ex} = 0.065$. $g_{in} = 0.4$ for both

changes in network activity as well as spatial inhomogeneities can be readily recognized.

When there is Gaussian white noise added to the Morris-Lecar equation (12.2), the sharp transition in activity as well as the complex behavior vanish. This is to be expected, since noise with variance σ^2 is added to each neuron independently and should disturb correlation and coherence between them. All considerations in the following part are therefore from data with $\sigma = 0$.

12.3.1 ER Random Network

In our network configuration (cf. Sect. 12.1) and with $g_{in} = 0.4$, the transition occurs between $g_{ex} = 0.06$ and $g_{ex} = 0.07$. Figure 12.5 shows two simulations with $g_{ex} = 0.062$ and $g_{ex} = 0.065$.

For these parameter values, there are two distinct states of the network. The state with low activity has a spike rate of approximately 3 Hz. This is the main frequency of the external forcing (cf. Sect. 12.1), so there is almost no self-sustained activity. The high-activity state has a spike rate of approximately 26 Hz. In this state, the neurons seem to be synchronized (cf. Fig. 12.6), which suggests that the interaction between the neurons drives the network.

Within one simulation run, the network switches between the two states irregularly. With increasing g_{ex}, the network is in the high-activity state for longer time intervals (Fig. 12.5). Interestingly, the spike rate changes rather abruptly within 50 ms, whereas between the switches it remains constant for typically a few seconds.

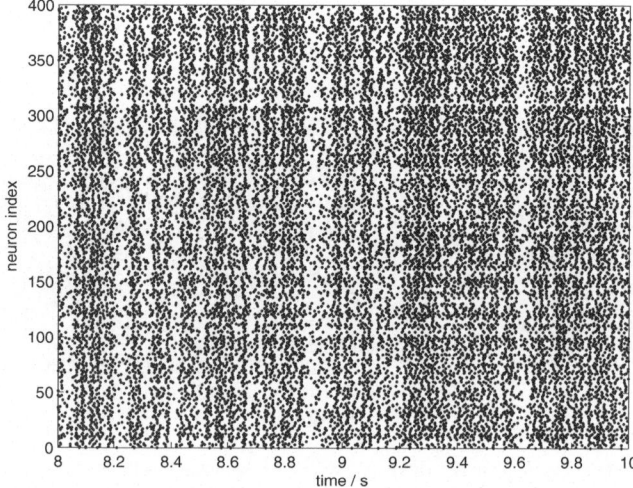

Fig. 12.6. Random network near critical coupling strength during high-activity phase. Enlargement of Fig. 12.5(b)

Figure 12.6 shows an enlargement of a high-activity phase of Fig. 12.5(b). There are irregular small-amplitude variations of activity on a time scale of a few milliseconds. The horizontal stripes are due to the rather strong activity variations between neurons: the average spike rate for a single neuron is 23 spikes per second, with a standard deviation of 9 spikes per second. We think that the deviations in activity are due to differences in the number of connections. Neurons with fewer connections have less input and therefore should be less active on average.

12.3.2 Small-world Network

Raster plots for small-world networks also show complex behavior near the critical coupling strength, but there are differences to the behavior of random networks, which are related to network topology. In a small-world network, neurons with neighboring indices are strongly connected, while there are only a few random connections between distant neurons (cf. Sect. 12.1).

The small-world topology influences the dynamics of the network in two ways:

(i) By the connection of neurons with neighboring indices, the concepts of neighborhood and distance between neurons are introduced. As a consequence, the spreading of activity is represented in the raster plots as non-orthogonal features, which almost certainly do not occur in random networks (compare Figs. 12.5 and 12.7).

(ii) Neighboring neurons can group into clusters that are strongly interconnected, but weakly connected to the rest of the network. The existence of inhibitory neurons supports this separation into clusters. In a raster plot, the clustering is indicated by broad horizontal stripes with distinct spiking behavior.

Figure 12.7 shows two examples of small-world networks with a close-to-critical coupling strength g_{ex}. As in the case of random networks, there is a state in which the network activity is only driven by the 3 Hz Poissonian input noise, and a state in which the coupling leads to a rather strong activity. For small-world networks, the whole network is not all in one single state; rather, the different clusters can have different states. For example, the cluster around neuron 300 in Fig. 12.7(a) is silent all the time, while the state of the neurons 100 to 250 changes irregularly.

Generally, with increasing g_{ex}, the high-activity state dominates the low-activity state, which is similar to the case of random networks. But the simulation run shown in Fig. 12.7(a) has a higher activity than the one shown in Fig. 12.7(b), although its coupling strength is lower. Bearing in mind that for each simulation run a new realization of the network setup was used, this gives a hint that the dynamics of near-critical small-world networks is more sensitive to the details of the network setup.

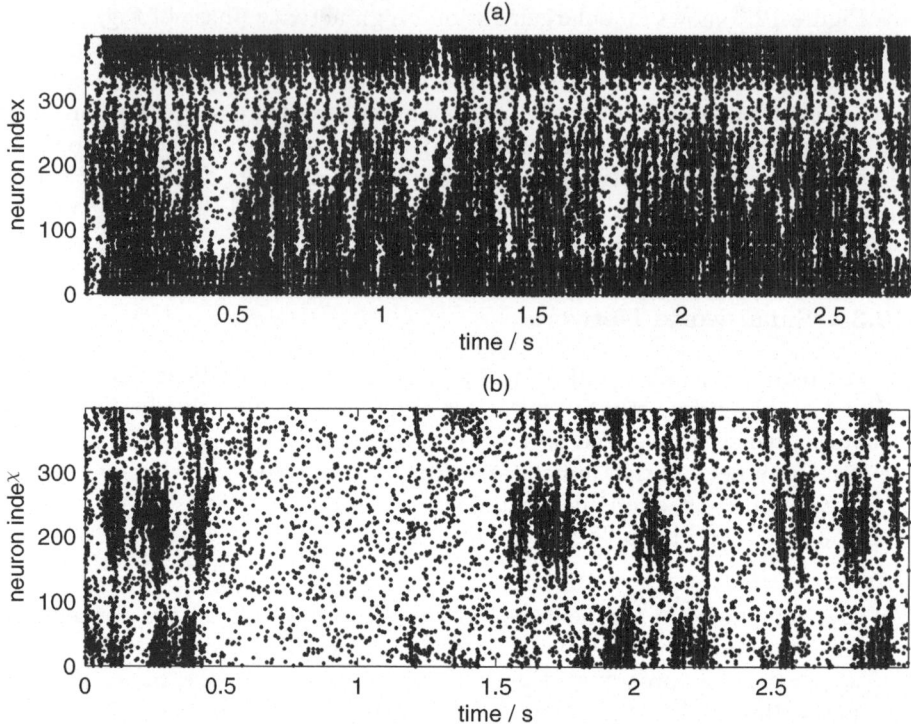

Fig. 12.7. Raster plots of small-world networks of Morris-Lecar neurons for two different coupling strengths that are in the critical range: (**a**) $g_{ex} = 0.065$; (**b**) $g_{ex} = 0.071$. $g_{in} = 0.4$ for both

In Fig. 12.8, which is an enlargement of Fig. 12.7(b), it can be seen that during high-activity phases, the neurons are strongly synchronized. They spike together after regular time intervals of approximately 20 ms. Another interesting feature is the spreading of activity that takes place between $t = 0.5$ s and $t = 0.6$ s; initially only neurons 0 to 50 are active, but then activity spreads up to neuron 250, where the propagation stops at the silent cluster around neuron 300.

12.4 Conclusion

Our study of networks of Morris-Lecar neurons shows that the network activity, measured as the average spike rate per neuron, strongly depends on the coupling strength between the excitatory neurons. There is a lower threshold coupling strength at which the activity abruptly rises, a maximal activity for intermediate coupling strengths and a saturation effect for high coupling strengths.

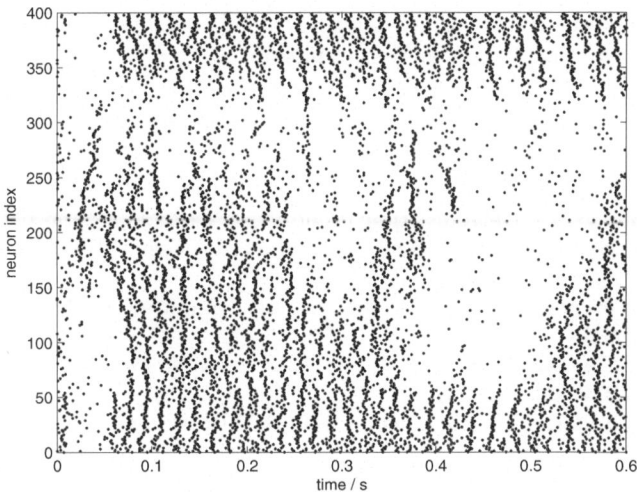

Fig. 12.8. Small-world network near critical coupling strength. Enlargement of Fig. 12.7(a)

Additive noise in the Morris-Lecar equations changes the shape of the activity function: for small coupling strengths, the activity increases smoothly — there is no threshold coupling strength. The maximum of activity is at higher coupling strengths. Our data is not sufficient to establish a saturation effect for the case of additive noise. Future studies should be carried out for larger parameter intervals.

For coupling strengths that are close to the lower threshold, complex dynamical patterns are possible. These patterns, visualized in raster plots, depend on the network topology and the coupling strength. Internal noise seems to inhibit the patterns. We suspect that the reason for this is the destruction of coherence between the neurons by the independent random signals fed to each neuron.

Our results suggest that the dependence of activity on coupling strength is similar for small-world networks and random networks. Nevertheless, the dynamical patterns inferred from the raster plots are rather different. Furthermore, the behavior of small-world networks seems to be more sensitive to connection set up. However, our results for small-world networks are only preliminary and require validation. Further studies should also try to systematically investigate the influence of the internal noise strength.

Two important questions remain open. First, how can one quantify the network behavior by different measures than the spike rate? For example, the detection of phase synchronization of the neurons would certainly give deeper insight into the network dynamics (compare also Chap. 6). Second, the character of the transition at the lower threshold of coupling strength should be explored. This regime may be particularly interesting for brain dynamics, since the intermediate degree of synchronisation allows for coherent but complex reaction to external stimulation.

References

1. J. Rinzel and B. Ermentrout: Analysis of neural excitability and oscillations. In: *Methods in Neuronal Modeling* ed. by C. Koch, I. Segev (MIT Press, Cambridge 1991).
2. M. St-Hilaire and A. Longtin: Journal of Computational Neuroscience **16**, 299–313 (2004).
3. N. Montejo, M. N. Lorenzo, V. Pérez-Villar, and V. Pérez-Muñuzuri: Physical Review E **72**, 011902 (2005).
4. E. Izhikevich, IEEE Transactions on Neural Networks, **15**, 1063–1070 (2004).

Traversing Scales:
Large Scale Simulation of the Cat Cortex
Using Single Neuron Models

Martin Vejmelka[1], Ingo Fründ[2] and Ajay Pillai[3]

[1] Department of Cybernetics, Faculty of Electrical Engineering, Czech Technical
University, Karlovo náměstí 13, 121 35 Praha 2, Czech Republic
Institute of Computer Science, Czech Academy of Sciences, Pod Vodárenskou
věží 4, Praha 8, Czech Republic
vejmelka@cs.cas.cz
[2] Department for Biological Psychology, Otto-von-Guericke University, Magdeburg
PO-Box 4120
ingo.fruend@nat.uni-Magdeburg.de
[3] Theoretical Neuroscience Group, Center for Complex Systems & Brain Sciences,
Florida Atlantic University, Boca Raton, Florida, USA
pillai@ccs.fau.edu

13.1 Introduction

The average adult human brain has about 100 billion neurons. Taken together,
these neurons form several trillions [1] of connections with each other. Though
we understand the functioning of single neurons in quite some detail, the same
cannot be said about large-scale neural network dynamics and the mechanisms
that generate them. Issues pertaining to large-scale neural interactions remain
open questions in the neurophysiological community. Even amongst computational neuroscientists who use neural modeling to investigate brain dynamics,
scale has remained a challenging issue.

To date, there are two major lines in modeling neural activity. On the one
hand, individual neurons are modeled in more or less detail [2] (cf. Chaps.1
and 2) and on the other hand, field equations that do not explicitly contain
individual neurons are derived for the propagation of activity in neural tissue
[3](cf. Chaps.1 and 8). Both these approaches have advantages but at the
same time they have serious shortcomings.

When modeling individual neurons, one has access not only to spike rates
but also to spike timings and the relations between the timings of individual spikes. Such timing aspects seem to play important roles for neural processing and coding [4–6]. It has been argued that only codes that rely on
spike timing could account for certain types of experimentally observed neuronal responses [7]. However, modeling individual neurons is usually limited

to relatively small numbers of neurons, typically several hundred, e.g. [8]. To overcome this limitation, several models that describe the propagation of statistical properties (such as spike rate or average postsynaptic potential) of neural tissue, so-called neural "field equations", have been developed [3,9,10]. In these types of models, individual cells are not modeled explicitly but instead the neural tissue by virtue of its high density of neurons is modeled as a continuum. Although these models capture dynamic aspects of large numbers of neurons, they do not capture all the details like an individual neuron's spikes and hence cannot provide information such as spike timing.

In this chapter, we explore the possibility of bringing the best of both worlds into one approach. We modeled a large-scale hierarchically organized neuronal network made up of individual neurons with a connection topology based on real physiological data. The connectivity of this large network mimicked the connectivity patterns found in the cat cortex by Scannel et al. [11] previously.

Simulating hundreds of thousands of neurons requires far more computing power than what can be supplied by even the most powerful desktop computers of today. In addition, it also presents some difficult numerical and computational challenges. We employed clusters of PCs operating in parallel to simulate our very large and detailed neuronal network. Thus, we were able to capture details of the network like spike timing and at the same time have the benefit of a large spatial scale and realistic connectivity. Using this approach, one set of results may be analyzed on multiple levels of detail at the same time.

13.2 Materials and Methods

We have modeled the distributed large-scale cortical activity of the cat using a neural network model with three hierarchically organized scales. On the lowest level of hierarchy, single neuron dynamics was modeled using point neuron models (cf. Chap. 1). These neuron models were connected to form local networks with a random topology (cf. Chap.3). The local networks were subsequently connected together to form a global network using a connection scheme respecting physiologically known information about long-range connectivity in the mammalian brain [11] (cf. Chaps. 4 and 9).

13.2.1 Neuron Model

Single neurons were modeled using the simple neuronal model proposed by Izhikevich [12] (cf. Chap. 1). This model consists of two state equations (13.2), representing the fast dynamics of the membrane potential and slow recovery effects due to activation of K^+ and inactivation of Na^+ currents.

$$\frac{dv}{dt} = 0.04v^2 + 5v + 140 - u + I \qquad (13.1)$$

$$\frac{du}{dt} = a(bv - u),$$

where u is a variable representing the Na$^+$ channel activation and K$^+$ channel inactivation and v is the membrane potential. If the membrane potential exceeds $30\,\mathrm{mV}$, the model is reset to $v = c$ and $u = u + d$ and a spike is emitted. The model can be tuned using the four parameters a, b, c, d to mimic the dynamical features of a wide variety of cortical and subcortical neurons. The variable I represents the total input current into the neuron. Motivated by the anatomy of the mammalian cortex [13], we chose the ratio of excitatory to inhibitory neurons to be 4 to 1. Excitatory neurons were chosen from a distribution containing the dynamical features of regular spiking, intrinsically bursting and chattering cells [14, 15]; typical activity of each type is shown in Fig. 13.1. For every excitatory neuron, the parameters were set to $(a, b) = (0.02, 0.2)$ and $(c, d) = (-65, 8) + (15, -6)r^2$ where $r \in \langle 0, 1 \rangle$ is a uniformly distributed variable. As regular spiking cells are more frequent in the cortex, the choice of excitatory neurons was biased towards this cell type. The term r^2 serves to bias the distribution.

Inhibitory neurons were chosen from a distribution containing fast spiking and low threshold firing cells, the parameters given by $(a, b) = (0.02, 0.25) + (0.08, -0.05)r$ and $(c, d) = (-65, 2)$, where again $r \in \langle 0, 1 \rangle$ is uniformly

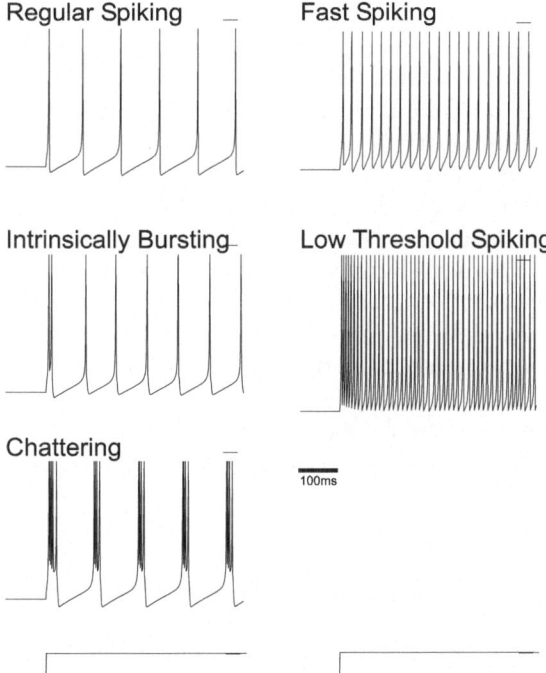

Fig. 13.1. Membrane potential evolution of samples from the distribution of excitatory (*left*) and inhibitory (*right*) cells in response to injected current (*bottom traces*)

distributed. Samples of the activity of both types are also in Fig. 13.1. Low threshold firing cells are characterized by strong spike frequency adaption during tonic stimulation [12].

13.2.2 Local Networks

We have decided to use a random connection topology for local connections because we found the current anatomical and physiological knowledge insufficient to support the adoption of a more specific topology. In order to mimic the sparsity of neural connections, only 10% of all possible connections were generated. Connections were randomly selected from the set of all possible connections. In total, 53 local networks were modeled, one for each area of the cortex as mapped by Scannel et al. [11]. If one neuron was connected to another neuron, spikes of the first neuron triggered postsynaptic potentials in the second neuron with a certain conduction delay. Postsynaptic potentials were modeled by the function

$$V_{\mathrm{PSP}}(\tau) = \frac{\sigma}{\tau_1 - \tau_2} \Big(\exp(-\tau/\tau_1) - \exp(-\tau/\tau_2) \Big) \qquad (13.2)$$

with a rising time constant $\tau_1 = 1\,\mathrm{ms}$ and a falling time constant of $\tau_2 = 3\,\mathrm{ms}$ added to the membrane potential of the postsynaptic cell, and where σ is the peak value of the exponential. The time constants of the postsynaptic potential were chosen to mimic the dynamics of AMPA and GABA$_\mathrm{A}$ receptors of excitatory and inhibitory synapses respectively [13]. Cortical conduction delays vary over a broad range between 0.1 ms to delays as long as 44 ms. Data by Swadlow [16] show two peaks in the distribution of conduction delays. One of these peaks is in the range of delays clearly below 10 ms, the other peak is in a range between 10 ms and slightly above 20 ms. We modeled local conduction delays in the range between 0.5 and 4 ms, which roughly corresponds to the first peak in the conduction delay distribution. In this way, conduction delays for long range connections could be modeled using conduction delays from the second peak. The strengths of the connections in the local networks were assigned to the lowest value for which the stimulation of 10% of the neurons in the network still lead to an overall response of that area.

13.2.3 Long Range Networks

The local networks were connected to form one single large-scale network. Very little is known about the exact strengths of long range corticocortical connections. However, Scannel et al. [11] rated connectivity data from the cat cortex according to whether a connection was strong, intermediate, weak or absent. Figure 13.2 shows the connections between areas in matrix form. See [11] for assignment of the numbers to anatomical names. A total of 826 connections have been identified.

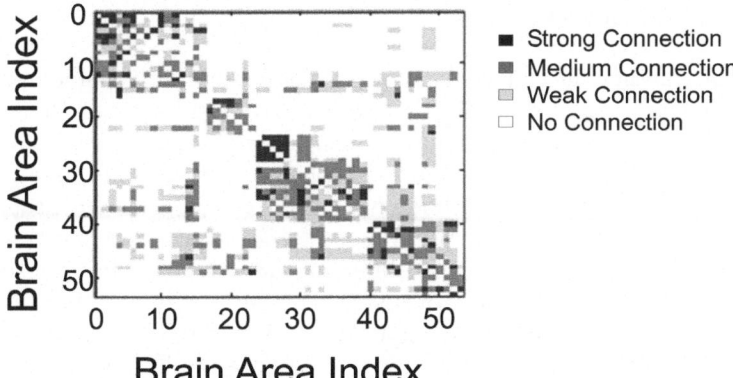

Fig. 13.2. The connectivity map used to set up the connections in the model. Adopted from [11]

We simulated long range connectivity by associating each of the 53 local networks with one cortex area. If the data by Scannel et al. [11] reported a connection between two areas, we randomly selected 5% of the excitatory cells from the first area and connected these cells to a randomly selected 5% cells from the other area. Connections were modeled as described in Sect. 13.2.2 (see Chaps. 3, 9). The alpha functions modeling the postsynaptic potential shape of these connections were scaled by a factor depending on the connection strengths reported by [11]. If a strong connection was reported, the alpha function was scaled by a factor of $\sigma = 3$, if an intermediate connection was reported the alpha function was scaled by $\sigma = 2$. For weak connections, the alpha function was not scaled, $\sigma = 1$. We have thus used a linear scale for the connection strengths. Other options include polynomial (e.g. quadratic) scaling or exponential scaling ($10^1, 10^2, 10^3$).

Conduction delays of long range connections were randomly assigned to a value between 10 and 20 ms in accordance with data provided by Swadlow [16]. Connection strengths between local networks were assigned the lowest value for which the stimulation of one local network still visibly spread to other local networks.

13.2.4 Stimulation Paradigms

During the simulation, the network was exposed to three different types of input.

(i) Unspecific thalamic input which was modeled by adding Gaussian white noise to the membrane potential of all cells.
(ii) Specific thalamic input, which was modeled by the injection of direct current into some of the cells belonging to the primary visual cortex (Brodmann's area 17).
(iii) Combined specific and unspecific thalamic input.

13.2.5 Rastergram Analysis

To find out whether the signal was propagating in an orderly fashion primarily along the paths suggested by the anatomical data, we analyzed the resulting spike time series to recover propagation delays of the signals from Brodmann's area 17 and related this to the shortest pathlength to all other areas.

Two different detection methods were used to estimate the propagation delay. The neural spike traces were preprocessed to generate one time series per local network, resulting in 53 different time traces. The preprocessing algorithm computed the number of spikes per time step in each local network. The first method was a threshold algorithm, which triggered a detection event when spiking activity in a local network crossed a minimum threshold. The propagation delay was defined as the difference of the time instance when activity was detected and the beginning of DC current injection. The second method was cross-correlation, where the propagation delay was derived from the lag at which the cross-correlation exhibited a maximum, zero lag coinciding with the time instant when the DC current injection started.

To provide further evidence that the synaptic connections are well adjusted and excitation is spreading primarily along the long range connectivity paths, we analyzed the local network cumulative spike traces resulting from the simulation to detect dominant connectivity patterns. Again, two different methods were used, cross-correlation and mutual information. The algorithm operated with prior information on how many connections were in the original input matrix. The problem of selecting a suitable detection threshold was thus circumvented. The question was, given the correct number of connections, will the detected connections between the area be similar to the anatomical data used as input to the simulation? Mutual information was computed for each pair of the time series. The computed values were sorted by magnitude from largest to smallest and only the 826 largest values (the number of connections in the data given by Scannel et al.) were considered to be detected connections.

13.3 Results

The model was run on 16 nodes of the Linux cluster "Peyote" of the Max Planck Institute for Gravitational Physics, Potsdam, Germany (cf. Chap. 11). The cluster is populated by 128 nodes with 2 CPUs (Intel Dual Xeon) and 2GB RAM each. The model was run under the three above mentioned conditions and a 10 second spike rastergram was obtained from each simulation. Here, we show results from simulations performed with 4096 neurons in each local network, that is, 217,088 neurons total. The rastergrams have simulation time on the horizontal axis and neuron index on the vertical axis (cf. Chaps. 12 and 14). When a neuron emits a spike, a dot is placed on the rastergram at the point corresponding to the time instant and neuron index. The network was simulated using Gaussian integration with a step size of 0.5 ms.

In the first experiments (specific thalamic input), we injected 50 nA of direct current into 10% of the neurons in Brodmann's area 17 (primary visual cortex) periodically every second for 100 ms. The rest of the network had no external excitatory input except incoming postsynaptic potentials from other neurons. The rastergram in Fig. 13.3 shows a one second segment of the simulation.

Figure 13.4 clearly shows that the propagation delays from Brodmann's area 17 to other local networks are well correlated with the minimum path-length between the local network and Brodmann's area 17. Both detection methods show similar results.

The plot showing connections detected using mutual information is shown in Fig. 13.5.

The computed connectivity matrix exhibits a reasonable degree of similarity to the original matrix.

In the second experiment (unspecific thalamic input), Gaussian white noise with mean 0 and standard deviation 5 nA was added to the membrane potential, v, in (13.2) of each neuron in each time step. The model network was run for 10 seconds and a trace of the spikes of each neuron was captured. Figure 13.6 shows the resulting trace image of the first two seconds. The network exhibits a synchronous rhythm with a frequency of approximately 5 Hz.

To simulate more realistic conditions, we applied the direct current injection in the presence of unspecific thalamic input. The resulting trace is shown in Fig. 13.7.

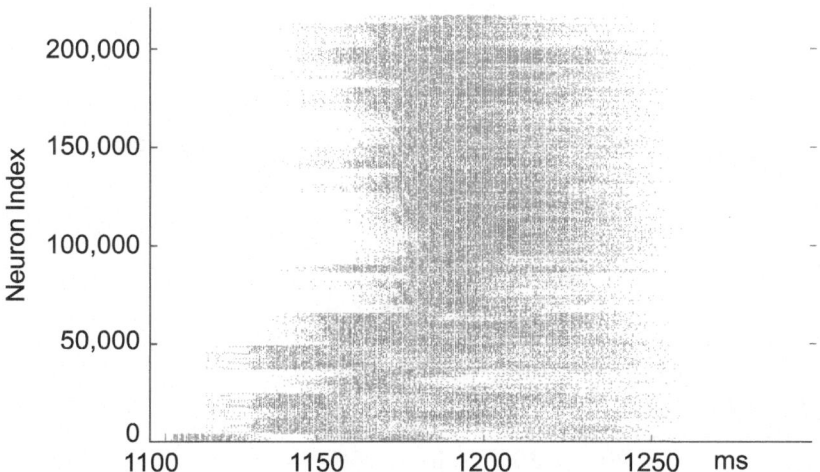

Fig. 13.3. The spike rastergram generated by the model when DC current injection was applied to 10% of the cells of Brodmann's area 17

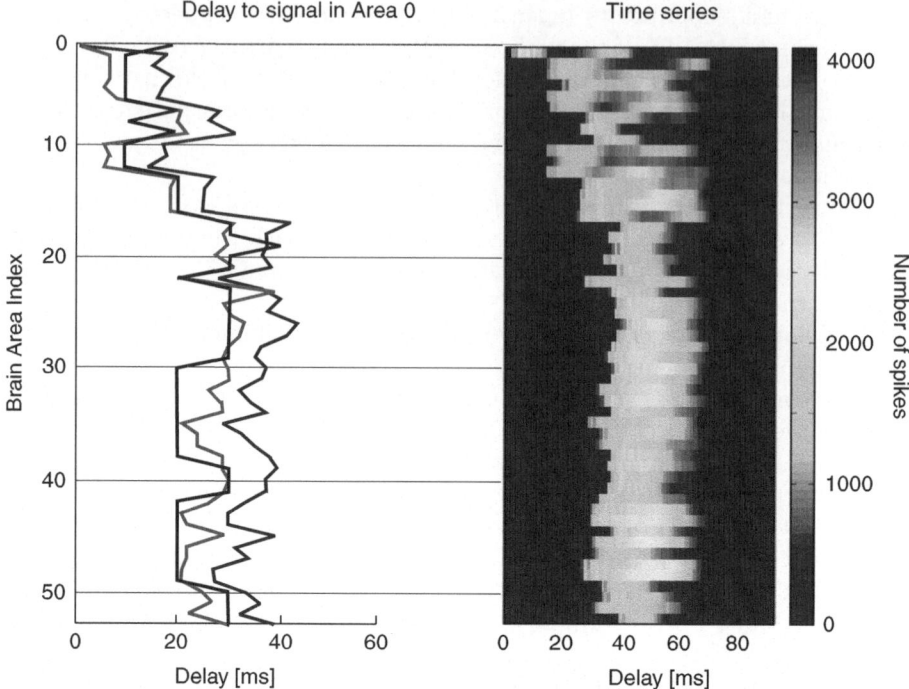

Fig. 13.4. Signal propagation delays from Brodmann's area 17 (*index 0*) to other areas. Black line is the delay corresponding to the minimum pathlength from area 17 to each area in turn. The blue line is the propagation delay computed by cross-correlation and the red line is the propagation delay computed by activity detection

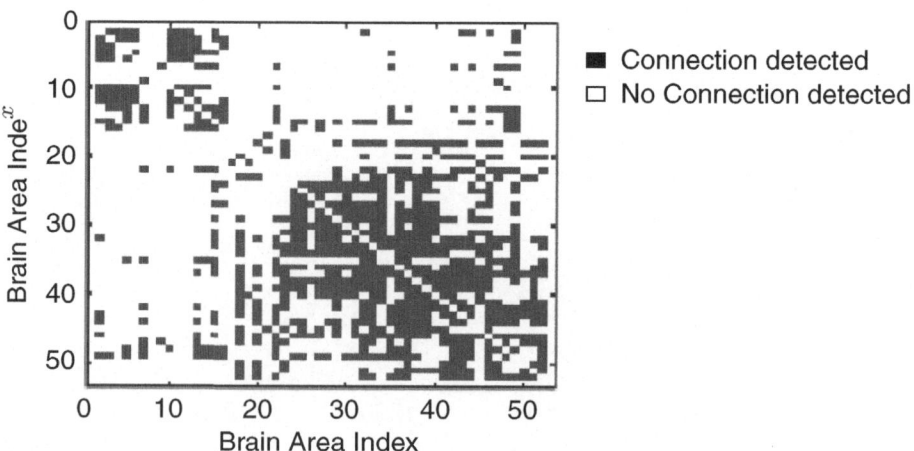

Fig. 13.5. Connectivity matrix computed from the time series generated by the model

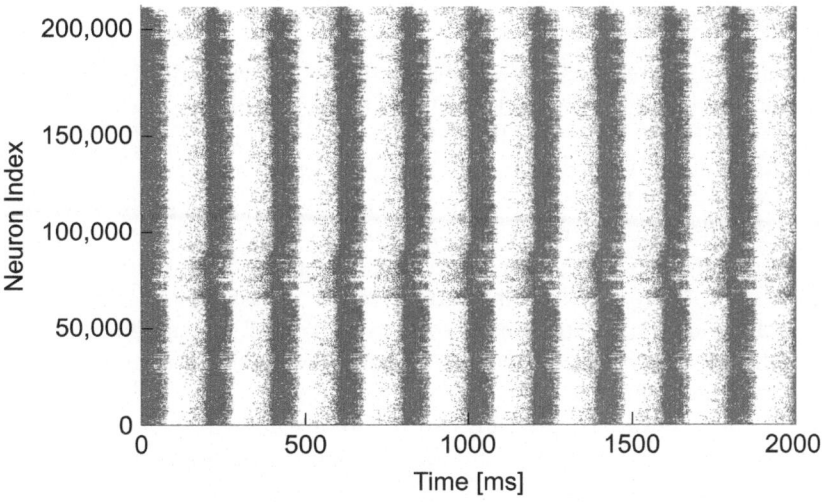

Fig. 13.6. Unspecific thalamic input rastergram

Qualitatively, the model exhibited the same behavior when unspecific thalamic input was present in addition to the direct currentinjection and again generated a 5 Hz rhythm. This could be explained by the fact that we only applied weak stimulation, which we verified would propagate from local network to local network, but was itself not strong enough to significantly alter the dynamics of the entire network.

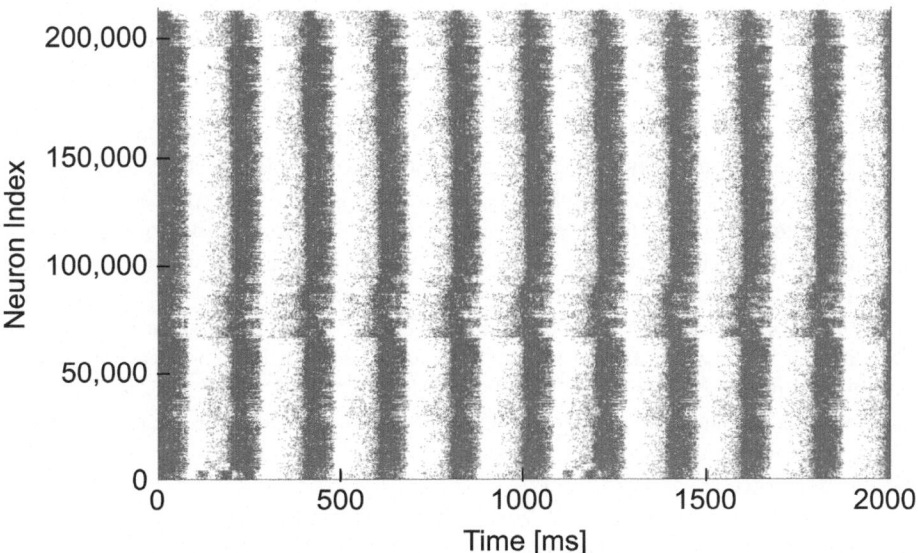

Fig. 13.7. Combined stimulation rastergram

13.4 Discussion

In the current investigation, we observed highly synchronized spike waves when the network was exposed to global unspecific noise. If one area of the network was directly activated by adding a fixed current to 10% of the neurons in that area, activity spread from this area to other areas.

We observed 5 Hz oscillations that were highly synchronized across the whole network. What is the origin of these oscillations? We can distinguish three different reasons that our network might show these oscillations:

(i) If excited by a constant current, most excitatory cells in our model spike with an interspike interval of roughly 200 ms [12]. It might thus be that an overall 5 Hz rhythm is due to the fact that every single neuron fires spikes with a frequency of 5 Hz.

(ii) If the overall 5 Hz rhythm can not be explained by the properties of the single neurons, this rhythm might evolve from the dynamics of the local randomly connected networks.

(iii) A third possible reason for the emergence of 5 Hz activity could be the large-scale connection properties that were adapted from Scannel et al. [11].

From these three mechanisms that might underlie the 5 Hz oscillations we observed in our simulation, the first mechanism is easily ruled out. In a simulation without any connections, only the properties of the single cells would play a role. In such a simulation, the first spike wave might be synchronized. However, subsequent spike waves will be increasingly scattered due to the noisy input. As long as there is no mechanism that counteracts the scattering of spike times due to noise, an overall rhythm will not emerge as a rule. The second and the third mechanisms are much harder to differentiate. Studies of networks with one level of hierarchy (i.e. local networks, in our simulation) have however not shown any such oscillations. In such simulations, usually much higher network frequencies (> 25 Hz) than 5 Hz are observed [2, 8]. Some authors also described lower frequency components in the alpha range (8–12 Hz, [12]). Therefore, we believe that our results can be accounted for by the large-scale connectivity that was adapted from Scannel et al. [11].

The frequency of the synchronized activity was an unexpected result. In general, the main frequency components of the brain can be found around 10 Hz and 40 Hz [17]. However, it could be shown in field equation simulations that under conditions with either very low external input [18] or weak connections [19], oscillation periods can also be in the range that was observed in the current investigation. Both these conditions might be present in the current simulation. However, the data from Fig. 13.4 demonstrate that the connections are at least strong enough to ensure a reliable propagation of activity between areas. A more rigorous investigation of this issue is required to clearly separate these two points.

Under what conditions do "real" brains show such highly synchronized low frequency activity? The electroencephalogram displays synchronized oscillations in the so-called theta range (3–7 Hz) during sleep stage 1 (drowsiness) [20] and most prominent during deep meditation [21, 22]. Interestingly, these brain states can be associated with reduced sensory input. During sleep stage 1, the eyes are closed and sounds are perceived to be damped. Persons performing deep meditation report that sensory stimulation is attenuated, which is also confirmed by event related potentials (Coromaldi, pers. comm.). Thus, the emergence of 5 Hz rhythmic activity during very weak stimulation seems to have a psycho-physiological counterpart.

If unspecific input to all areas was combined with direct excitation of one area, the well defined propagation pattern observed in Figs. 13.3 and 13.4 was lost. A plausible reason for this might be that the specific stimulus was too weak to trigger an overall change of the dynamics of the network. However, as we did not analyze the spike patterns at the level of single neurons, we cannot exclude the possibility that there were stimulus specific patterns in the spike responses of the single neurons. The importance of such patterns has been highlighted by several authors [2,4,7,23].

13.5 Conclusion

The current results indicate that spiking neuron models can display dynamics that are comparable to those obtained from neural field equations. If the number of simulated spiking neurons is sufficiently large, weak noisy input can drive activity patterns that are observed in field equations and neural mass recordings. Further investigations are required to identify the precise relations between spike timing and spike rates in such large scale simulations.

Acknowledgments

We would like to extend our grateful thanks to the scientific directors of the Helmholtz International Summer School Dr. Changsong Zhou, Dr. Marco Thiel and Dr. Peter beim Graben for giving us the chance to work in an exciting area of today's research and for providing us with state-of-the-art hardware to run the simulations. We would also like to thank the lecturers for their engaging talks and interesting discussions. Special thanks are due to Werner von Bloh for providing us with the original version of the simulation code and for helping with the modifications. Finally, we would like to thank all the students who participated in the Summer School for the stimulating discussions and inspiring time we all had in Potsdam in September 2005.

References

1. Christof Koch. *Biophysics of Computation: Information Processing in Single Neurons.* Oxford University Press, New York, 1999.
2. E M Izhikevich. Polychronization: computation with spikes. *Neural Computation*, 18:245–282, 2006.

3. V K Jirsa and H Haken. A derivation of a macroscopic field theory of the brain from the quasi-microscopic neural dynamics. *Physica D*, 99:503–526, 1997.
4. M Abeles. Time is precious. *Science*, 304:523–524, 2004.
5. E Ahissar and A Arieli. Figuring space by time. *Neuron*, 32:185–201, 2001.
6. E M Izhikevich, N S Desai, E C Walcott, and F C Hopensteadt. Bursts as a unit of neural informations: selective communication via resonance. *Trends in Neuroscience*, 26(3):161–167, 2003.
7. S Thorpe, A Delorme, and R Van Rullen. Spike-based strategies for rapid processing. *Neural Networks*, 6–7(14):715–725, 2001.
8. N Kopell, G B Ermentrout, M A Whittington, and R D Traub. Gamma rhythms and beta rhythms have different synchronization properties. *Proc Natl Acad Sci USA*, 97(4):1867–1872, 2000.
9. W J Freeman. *Mass Action in the Nervous System. Examination of the Neurophysiological Basis of Adaptive Behavior through the EEG*. Academic Press, New York, 1975.
10. H R Wilson and J D Cowan. Excitatory and inhibitory interactions in localized populations of model neurons. *Biophys J*, 12(1):1–24, 1972.
11. J W Scannel, G A P C Burns, C C Hilgetag, M A O'Neil, and M P Young. The connectional organization of the cortico-thalamic system of the cat. *Cerebral Cortex*, 9(3):277–299, 1999.
12. E M Izhikevich. Simple model of spiking neurons. *IEEE Transactions on neural networks*, 14(6):1569–1572, 2003.
13. E M Izhikevich, J A Gally, and G M Edelmann. Spike-timing dynamics of neuronal groups. *Cereb Cortex*, 14(8):933–944, 2004.
14. B W Connors and M J Gutnick. Intrinsic firing patterns of diverse neocortical neurons. *Trends in Neuroscience*, 13(9):365–366, 1990.
15. C M Gray and D A McCormick. Chattering cells: superficial pyramidal neurons contributing to the generation of synchronous osillations in the visual cortex. *Science*, 274(5284):109–113, 1996.
16. H A Swadlow. Physiological properties of individual cerebral axons studied in vivo for as long as one year. *J Neurophysiol*, 54(5):1346–1362, 1985.
17. O David and K J Friston. A neural mass model for MEG/EEG: coupling and neuronal dynamics. *NeuroImage*, 20:1743–1755, 2003.
18. L H A Monteirony, M A Bussaby, and J G Chaui Berlinckz. Analytical results on a Wilson-Cowan neuronal network modified model. *J Theor Biol*, 219:83–91, 2002.
19. R M Borisyuk and A B. Kirillov Bifurcation analysis of a neural network model. *Biol Cybern*, 66(4):319–325, 1992.
20. J S Barlow. *The Electroencephalogram. Its Patterns and Origins*. MIT Press, 1993.
21. L I Aftanas and S A Golocheikine. Human anterior and frontal midline theta and lower alpha reflect emotionally positive state and internalized attention: High-resolution EEG investigation of meditation. *Neurosci Lett*, 310:57–60, 2001.
22. E Coromaldi, C Başar-Eroglu, and M A Stadler. Langanhaltende Theta-Aktivität während tiefer Meditation: Eine Einzelfallstudie mit einem Zen-Meister. *Hypnose und Kognition*, 21:61–77, 2004.
23. E Körner, M-O Gewaltig, U Körner, A Richter, and T Rodemann. A model of computation in neocortical architecture. *Neural Networks*, 12:989–1005, 1999.

Parallel Computation of Large Neuronal Networks with Structured Connectivity

Marconi Barbosa[1], Karl Dockendorf[2], Miguel Escalona[3], and Borja Ibarz[4]
Aris Miliotis[2], Irene Sendiña-Nadal[4], Gorka Zamora-López[5]
and Lucia Zemanová[5]

[1] University of São Paulo, Brazil
marconi@if.sc.usp.br
[2] University of Florida, USA
am397@ufl.edu
[3] Universidad de Los Andes, Venezuela
angele@ula.ve
[4] Universidad Rey Juan Carlos, Spain
borja.ibarz@urjc.es
[5] University of Potsdam, Germany
gorka@agnld.uni-potsdam.de

14.1 Introduction

One does not need to delve into complex modern physical phenomena to realize that laws of physical nature are vastly employed and exploited by nature. We can see in an object as ubiquitous as a flower, among other striking properties related to its form, that the anther is isolated and the stigma is grounded, readily providing an electrostatic mechanism for charged bees to carry the pollen [1]. This phenomenon is so basic, yet shows what years of evolution manage with one charged particle, its surplus and absence.

The quest to understand and somehow control biological systems, using fundamentals and even by-products from physics and mathematics, has a long history. Only to mention a very few examples and recent observations: an original model of rhythmic waves coordinated by a central pattern generator in multi-legged animal locomotion [2,3], models dealing with the mechanism of pattern formation (a perspective in [4]); a description of the formation of sunflowers' spirals and their relationship to Fibonacci series [5]; and new phenomena and nonlinear mechanical models of hearing [6,7]. Most of these works require multidisciplinary thinking and address a larger community at various levels of mathematical sophistication, see [8] for a recent selection of biologically related material.

Among biological systems in general, the mammalian nervous system is arguably the richest example of the interplay between biology, physics, chemistry and geometry. In the brain, vastly complex physical phenomena occur, but not only because of the raw number of interacting units and their hierarchical organization. The units themselves [9–12] as well as their effective connectivity [13–15] are still a major challenge to comprehension.

Various reports attempting to model aspects of neuronal activity with tools from nonlinear dynamics have aroused a growing acceptance of the fact that the diverse biological reactions of a cell to a stimulus can be explained by bifurcation theory. While many types of receptors and channels (which need to be taken into account in a description of the cell's intrinsic properties) might be present in a specific cell, they only determine the type of bifurcation of the neural dynamical behavior [16].

Of more interest to us is the thriving activity in modeling networks of neurons with diverse levels of biological realism, focused on explaining a few features of the collective behavior. Using tools related to nonlinear dynamics, such models come in various forms like traditional dynamical systems [17–19], continuous media [20–22], mean field approximations [23], maps [24,25]; models incorporating morphology and structure [26–31]; models with competition both at the network level [19] or at the synaptic level via plasticity [15] (see Chap. 1 for a survey). As usual, a trade-off is necessary to balance the level of detail and the scope of any attempt to come out with a useful model, i.e. to predict behavior or a particular trait accurately and reproducibly.

In [16], a thorough exposition of the basic mechanisms by which neuronal dynamical features can be understood is presented and the common idea of the existence of a threshold is challenged. In [32], a review of the state of affairs of neuronal network modeling surveying a wide range of techniques and information of the present state of phenomenological advances is available. The frontier between complex network structures and associated dynamics is extensively detailed in [33]. A bold perspective appearing in [34] argues that the level of knowledge of the biological intricacies of brain areas and layers has reached such a mature level that time is ripe to attempt a larger scale brain simulation of a microcolumn, calibrated to be indistinguishable from a real one.

The real lack of detailed knowledge of connectivities in the mammalian brain suggests that a wide range of possibilities should be tried when simulating networks of many neurons. This poses a computational challenge. The main goal of this chapter is to show the simulation work we produced during the Summer School using and modifying the code presented in Chap. 11 using the general scheme of Chap. 9 that tries to take on account neural structures and connectivities. It is important to remember that other efforts to simulate computationally large communities of neuronal cells have been done with different objectives in mind, for example [35,36] or the various references in [34]. Although interesting and efficient, those initiatives do not permit a straightforward extension to allow for structured areas with their natural connectivity patterns.

This chapter reports on a number of activities developed at the 5th summer school where we implemented a working framework for the simulation of large populations of neurons. Those neurons were treated as dynamical systems near a bifurcation so that variations of critical parameters would convert an otherwise quiescent state into a state of firing/bursting activity. Keeping track of all the information describing the network state in files would of course become prohibitive, so strategies for dealing with the data also were envisaged.

We chose the Morris-Lecar model as the unit in our simulations. This choice is based on the fact that its dynamics is well understood for the sake of the computational efficiency needed to produce data on a fine time scale and long simulation range. Its efficiency could be one order of magnitude better than the full Hodgkin-Huxley prototype neuron. Morris-Lecar is a conductance-based model like Hodgkin-Huxley but with only two persistent channels, one fast (Ca^{2+}) and one slow (K^+). Another group of participants dealing with the cat map (Chap. 13) implemented the Izhikevich model, see [17]. Any other dynamical model can be easily implemented as an additional module to the code (cf. Chap. 1).

The choice of the Morris-Lecar model could seem odd to a biologist, for this model was originally proposed to simulate features of a muscle cell, but the trade-off of biological realism for the possibilities of investigating a longer simulation and actually observing size dependent phenomena has paid off, especially when considering that many other biological facts were already left aside or oversimplified (e.g. synaptic dynamics, morphology, delay, etc.) for this particular set of studies. This is in full accordance with the original proposal of building a simple framework from scratch, avoiding black boxes, in a bid to better understand the role played by connectivity in large networked dynamical systems.

The idea of building a framework as general as we implemented is to start probing the functioning of the cat's brain. A vast amount of knowledge of its structures and connectivities has been collected during the last few years [37]. An important feature of the cat's brain is its subdivision into 53 functioning areas. The connection strength of these areas is assumed to be proportional to the thickness of nerves connecting them and is implemented in our code as well. One question that comes to mind is to what extent stimulating one area would spread activity to other areas. This main task is what we set about to address after polishing the code. Our results, while still too preliminary, look promising.

The chapter begins with a brief description, in Sect. 14.2, of the Morris-Lecar prototype neuron dynamics and its coupling to other neurons and the noisy environment. In Sect. 14.3, the idea of one neuronal area is developed and the conjecture of small world connectivity is implemented. This section also addresses the tuning of the network to a natural baseline behavior through parameter search. In Sect. 14.4, the procedure of tuning parameters to baseline behavior is re-introduced and preliminary results of our simulations, mainly the stimulation/ablation of areas and the effect of the size of the network, are presented. Section 14.5 describes the relation of the known connectivity structure of cortex with the observed activity in the simulations. Last, Sect. 14.6 provides discussions on our results and poses important perspectives to be considered in the future using the general framework presented in Chap. 9.

14.2 The Model

14.2.1 The Morris-Lecar Neuron Model

The neuron model used in our simulations is the Morris-Lecar model. It is a simplified conductance model of the barnacle muscle fiber, with two variables obeying the equations:

$$\dot{v} = -g_{\text{Ca}}(v)(v - V_{\text{Ca}}) - g_K w(v - V_K) - g_L(v - V_L) + I_{\text{ext}} \quad (14.1)$$

$$\dot{w} = \frac{\phi}{\tau_w(v)}(W_\infty(v) - w) \quad (14.2)$$

This is the dimensionless form of the model as presented by Rinzel and Ermentrout in their classic exposition [38]. Voltage v is normalized to the reversal potential of the excitatory ion Ca^{2+}, so voltage parameters are $V_{\text{Ca}} = 1.0$, $V_K = -0.7$ and $V_L = -0.5$. Conductances have been normalized by a reference conductance $G_{\text{ref}} = 4$ mS/cm^2 and time by the time constant $\tau = C/G_{\text{ref}} = 5$ ms, where $C = 20\ \mu\text{F/cm}^2$. Thus we arrive at values $g_K = 2.0$, $g_{\text{Ca}} = 1.0$ and $g_L = 0.5$. In order to better match the model to the simulation of mammal cortex rhythms, one (dimensionless) time unit will henceforth be equivalent to 1 ms. Finally, $\phi = \frac{1.0}{3.0}$, and:

$$g_{\text{Ca}}(v) = 0.5\left[1 + \tanh\left(\frac{v + 0.01}{0.15}\right)\right] \quad (14.3)$$

$$\tau_w(v) = \frac{1}{\cosh\left(\frac{v - 0.1}{0.145}\right)} \quad (14.4)$$

$$W_\infty(v) = 0.5\left[1 + \tanh\left(\frac{v - 0.1}{0.29}\right)\right] \quad (14.5)$$

This leaves the external current I_{ext} as the only free parameter. For low values ($I_{\text{ext}} < I_{\text{SN}}$), all orbits of the system are attracted to the unique stable node with most potassium channels closed and a low, polarized voltage (between V_L and V_K). As the external current grows beyond $I_{\text{SN}} = 0.0833$, the stable node disappears via a saddle-node bifurcation in an invariant circle, giving way to an attracting limit cycle. In this cycle, the voltage jumps towards V_{Ca} and triggers the opening of potassium channels, which in turn pull the voltage back to polarized values; potassium channels then close and the cycle begins again. The result is rhythmic spiking beginning at infinitely low frequencies,

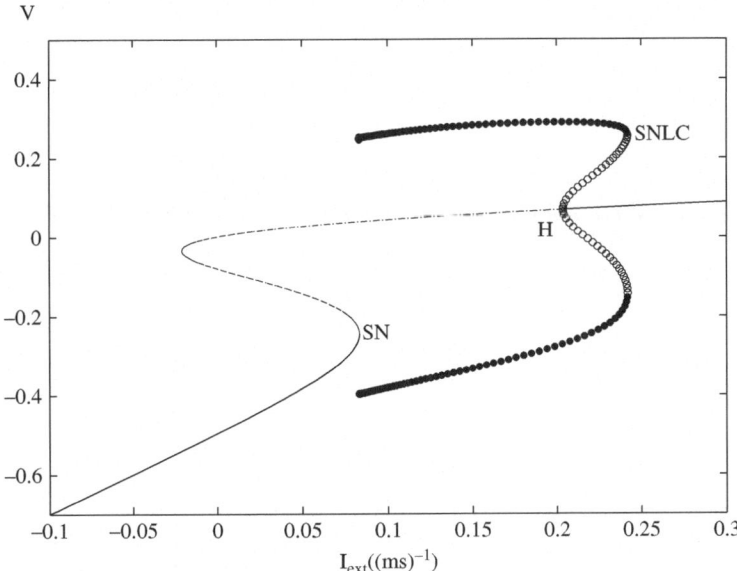

Fig. 14.1. Bifurcation diagram of the Morris-Lecar model used in this chapter. SN, saddle-node on invariant circle. H, Hopf. $SNLC$, saddle-node of limit cycles

and therefore, for the values of parameters chosen, Morris-Lecar is a class I model (for other choices of parameters, it may be a class II model ; see [38]).

Figure 14.1 completes the picture of bifurcations with external current. When $I_{ext} = I_{SNLC} = 0.242$, the limit cycle disappears through collision with an unstable limit cycle (born at a subcritical Hopf bifurcation of the hitherto unstable focus of the system) and the neuron becomes silent again, this time at a depolarized value of voltage. In this state, the external current is so strong that it effectively offsets the injection of potassium ions into the cell, preventing the firing of action potentials.

The parameter region of interest for our networks is the so-called excitable regime found at values of I_{ext} just below the threshold I_{SN} of rhythmic spiking. In this regime, if the voltage is pushed by the arrival of a synaptic impulse or by random noise out of equilibrium and across the unstable manifold of the saddle point, the system will make a long excursion in the phase plane, producing a single spike. This is illustrated in Fig. 14.2.

14.2.2 Coupling Between Neurons and External Stimulation

In the previous section, external current I_{ext} has been treated as a constant parameter. In network simulations, I_{ext} is a time-varying current coming from three sources:

$$I_{ext}(t) = I_{bias} + I_{syn}(t) + I_{Poiss}(t)$$

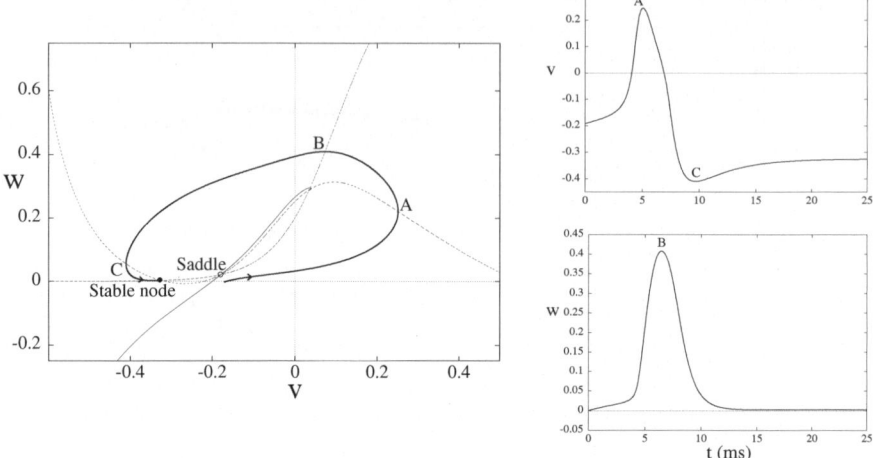

Fig. 14.2. Phase plane representation (**left**) and time evolution of variables v and w (**right**) for a single spike in the excitable regime ($I_{ext} = 0.07$) of the Morris-Lecar model. In the phase plane diagram, the thin continuous line is the unstable manifold of the saddle, dashed lines are the v and w nullclines, and the thick continuous line is the system trajectory. Point A, both in the phase plane and in the $v - t$ diagram, is the maximum of the action potential, where the trajectory crosses the v-nullcline. At point B, where the trajectory intersects the w-nullcline, there is maximal opening of potassium channels. Point C is the after-hyperpolarization peak due to remaining open potassium channels when the voltage first returns to the equilibrium level

- I_{bias} is a constant external bias current which will set the neurons in the excitable regime, as described in the previous section. It is common for all neurons.
- I_{syn} is the sum of synaptic currents arising from connections with other neurons in the network.
- I_{Poiss} also comes in the form of a synaptic current, but its origin lies outside the network; the presynaptic spikes that generate this current do not come from neurons in the network, but are instead randomly generated according to a Poisson process. These currents allow us to inject external stimulation.

We now describe the synaptic current $I_{syn}(t)$. It is a sum of currents due to both excitatory and inhibitory chemical synapses. Indeed, neurons in the network are classified as excitatory or inhibitory. If an excitatory presynaptic neuron fires at time t_{sp} (i.e. its voltage crosses zero with positive derivative), it adds to the term $I_{syn}(t)$ of the postsynaptic neurons an ohmic current $I_{syn,exc}(t)$ with reversal potential $V_{exc} = 0.05$ and time-varying, alpha function shaped conductance, thus:

$$I_{syn,exc}(t) = -g_{exc}\alpha_{exc}(t - t_{sp} - t_{del}) \cdot \Theta(t - t_{sp} - t_{del}) \cdot (v_{post}(t) - V_{exc}) \quad (14.6)$$

Here, g_{exc} is the strength of connection between pre- and postsynaptic neurons, t_{del} is the time delay of this connection, $\Theta(t)$ is the Heaviside step function, $v_{\mathrm{post}}(t)$ is the postsynaptic neuron voltage and $\alpha_{\mathrm{exc}}(t)$ is the alpha function:

$$\alpha_{\mathrm{exc}}(t) = \frac{1}{\tau_{1,\mathrm{exc}} - \tau_{2,\mathrm{exc}}}(e^{-\frac{t}{\tau_{1,\mathrm{exc}}}} - e^{-\frac{t}{\tau_{2,\mathrm{exc}}}})$$

The smaller of the two time constants $\tau_{1,\mathrm{exc}}$ and $\tau_{2,\mathrm{exc}}$ is the rise time and the larger one is the decay time of the function. If, instead, the presynaptic neuron is inhibitory, the reversal potential is $V_{\mathrm{inh}} = -0.50$ (i.e. equal to V_L) and the synaptic current added to $I_{\mathrm{syn}}(t)$ is similarly:

$$I_{\mathrm{syn,inh}}(t) = -g_{\mathrm{inh}}\alpha_{\mathrm{inh}}(t - t_{\mathrm{sp}} - t_{\mathrm{del}}) \cdot \Theta(t - t_{\mathrm{sp}} - t_{\mathrm{del}}) \cdot (v_{\mathrm{post}}(t) - V_{\mathrm{inh}}),$$

where $\alpha_{\mathrm{inh}}(t)$ is now timed according to (possibly different) constants $\tau_{1,\mathrm{inh}}$ and $\tau_{2,\mathrm{inh}}$.

Finally, the $I_{\mathrm{Poiss}}(t)$ term is very similar to $I_{\mathrm{syn}}(t)$. The only difference is that it is made up exclusively of excitatory currents of the form of 14.6, and the times t_{sp} do not correspond to spikes of presynaptic neurons but are instead generated by a Poisson process. By varying the rate λ of this process, the amount of external stimulation injected into the different areas of our network may be chosen. Baseline values for non-stimulated areas are around $\lambda = 3$ Hz (mean period $T_{\mathrm{Poiss}} = 333$ ms).

14.3 Setting Proper Parameters

The two-level network described in Chap. 9 gives rise to a moderately large number of parameters (connection numbers and strengths) that have to be tuned if we want our model to mimic cortex behavior. In this section, we describe the tuning procedure in two steps: first for intra-area parameters, and then for the whole 53 area network. Table 14.1 summarizes all the relevant parameters of the model and gives default values for them.

14.3.1 Optimal Inhibitory and Excitatory Coupling Strength for one area

In the absence of specific external stimulation, we would like neurons in our network to receive balanced excitatory and inhibitory input, so as to maintain a baseline activity corresponding to the non-specific Poissonian stimulation of 3 Hz. If this balance is achieved, scaling of connection strength with the square root of the degree (see Chap. 9) will ensure that input amplitude is independent of the number of neurons in the network (which is bounded by computational constraints). Balance of excitation and inhibition in one area depends on coupling parameters $g_{1,\mathrm{inh}}$ and $g_{1,\mathrm{exc}}$. In order to find appropriate values for these parameters, we did the following:

Table 14.1. Parameters of the network model and their default values grouped as: neuronal model parameters, network topology, connectivity strength and delays

Parameter	Description	Default value
I_{bias}	Constant bias current	0.08
V_{exc}	Reversal potential for excitatory synapses	0.05
V_{inh}	Reversal potential for inhibitory synapses	−0.5
n	neurons per area	512
p_{inh}	Ratio of inhibitory neurons	0.2
p_{ring}	Ratio of connections inside one area	0.1
p_{rew}	Probability of rewiring	0.3
p_3	Ratio of neurons receiving synapses from a connected area	0.05
p_4	Ratio of neurons with synapses towards a connected area	0.05
$g_{1,exc}$	Non-normalized strength of intra-area excitatory synapses	0.075
$g_{1,inh}$	Non-normalized strength of intra-area inhibitory synapses	2.5
$g_{2,exc}$	Non-normalized strength of inter-area excitatory synapses	0.075
$g_{2,inh}$	Non-normalized strength of inter-area inhibitory synapses	0
g_{ext}	Strength of connection for Poissonian currents	0.1
$\tau_{1,exc}, \tau_{2,exc}$	Rise and delay times of excitatory synaptic current	1 ms, 3 ms
$\tau_{1,inh}, \tau_{2,inh}$	Rise and delay times of inhibitory synaptic current	1 ms, 3 ms
$t_{del,1,exc}$	Delay of intra-area excitatory synapses	1 ms
$t_{del,1,inh}$	Delay of intra-area inhibitory synapses	3 ms
$t_{del,2,exc}$	Delay of inter-area excitatory synapses	3 ms
$t_{del,2,inh}$	Delay of inter-area inhibitory synapses	9 ms
$T_{Poisson}$	Mean period of Poisson excitation	333 ms

- In the absence of inhibition ($g_{1,inh} = 0$), we measured the mean firing rate (MFR) of the network (total number of spikes per second per neuron) as a function of the strength of excitatory synapses $g_{1,exc}$. Poissonian stimulation at a rate of 3 Hz is added to elicit baseline network activity. As is shown in Fig. 14.3(left), for $g_{1,exc} < 0.05$, the MFR is close to the externally imposed Poisson rate. At around $g_{1,exc} \approx 0.05$, activity blows up to a state of higher MFR where spiking is self-sustained and independent of the external input.
- Selecting a value for $g_{1,exc}$ barely above the threshold of sustained MFR ($g_{1,exc} = 0.075$), we increased the strength of inhibitory connections until the activity turned back to the background level as shown in

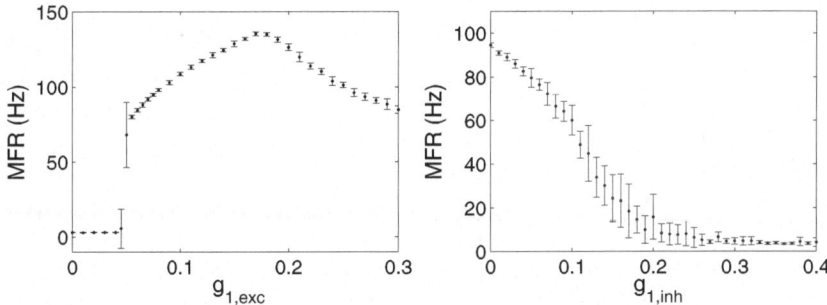

Fig. 14.3. (**Left**) Mean firing rate in one area of $n = 512$ neurons as a function of the strength of excitatory synapses for $g_{1,inh} = 0$. Each point is an average of 25 simulations; (**Right**) Mean firing rate in one area of $n = 512$ neurons as a function of the strength of inhibitory synapses for $g_{1,exc} = 0.075$. Each point is an average of 25 simulations

Fig. 14.3(right). At this point, excitatory and inhibitory forces within one area are balanced.

In order to check if both forces are balanced, we have measured the MFR as a function of the number of neurons n in one area for two sets of parameters. The first one corresponds to the pair ($g_{1,exc} = 0.075$, $g_{1,inh} = 0.25$), which produces a 3 Hz firing rate (see Fig. 14.3) and the second one, ($g_{1,exc} = 0.075$, $g_{1,inh} = 0.1$), is chosen such that there is more excitation than inhibition. This imbalance is going to be dependent on area size as shown in Fig. 14.4, while for the right selection of the excitatory and inhibitory strengths, the MFR remains constant at background level.

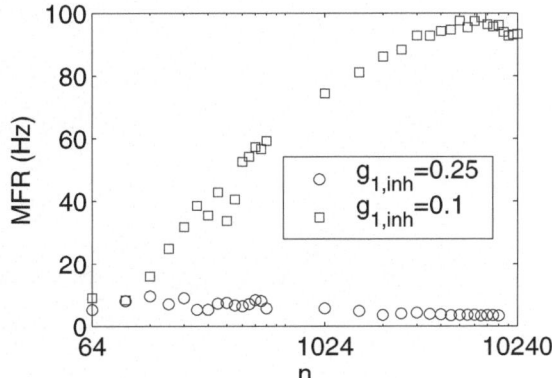

Fig. 14.4. Comparison of the MFR as a function of the number of neurons n per area between two sets of coupling strengths. When inhibition is not well balanced, the MFR increases as n becomes larger

14.3.2 Tuning Inter-area Parameters

When the 53 areas are coupled with the choice of parameters of the previous section, making $g_{2,\text{exc}} = g_{1,\text{exc}}$ and $g_{2,\text{inh}} = 0$, a firing pattern that we will call "generic" was observed, see Fig. 14.5. All areas show a similar homogeneous pattern of activity, only the firing rate differs. Overexcitation can be determined to be responsible for this result. Therefore, a more extensive search for parameters that yielded spontaneous bursting (the desired physiologic phenomenon) was performed to elucidate suitable ranges and ratios of parameters in the model. In addition, trends were noted on how such parameters affect general behavior, including bursting, spike rates, and propagation through the systems of the simulated cortical structures.

Even though the parameter values chosen for a single area produced activity patterns and firing rates analogous to "natural" activity, the extension to 53 areas, through incorporation of the connectivity matrix, affected the behavior of each area. The primary effect was the introduction of additional activity in each area from all the areas that it is connected to with afferent connections. This extra activity increased the mean firing rate of each area to frequencies higher than desired, higher than the 10–40 Hz range. Since the parameters for excitatory and inhibitory connections within each area have the greatest influence on the mean firing rate, we performed simulations of the whole model with different combinations of these values. Figure 14.6 summarizes the results of all these simulations. From these plots, we chose the parameters $g_{1,\text{inh}} = 0.4$ and $g_{2,\text{exc}} = 0.075$ as the ones providing the most appropriate firing rates.

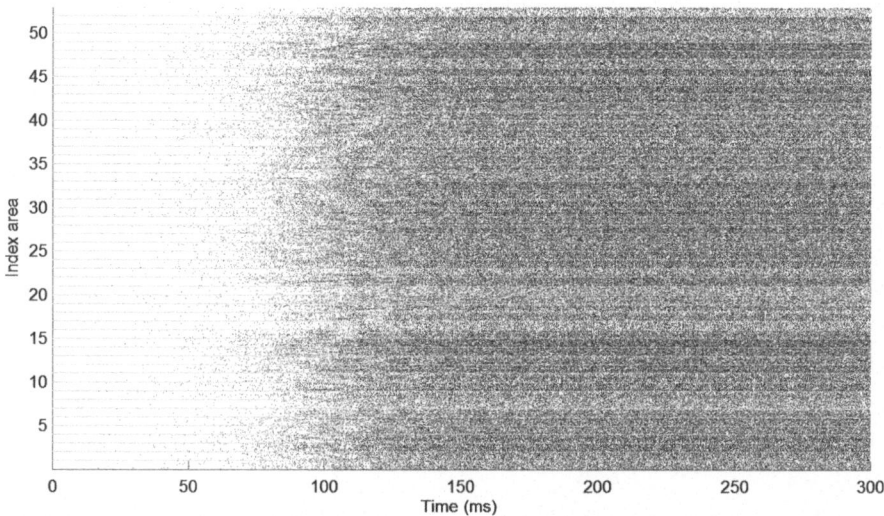

Fig. 14.5. Raster plot of the behavior of the whole cortex characterized as "generic". $g_{1,\text{inh}} = 0.4$, $g_{2,\text{inh}} = 0$, $g_{1,\text{exc}} = g_{2,\text{exc}} = 0.075$, $p_{\text{ring}} = 0.05$, $p_{\text{rew}} = 0.2$

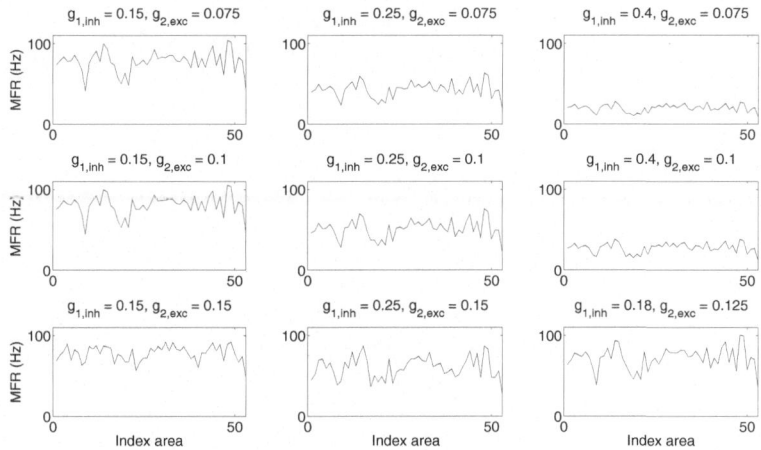

Fig. 14.6. Mean firing rate of each area, under nine different combinations of $g_{1,\text{inh}}$ and $g_{2,\text{exc}}$. $p_{\text{ring}} = 0.05$, $p_{\text{rew}} = 0.2$. Chosen parameters for further simulations: $g_{1,\text{inh}} = 0.4$ and $g_{2,\text{exc}} = 0.075$

Despite the fact that our model was now able to produce activity within the desired firing rate range, the overall behavior of areas continued to be rather homogeneous. We interpreted this result as not being "natural" behavior and called it "generic".

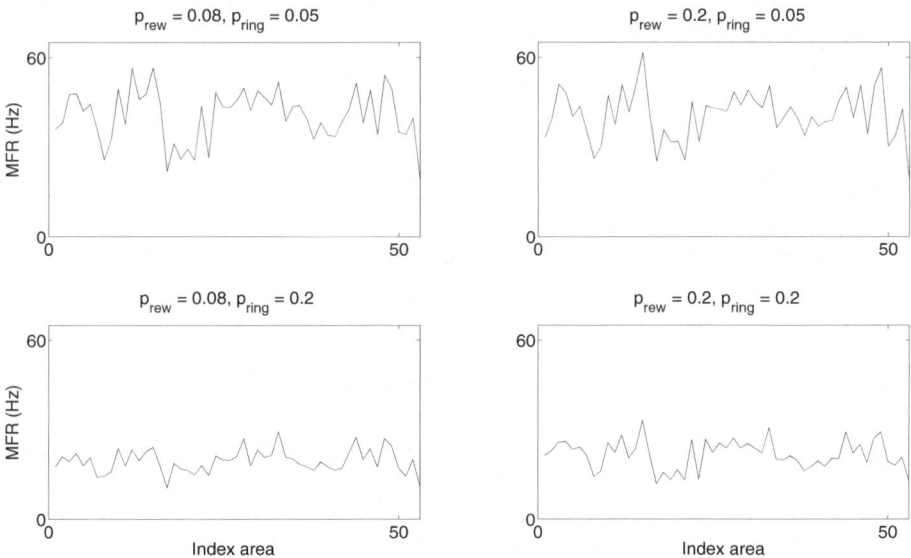

Fig. 14.7. Mean firing rate of each area, under four different combinations of p_{ring} and p_{rew}. Chosen parameters for later simulations: $p_{\text{ring}} = p_{\text{rew}} = 0.2$

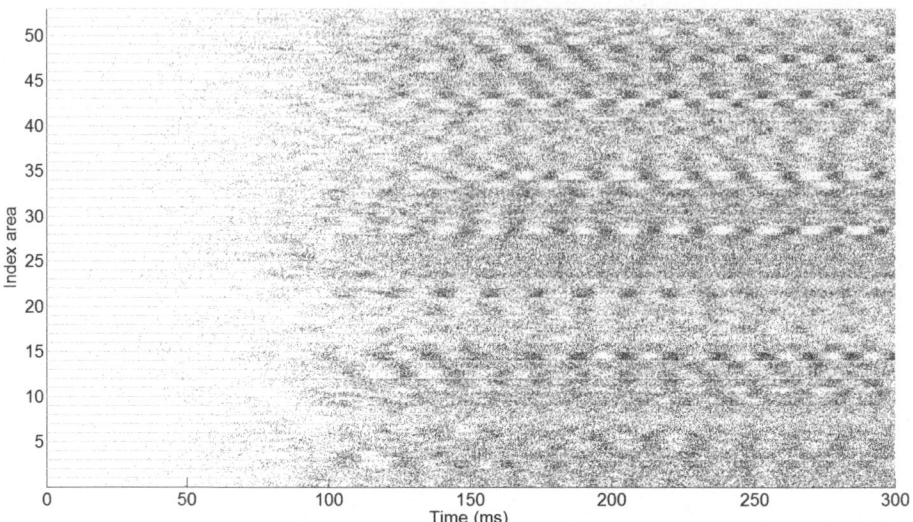

Fig. 14.8. Raster plot of the behavior of the whole cortex characterized as "natural". $g_{1,\text{inh}} = 0.4$, $g_{1,\text{exc}} = 0.075$, $p_{\text{ring}} = p_{\text{rew}} = 0.2$

Therefore, a set of further parameter search simulations were carried out where different combinations of the connectivity within each area were investigated, see Fig. 14.7. In this set of simulations, as before, our primary determining factor for "natural" behavior was to conserve the mean firing rate within the desired ranges. The additional condition was to obtain raster plots that presented different patterns of activity. Considering that the highest number of connections between areas is within each system, we desired that the areas of each system behave in a similar manner, and the behavioral pattern of each system be different from each other, hence representing each system's different function. This second condition was satisfied by setting parameters p_{ring} and p_{rew} to 0.2 (each area is modeled by a "small-world" subnetwork composed of 512 neurons. The number of connections is controlled by parameter p_{ring} and the deviation from the initial ring by p_{rew} (see Chaps. 3 and 9 for detailed descriptions)). Figure 14.8 presents the behavior obtained with these parameters.

14.4 Simulation of the Cat Cerebral Cortex

Once we managed to configure our model to behave in the manner of Fig. 14.8, we were more confident that the effects of the connectivity matrix were significant and that the behavior of the model could be described as "natural". We remind the reader that, as discussed in Chap. 3, the given corticocortical network has been found to be divided into four major clusters, corresponding to sensorial systems. Areas indexed 1–16 correspond to *visual cortex*, 17–23 represents the *auditory cortex*, indices 24–39 *somatosensory-motor cortex* and 40 − 53 *frontolimbic* cortical areas. Next, we aimed to simulate different conditions of the neural network:

- The already achieved natural behavior, under the conditions of background noise (simulated by low frequency Poisson noise). Area coupling propels connected areas into correlated patterns of bursting, behavior remaining similar within each of the 4 cortical systems. Due to the small number of connections between systems, each system is to follow its own independent behavior.
- Stimulation of a single area, and of a whole system. From the sparseness of connection weights in the matrix beyond each system, and the existence of areas operating as "communication hubs", a preferential pathway of inter-system communication is implied. Stimulation of an area, or system, with a simulated external signal was expected to propagate to other systems of the network along this preferential pathway.
- The effects of ablating an area. The existence of only a small number of areas preferential for inter system communication implies a greater significance for these areas for network operation. Removal of these areas from the model should contribute to a great change of the overall activity of the network. To further confirm that our model was adequately representative of cat cerebral cortex and to observe the effects of the removal of such areas, we carried out simulations where such areas were ablated.

We thus proceeded to simulate stimulation of a part of the cortex to observe the propagation of the stimulation and the effects on the non-stimulated parts. Considering that the visual system is one of the most important, if not the primary, stimulus receiving part of the cortex, we simulated our model with stimulation of pulses at 25 Hz on the whole visual system. It is clear from Fig. 14.9, as expected, that the independent behavior of the other systems and areas of the cortex was dominated by the stimulation. Specifically, we can observe that the visual system transfers the introduced activity into the other systems and functions as the driving system for the whole cortex.

Our final simulations considered the effects of damaged tissue. To observe these effects, we simulated our model with stimulation and in addition, we inactivated areas that operated as midpoints in the pathway from the stimulation area to other areas. In these simulations, to represent area "death" or "ablation", we lowered the excitability parameter, I_{bias}, of the neurons within the target area slightly below their excitable regime. In addition, since the effects of stimulation of the whole visual cortex were so dominating that subtle changes of the network ("ablation" of an area) were not affecting the overall behavior, we limited ourselves to stimulating only "area 17" (*primary visual cortex*, index: 1) with pulses at 25 Hz. Areas "Ia", "35", and "36" (indexes: 43, 48 and 49; frontolimbic areas) were active (Fig. 14.10) or inactive (Fig. 14.11). Inactivation of these frontolimbic areas can be seen to

Fig. 14.9. Raster plot of the behavior of the whole cortex characterized under stimulation of the whole visual cortex

affect the auditory system (indices: 17–23) lowering its activity and synchronization as expected, since part of the functions of the limbic system is to connect the visual system with the auditory system.

Aside from configuring and simulating our model to observe the above behaviors, we also attempted a simulation with a large number of neurons ($> 10^6$). This was done mostly as a computational task and the parameters for this simulation were not in accordance to the other simulations. Regardless, the behavior exhibited by our model with this large number of neurons is

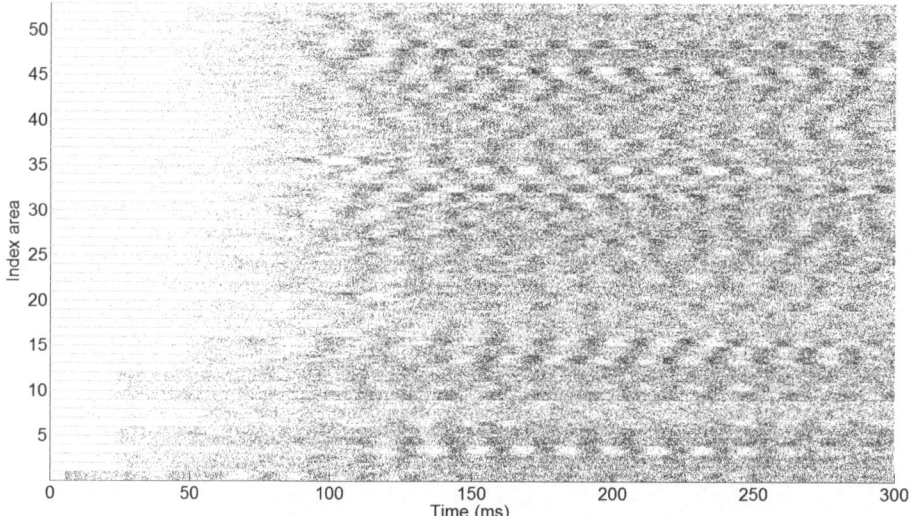

Fig. 14.10. Raster plot of stimulation of area "17" (primary visual cortex, index: 0)

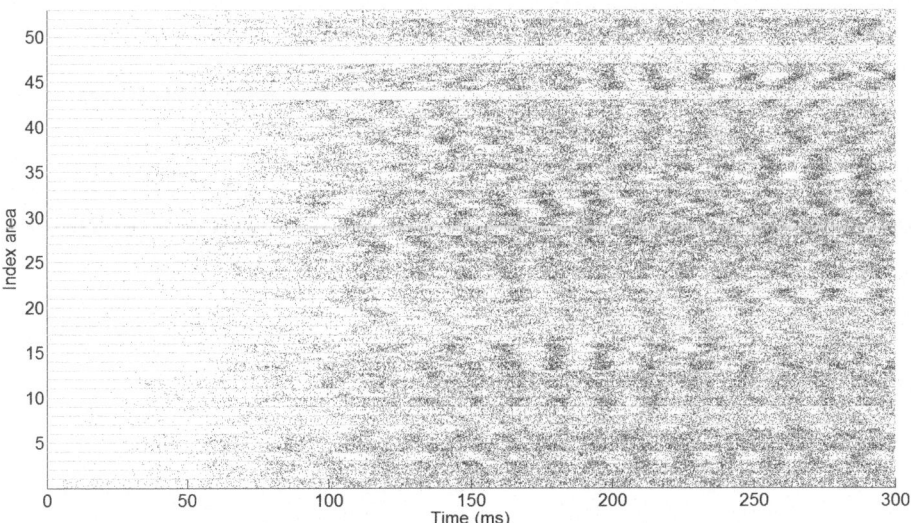

Fig. 14.11. Stimulation of area "17", while areas "LA", "35", and "36" (indices: 43, 48 and 49; frontolimbic cortex) are inactive

presented in Fig. 14.12. Considering that the parameters were different from our other simulations, the result is obviously misleading. We believe the strong oscillatory pattern observed to be related to the dynamics of the Morris-Lecar neuronal model. Looking at its bifurcation diagram, shown in Fig. 14.1, we argue that for the given parameters, all neurons must have collectively followed similar changes in their dynamical states. When so many neurons are present

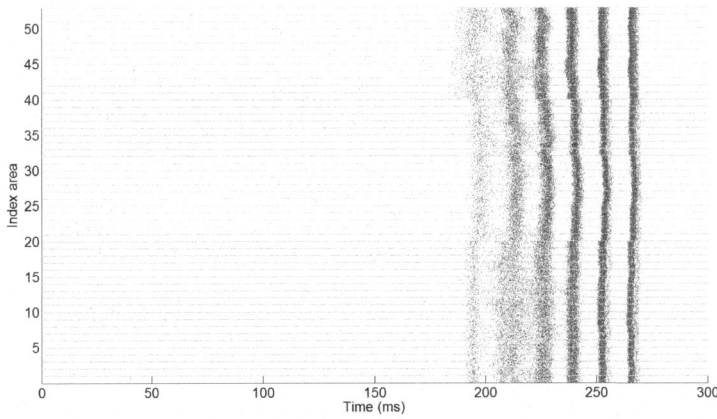

Fig. 14.12. Raster plot of the behavior of a cortex of $1,085,440$ neurons, which corresponds to $20,480$ neurons per area. $g_{1,\mathrm{inh}} = 0.4$, $g_{1,\mathrm{exc}} = 0.075$, $g_{2,\mathrm{exc}} = 0.15$, $p_{\mathrm{ring}} = 0.05$, $p_{\mathrm{rew}} = 0.1$

within each cortical area, each neuron receives an extremely large amount of input, which, after a brief period of intense firing, raises their I_{bias} triggering all neurons beyond the *SNLC* point into the "silent regime". Once all neurons are silent, only noise is present in the system, allowing neurons to recover their "excitatory state" and start firing again after a brief pause. Unfortunately, we could not perform further simulations of this size due to computer time limitations.

14.5 Dependence of MFR on Anatomical Connectivity

In this section, we present a brief attempt to explore the relationship between the observed behavior of the simulated system and the structural properties of the network. We will look for correlations between the characteristic firing rate of each cortical area obtained in the simulations with its degree and intensity. We will also try to find a simple analytical solution to explain the observed dynamics.

In Sect. 14.4, firing rates of individual cortical areas were estimated from the simulations. Frequency is observed to be modulated within about 10–20 ms (look at the fine structure of Fig. 14.8). On the other hand, under stationary conditions it varies significantly from area to area (Fig. 14.13). The variation of $g_{2,exc}$ (inter-areal excitatory coupling strength) and $g_{1,inh}$ (intra-areal inhibitory coupling strength) contributes to the absolute value of the mean firing rate. In the chosen parameter range, $g_{2,exc} \in [0.075, 0.15]$ and

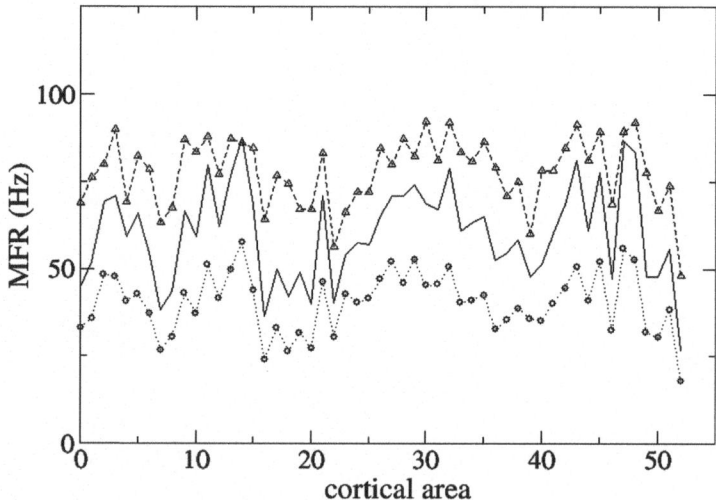

Fig. 14.13. Dependence of the firing rate on the internal and external coupling strength. $g_{2,exc} = 0.15$ in all cases. Dashed line: $g_{1,inh} = 0.15$, solid: $g_{1,inh} = 0.25$, dotted: $g_{1,inh} = 0.4$

$g_{1,\text{inh}} \in [0.15, 0.4]$, the mean firing rate of the areas was found to lie between 25 to 100 Hz. The main curve profile remains unchanged for different values of $g_{1,\text{inh}}$ and $g_{2,\text{exc}}$. Modification of the $g_{1,\text{inh}}$ coupling shifts the curve in the vertical direction, i.e. higher frequencies are achieved with lower inhibition. This fact indicates the importance of inhibitory connections in the modulation of brain dynamics. Indeed, inhibitory coupling has been shown to suppress oscillations induced by the excitatory coupling as is known to happen in pathology such as epileptic seizures caused by excessive synchronization of neuronal activity [39].

14.5.1 Correlation to k^{in} and S^{in}

The input degree of a node k^{in} refers to the number of connections a node receives. Its natural extension for weighted networks, the input intensity of a node S^{in}, is the sum of the strength of its input connections. Although there are many existing network measures (see Chap. 3 for descriptions of network characterization and properties of the cat cortex), here, we will only explore the relationship between the mean firing rate and k^{in} and S^{in}.

The average response of a cortical area depends directly on the amount of input signal received, thus we will correlate the MFR to k^{in} and S^{in} of the cortical areas. Linear correlation of both measures with the MFR obtained from simulations is depicted on Fig. 14.14. As expected, it is a monotonously increasing function of k^{in} and S^{in}. The more input a cortical area receives, the more often its neurons will fire. The linear fit is better in the case of intensity, since the relation between the MFR and k^{in} slightly saturates at high degrees ($k^{\text{in}} \geq 20$). This saturation is more pronounced in other parameter sets (not shown). Figure 14.17 shows that correlation between the MFR and S^{in} is higher for most of the parameter sets.

14.5.2 Analytical Estimation

In the following, our modest effort to model the observed MFRs for each cortical area is presented. A commonly used approach in artificial neural networks

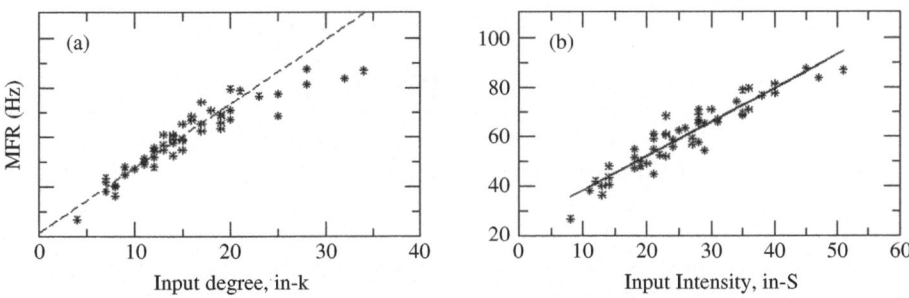

Fig. 14.14. Correlation of average simulated Mean Firing Rates per area (MFR) with: (a) degree and; (b) intensity

is to define an activation function describing the average response of neurons to input received from its neighbors. In a general form, the equations are written as:

$$r_i(t+1) = F(h_i)$$

$$h_i = a \sum_{j=1}^{N} W_{ij} r_j(t) + \xi, \quad i = 1, ..., N,$$

r_i being the activity of the neuron, W_{ij} the weighted adjacency matrix of the network and ξ some external input, i.e. noise. $F(h_i)$ is usually some sigmoidal saturation function and is normalized either to $[0,1]$ or $[-1,1]$. Parameter a controls the slope of the saturation function tuning the scale of the response. Such a function sums up all inputs the neuron receives and returns a normalized output representing its average activity response.

Similar approaches have already been used for cortical models of the cat [40, 41]. Here, we simulate cortical areas instead of individual neurons. This approach is reasonable since the mean activation level of a cortical area strongly depends on the amount of input received from its neighbors in a cumulative and smooth manner, which could be highly arguable in the case of individual neurons [42, 43]. As said above, the number of firing neurons scales with the input an area receives. Here, mean activity level will be considered to be equivalent to MFR. The simplest choice is to assume a linear approximation

$$F(h_i) = \alpha h_i + \beta, \text{ where } \beta = 0$$

(the saturation function crosses the origin). Our study is then limited to estimating the slope parameter a and noise level ξ to provide an optimal approximation to the results obtained from the simulations.

In the steady state $r_i(t+1) = r_i(t)$ and after taking $F(h_i)$ to be linear, equations reduce to:

$$r_i = \alpha \left(a \sum_{j}^{N} W_{ij} r_j + \xi \right).$$

After rescaling, we arrive at the following equation in matrix form:

$$r' = (I - a'W)^{-1} \xi', \tag{14.7}$$

where r' is the MFR vector of the 53 areas, I is the identity matrix, W is the adjacency matrix of the cortical network, $a' = \alpha a$, and ξ' is a column vector where all elements are $\alpha \xi$.

Our main purpose is to find the coupling strengths $g_{1,\text{inh}}$ and $g_{1,\text{exc}}$ that produce estimated MFRs as closely correlated to the MFRs from our simulations as possible. We face the problem of setting both slope a' and noise

ξ' parameters satisfactorily. There is a limited range of a's we can examine
because the maximum eigenvalue of our dynamical system (given by the adja-
cency matrix W) is $\lambda_{\max} = 29.08$, and thus $a' \approx 1/\lambda_{\max} = 0.034385$. At this
value, the system has a singularity and solutions with larger a' are unstable.
Correlation is "blind" to this singularity, and thus, in order to look for proper
a' and ξ', we will calculate the Euclidean distance between the estimated r'
vectors and the vectors of MFRs for different $g_{1,\text{inh}}$ and $g_{2,\text{exc}}$ obtained from
our simulations. Figure 14.15a shows how estimated MFRs differ from simu-
lated MFRs for different values of a'. In this representation, the singularity is
clearly observed as the distance between r' and MFR vectors grows to infinity
(Fig. 14.15b).

Importantly, Fig. 14.15(b) also shows the presence of a minimal distance,
so in the following, our optimization problem is to find the closest r' solutions
to the MFR from simulations. Among the stable solutions ($a' < 0.034385$),
the shortest distance depends both on a' and ξ' for each coupling strength
combination. After solving (14.7) for different parameters a' and ξ', optimal
values were found for each combination of $g_{1,\text{inh}}$ and $g_{2,\text{exc}}$ as summarized in
Table 14.2.

We are now ready to look for the optimal coupling strengths. For each pair
of $g_{1,\text{inh}}$ and $g_{2,\text{exc}}$, the simulated MFRs and the vector r' estimated for corresponding optimal a' and ξ' (see Table 14.2) are correlated. Results are shown
in Fig. 14.16. Note that the best correlation is for the values of the coupling
parameters that produce lower MFRs. Interestingly, the properly balanced inhibitory coupling allows us to achieve both "natural behavior" and maximal
correlation to the linear approximation. Too low as well as too high inhibition
gives rise to a marked decrease of this correlation for all excitatory values.

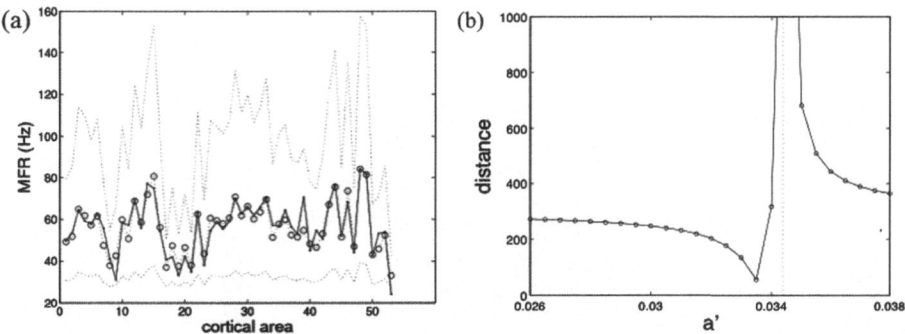

Fig. 14.15. a) Simulated MFR of the 53 cortical areas (*solid*) and estimated MFRs
with $\xi = 24.5$ and different a' values (*dotted lines*). Distance between simulated and
estimated MFRs varies significantly for different parameters; b) Setting $\xi' = 1.0$,
the singularity of the dynamical system appears as distance going to infinity at
$a' = 0.034385$. Larger values of a' represent unstable solutions

Table 14.2. Optimal values of slope a' and noise ξ' for different coupling strengths

$g_{2,exc}$	$g_{1,inh}$	opt a'	opt ξ'	$g_{2,exc}$	$g_{1,inh}$	opt a'	opt ξ'
0.075	0.15	0.0145	47.0	0.125	0.15	0.0135	50.0
	0.18	0.0175	33.5		0.18	0.0175	37.5
	0.2	0.0185	28.0		0.2	0.0190	32.0
	0.25	0.0200	19.0		0.25	0.0205	24.5
	0.3	0.0200	14.5		0.3	0.0210	20.0
	0.4	0.0195	9.0		0.4	0.0195	16.0
0.1	0.15	0.0140	50.0	0.15	0.15	0.0130	50.0
	0.18	0.0175	36.0		0.18	0.0165	39.5
	0.2	0.0190	30.0		0.2	0.0190	32.5
	0.25	0.0205	22.0		0.25	0.0220	23.5
	0.3	0.0200	18.5		0.3	0.0215	21.5
	0.4	0.0210	11.5		0.4	0.0205	17.5

Increasing the excitatory coupling $g_{2,exc}$ produces, in general, a monotonous increase of correlation.

Finally, we compare the results from this analytical linear estimation to the effects of degree and intensity distribution. Vectors of input degrees k^{in} and input intensities S^{in} of cortical areas are correlated to the MFRs from the simulation as shown in Fig. 14.17. As expected from the weighted nature of the adjacency matrix, intensities do correlate better than degrees. After all the optimization effort, high correlation values suggest that our model behaves as a linear system for a certain range of coupling strengths (see Fig. 14.17). This is also supported by the high correlation between simulated MFRs and intensities S^{in}. Indeed, the analytical estimations show only slightly better correlation than the S^{in}.

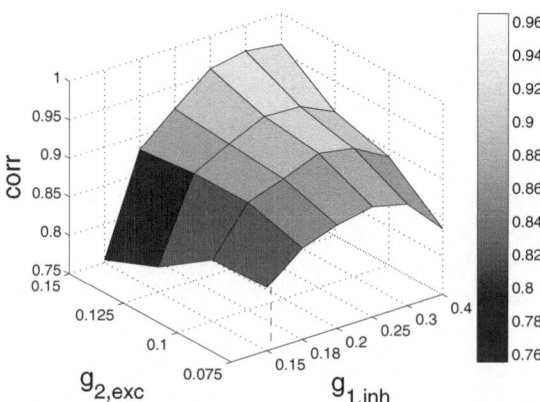

Fig. 14.16. Correlation between MFR from simulations and estimated r' using optimal a' and ξ' for each set of $g_{1,inh}$ and $g_{2,exc}$

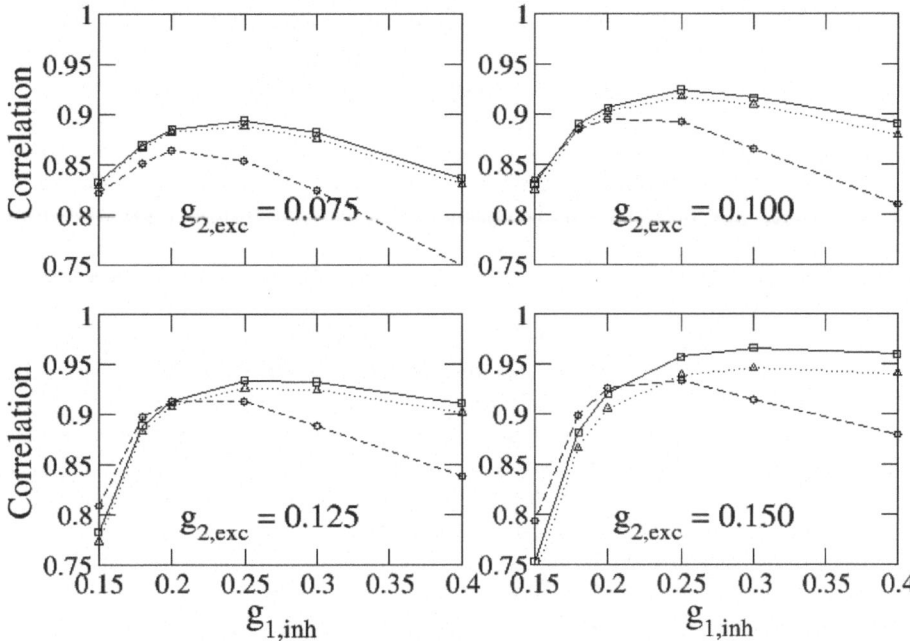

Fig. 14.17. Comparison of the correlation between computed MFR and estimated MFR at parameter a' (*solid line*), degree (*dashed*), intensity (*dotted*)

14.6 Conclusions and Outlook

During this set of computational exercises, we learned principles of large-scale neuronal simulations using the parallel code initially given to us. We defined a method to look for parameters that would provide a *realistic behavior*. We observed that finding suitable parameters and obtaining robust behavior is difficult. First, for the internal sub-network representing each cortical area, a set of excitatory/inhibitory strengths was found that would provide stable response with increasing size of the sub-network (see Fig. 14.4). The requirement was to keep average firing rate of individual neurons around 3 Hz. Then, after connecting the 53 subnetworks and assuming equal $g_{1,\text{exc}}$ and $g_{2,\text{exc}}$, the network showed too high a MFR, which was re-balanced by increasing the strength of inhibitory connections (Fig. 14.7). But this change happened to break the balance again as seen in the simulation with a million neurons, where the effect of scaling is evident.

However, raster plots displayed rather homogeneous and similar behavior of all cortical areas (Fig. 14.5), a behavior we called "generic". Looking for different, more realistic behaviors, we performed a parameter search on p_{ring}

and p_rew in order to change the network structure and hopefully also the behavior. A regime in which cortical areas exhibit bursting and silent epochs was found providing more interesting dynamics that we called "natural behavior" (see Fig. 14.8).

Finally, as discussed in Sect. 14.5, our analytical estimations show that the model behaves on average as a linear network model where, apart from the bursting dynamics, the MFR of each cortical area is highly proportional to the total input received (see Fig. 14.14).

Several open questions remain and a large set of possible implementations can be tried out:

First, we are aware of the arbitrary manner in which "natural behavior" was characterized: based exclusively on keeping the MFR at biologically reliable levels observed experimentally and on visual inspection of raster plots to avoid homogeneous dynamical responses. Thus, more convenient methodology founded on different measures would be desirable in order to characterize and classify the observed dynamics. Such measures could depend on, e.g., temporal correlations, frequency content, information transfer, etc.

Second, in our simulations, the small-world network topology following the Watts-Strogatz model was used for the internal neural connections within one cortical area. This topology has already been shown to enhance signal propagation and network synchronization which are so important for exchange of information.

It would be desirable, however, to introduce more realistic internal connectivities modeling finer cortical structures like layers, columns and if possible, the morphology of cortical neurons. Current sparse knowledge of detailed connectivity at the neuronal level makes such an implementation improbable in the nearest future. An intermediate solution might rely on taking just a small set of cat's cortical neurons, extracting their approximate local topology and randomly replicating it in order to mimic the internal structure of a cortical area. On the other hand, representing cortical layers and all the available experimental data about their interconnectivity offers an interesting opportunity to improve the internal architecture of the model. A first initial step should be the modeling of hierarchical organization of the cortex introducing hierarchical subnetworks for each cortical area rather than the small-worlds used here.

And finally, the linear behavior described by the model is not expected in real brains. As a complex system per excellence, the brain does not perform only such trivial behavior. Further modifications might include the introduction of delays and improved simulation strategies to limit the continuous spread of activity typical of pathological situations. For future work, we should remark that the dependence of the mean firing rates on the intra-areal connectivity among neurons is yet to be tested. Observing and characterizing brain activity under external stimuli is also of high interest.

References

1. W. S. Armbruster, Evolution of floral form: electrostatic forces, pollination, and adaptive compromise, New Phytol., 152(2):181–183, 2001.
2. M. Golubitsky, I. Stewart, P.-L. Buono and J. J. Collins, A modular network for legged locomotion, Physica D, 115(1):56–72, 1998.
3. M. Golubitsky, M. Pivato and I. Stewart, Interior symmetry and local bifurcation in coupled cell networks, Dyn. Sys., 19(4):389–407, 2004.
4. H. Levine and E. Ben-Jacob, Physical schemata underlying biological pattern formation—examples, issues and strategies, Phys. Biol., 1:14–22, 2004.
5. S. Douady and Y. Couder, Phyllotaxis as a physical self-organized growth process, Phys. Rev. Lett., 68(13):2098–2101, 1992.
6. V. M. Eguíluz, M. Ospeck, Y. Choe, A. J. Hudspeth and M. O. Magnasco, Essential nonlinearities in hearing, Phys. Rev. Lett., 84(22):5232–5235, 2000.
7. P. Martin and A. J. Hudspeth, Compressive nonlinearity in the hair bundle's active response to mechanical stimulation, Proc. Natl. Acad. Sci. USA, 98(25):14386–14391, 2001.
8. PhysicsWeb, Best of physicsweb, Best of Physics in Biology, http://physicsweb.org/bestof/biology, 2006.
9. C. Koch and I. Segev, The role of single neurons in information processing, Nat. Neurosci., 3:1171–1177, 2000.
10. B. J. O'Brien, T. Isayama, R. Richardson and D. H. Berson, Intrinsic physiological properties of cat retinal ganglion cells, J. Physiol., 538(3):787–802, 2002.
11. R. H. Masland, The fundamental plan of the retina, Nat. Neurosci., 4(9): 877–886, 2001.
12. C. F. Stevens, Models are common; good theories are scarce, Nat. Neurosci., 3:1177, 2000.
13. L. C. Jia, M. Sano, P.-Y. Lai and C. K. Chan, Connectivities and synchronous firing in cortical neuronal networks, Phys. Rev. Lett., 93:088101, 2004.
14. J. van Pelt, I. Vajda, P. S. Wolters, M. A. Corner, W. L. C. Rutten and G. J. A. Ramakers, Dynamics and plasticity in developing neuronal networks in vitro, Prog. Brain Res., 147:173–188, 2005.
15. A. Van Ooyen, Competition in neurite outgrowth and the development of nerve connections, Prog. Brain Res., 147:81–99, 2005.
16. E. M. Izhikevich, Dynamical systems in neuroscience: the geometry of excitability and bursting, MIT Press, 2007.
17. E. M. Izhikevich, Which model to use for cortical spiking neurons, IEEE Trans. Neural Netw., 15(5):1063–1070, 2004.
18. R. C. Elson, A. I. Selverston, R. Huerta, N. F. Rulkov, M. I. Rabinovich and H. D. I. Abarbanel, Synchronous behaviour of two coupled biological neurons, Phys. Rev. Lett., 81(25):5692–5695, 1998.
19. M. Rabinovich, A. Volkovskii, P. Lecanda, R. Huerta, H. D. I. Abarbanel and G. Laurent, Dynamical encoding by networks of competing neuron groups: winnerless competition, Phys. Rev. Lett., 87(6):068102, 2001.
20. C. J. Rennie, P. A. Robinson and J. J. Wright, Unified neruophysical model of EEG spectra and evoked potentials, Biol. Cybern., 86:457–471, 2002.
21. J. J. Wright, C. J. Rennie, G. J. Lees, P. A. Robinson, P. D. Bourke, C. L. Chapman, E. Gordon and D. L. Rowe, Simulated electrocortical activity at microscopic, macroscopic and global scales, Neuropsychopharmacology, 28: 80–93, 2003.

22. P. A. Robinson, C. J. Rennie, D. L. Rowe, S. C. O'Connor, J. J. Wright, E. Gordon and R. W. Whitehouse, Neurophysical modeling of brain dynamics, Neuropsychopharmacology, 28:74–79, 2003.

23. H. R. Wilson and J. D. Cowan, A mathematical theory of the functional dynamics of cortical and thalamic neuron tissue, Kybernetik, 13:55–80, 1973.

24. M. Bazhenov, N. F. Rulkov, J.-M. Fellous and I. Timofeev, Role of network dynamics in shaping spike timing reliability, Phys. Rev. E, 72:041903, 2005.

25. G. Tanaka, B. Ibarz, M. A. F. Sanjuan and K. Aihara, Synchronization and propagation of bursts in networks of coupled map neurons, Chaos, 16:013113, 2006.

26. G. A. Ascoli, Progress and perspectives in computational neuroanatomy, Anat. Rec. (New Anat.), 257(6):195–207, 1999.

27. P. C. Bressloff, Resonantlike synchronization and bursting in a model of pulse-coupled neurons with active dendrites, J. Comput. Neurosci., 6:237–249, 1999.

28. S. M. Korogod, I. B. Kulagina, V. I. Kukushka, P. Gogan and S. Tyc-Dumont, Spatial reconfiguration of charge transfer effectiveness in active bistable dendritic arborizations, Eur. J. Neurosci; 16:2260–2270, 2002.

29. P. C. Bressloff and S. Coombes, Synchrony in an array of integrate-and-fire neurons with dendritic structure, Phys. Rev. Lett., 78(24):4665–4668, 1997.

30. L. da F. Costa, Morphological complex networks: can individual morphology determine the general connectivity and dynamics of networks?, oai:arXiv.org:q-bio/0503041, 2005.

31. L. F. Lago-Fernández, R. Huerta, F. Corbacho and J. A. Sigüenza, Fast response and temporal coherent oscillations in small-world networks, Phys. Rev. Lett., 84(12):2758–2761, 2000.

32. M. I. Rabinovich, P. Varona, A. I. Selverston and H. D. I. Abarbanel, Dynamical principles in neuroscience, Rev. Mod. Phys., 78:1213–1265, 2006.

33. S. Boccaletti, V. Latora, Y. Moreno, M. Chavez and D.-U. Hwang, Complex networks: structure and dynamics, Phys. Rep., 424:175–308, 2006.

34. H. Markram, The blue brain project, Nat. Rev. Neurosci., 7:153–160, 2006.

35. R. D. Traub and R. K. Wong, Cellular mechanism of neuronal synchronization in epilepsy, Science, 216:745–747, 1982.

36. A. Morrison, C. Mehring, T. Geisel, A. Aertsen and M. Diesmann, Advancing the boundaries of high-connectivity network simulation with distributed computing, Neural Comput., 17:1776–1801, 2005.

37. J. W. Scannell, G. A. P. C. Burns, C. C. Hilgetag, M. A. O'Neill and M. P. Young, The connectional organization of the cortico-thalamic system of the cat, Cereb. Cortex, 9:277–299, 1999.

38. J. Rinzel and G. B. Ermentrout, Analysis of neural excitability and oscillations, in Methods in neuronal modeling: from synapses to networks, ed. C. Koch and I. Segev, 135–169, MIT Press, Cambridge, MA, 1989.

39. P. Kudela, P. J. Franaszczuk and G. K. Bergey, Changing excitation and inhibition in simulated neural networks: effects on induced bursting behavior, Biol. Cybern., 88:276–285, 2003.

40. R. Kötter and F. T. Sommer, Global relationship between anatomical connectivity and activity propagation in the cerebral cortex, Phil. Trans. R. Soc. Lond. B, 355:127–134, 2000.

41. M. P. Young, C. C. Hilgetag and J. W. Scannell, On imputing function to structure from the behavioural effects of brain lesions, Phil. Trans. R. Soc. Lond. B, 355:147–161, 2000.

42. W. J. Freeman, Tutorial on neurobiology: From single neurons to brain chaos, Int. J. Bifurcation Chaos, 2(3):451–482, 1992.
43. P. beim Graben and J. Kurths, Simulating global properties of electroencephalograms with minimal random neural networks, Neurocomputing, doi: 10.1016/j.neucom.2007.02.007, 2007.

Index

Understanding Complex Systems

Jirsa, V.K.; Kelso, J.A.S. (Eds.)
Coordination Dynamics: Issues and Trends
XIV, 272 p. 2004 [978-3-540-20323-0]

Kerner, B.S.
The Physics of Traffic:
Empirical Freeway Pattern Features,
Engineering Applications, and Theory
XXIII, 682 p. 2004 [978-3-540-20716-0]

Kleidon, A.; Lorenz, R.D. (Eds.),
Non-equilibrium Thermodynamics
and the Production of Entropy
XIX, 260 p. 2005 [978-3-540-22495-2]

Kocarev, L.; Vattay, G. (Eds.)
Complex Dynamics in Communication
Networks
X, 361 p. 2005 [978-3-540-24305-2]

McDaniel, R.R.Jr.; Driebe, D.J. (Eds.)
Uncertainty and Surprise in Complex Systems:
Questions on Working with the Unexpected
X, 200 p. 2005 [978-3-540-23773-0]

Ausloos, M.; Dirickx, M. (Eds.)
The Logistic Map and the Route to Chaos –
From the Beginnings to Modern Applications
XX, 413 p. 2006 [978-3-540-28366-9]

Kaneko, K.
Life: An Introduction to Complex Systems
Biology
XIV, 369 p. 2006 [978-3-540-32666-3]

Braha, D.; Minai, A.A.; Bar-Yam, Y. (Eds.)
Complex Engineered Systems – Science Meets
Technology
X, 384 p. 2006 [978-3-540-32831-5]

Fradkov, A.L.
Cybernetical Physics – From Control of Chaos
to Quantum Control
XII, 241 p. 2007 [978-3-540-46275-0]

Aziz-Alaoui, M.A.; Bertelle, C. (Eds.)
Emergent Properties in Natural
and Artificial Dynamical Systems
X, 280 p. 2006 [978-3-540-34822-1]

Baglio, S.; Bulsara, A. (Eds.)
Device Applications of Nonlinear Dynamics
XI, 259 p. 2006 [978-3-540-33877-2]

Jirsa, V.K.; McIntosh, A.R. (Eds.)
Handbook of Brain Connectivity
X, 528 p. 2007 [978-3-540-71462-0]

Krauskopf, B.; Osinga, H.M.;
Galan-Vioque, J. (Eds.)
Numerical Continuation Methods
for Dynamical Systems
IV, 412 p. 2007 [978-1-4020-6355-8]

Perlovsky, L.I.; Kozma, R. (Eds.)
Neurodynamics of Cognition and Consciousness
XI, 366 p. 2007 [978-3-540-73266-2]

Qudrat-Ullah, H.; Spector, J.M.; Davidsen, P. (Eds.)
Complex Decision Making – Theory and Practice
XII, 337 p. 2008 [978-3-540-73664-6]

beim Graben, P.; Zhou, C.; Thiel, M.;
Kurths, J. (Eds.)
Lectures in Supercomputational Neuroscience –
Dynamics in Complex Brain Networks
X, 378 p. 2008 [978-3-540-73158-0]